Hilbert–Courant

Elizabeth Reidemeister

David Hilbert, 1937

Constance Reid

Hilbert–Courant

With 65 Photographs

Springer-Verlag
New York Berlin Heidelberg Tokyo

Constance Reid
70 Piedmont Street
San Francisco, CA 94117
U.S.A.

AMS Classifications: 01A60, 01A70, 01A72

Library of Congress Cataloging in Publication Data
Reid, Constance.
  Hilbert–Courant,
  Previously published as 2 separate vols.: Hilbert.
1970, and Courant in Göttingen and New York. 1976.
  Hermann Weyl's appreciation of Hilbert's mathematical
work, originally in "Hilbert" has been omitted in this
new ed.
  Includes bibliographical references and index.
  1. Hilbert, David, 1862–1943. 2. Courant, Richard,
1888–1972. 3. Mathematicians—Biography. I. Title.
QA29.H5R42 1986   510'.92'2 [B]   86-3721

Previously published as two separate volumes: *Hilbert* by Constance Reid, Springer-Verlag, 1970;
Courant in Göttingen and New York: The Story of an Improbable Mathematician by Constance Reid,
Springer-Verlag, 1976.

© 1986 by Springer-Verlag New York Inc.
All rights reserved. No part of this book may be translated or reproduced in any form without
written permission from Springer-Verlag, 175 Fifth Avenue, New York, New York 10010,
U.S.A.

Printed and bound by Arcata Graphics/Halliday, West Hanover, Massachusetts.
Printed in the United States of America.

9 8 7 6 5 4 3 2 1

ISBN 0-387-96256-5 Springer-Verlag New York Berlin Heidelberg Tokyo
ISBN 3-540-96256-5 Springer-Verlag Berlin Heidelberg New York Tokyo

Preface

I am very pleased that my books about David Hilbert, published in 1970, and Richard Courant, published in 1976, are now being issued by Springer-Verlag in a single volume. I have always felt that they belonged together, *Courant* being, as I have written, the natural and necessary sequel to *Hilbert*—the rest of the story.

To make the two volumes more compatible when published as one, we have combined and brought up to date the indexes of names and dates. Unfortunately we have had to omit Hermann Weyl's article on "David Hilbert and his mathematical work," but the interested reader can always find it in the hardback edition of *Hilbert* and in Weyl's collected papers. At the request of a number of readers we have included a listing of all of Hilbert's famous Paris problems.

It was, of course, inevitable that we would give the resulting joint volume the title *Hilbert–Courant*.

San Francisco, California CONSTANCE REID
September 30, 1985

Hilbert 1
 Königsberg and Göttingen: An Album (following page 220)

Courant 221
 Göttingen and New York: An Album (following page 533)

Index of Names 535

Hilbert

To the memory of

Otto Blumenthal
1876—1944

Foreword

David Hilbert was one of the truly great mathematicians of his time. His work and his inspiring scientific personality have profoundly influenced the development of the mathematical sciences up to the present time. His vision, his productive power and independent originality as a mathematical thinker, his versatility and breadth of interest made him a pioneer in many different mathematical fields. He was a unique personality, profoundly immersed in his work and totally dedicated to his science, a teacher and leader of the very highest order, inspiring and most generous, tireless and persistent in all of his efforts.

To me, one of the few survivors of Hilbert's inner circle, it always has appeared most desirable that a biography should be published. Considering, however, the enormous scientific scope of Hilbert's work, it seemed to me humanly impossible that a single biographer could do justice to all the aspects of Hilbert as a productive scientist and to the impact of his radiant personality. Thus, when I learned of Mrs. Reid's plan for the present book I was at first skeptical whether somebody not thoroughly familiar with mathematics could possibly write an acceptable book. Yet, when I saw the manuscript my skepticism faded, and I became more and more enthusiastic about the author's achievement. I trust that the book will fascinate not only mathematicians but everybody who is interested in the mystery of the origin of great scientists in our society.

New Rochelle,
November 23, 1969   Richard Courant

Preface

To a large extent this book has been written from memory.

I received much friendly assistance from men and women who took their doctoral degrees from Hilbert: Vera Lebedeff-Myller (1906), Robert König (1907), Andreas Speiser (1909), Richard Courant (1910), Hugo Steinhaus (1911), Paul Funk (1911), Ludwig Föppl (1912), Hellmuth Kneser (1921), Haskell Curry (1930), Arnold Schmidt (1932), Kurt Schütte (1934).

Written recollections of other former students, no longer living, were also of great assistance. I would like to acknowledge here my special debt to Otto Blumenthal (1898), who wrote the biographical sketch for Hilbert's collected works and the one for the special edition of *Naturwissenschaften* honoring Hilbert's sixtieth birthday; and to Hermann Weyl (1908) for the obituary notice for the Royal Society and the article "David Hilbert and his mathematical work," which is reprinted in this book.

Perhaps most helpful to me because they had the longest, closest acquaintance with Hilbert were Richard Courant, who was his colleague from 1919 to 1933, most of that time as head of the Mathematical Institute; and Paul Bernays, who was from 1917 to 1934 his assistant and his collaborator in his work on logic and the foundations of mathematics.

Among Hilbert's former physics assistants, Alfred Landé, Paul Ewald, Adolf Kratzer and Lothar Nordheim were most generous with their time and knowledge. I would especially like to thank Professor Ewald for his suggestions on the literary treatment of Hilbert's life.

I was also able to obtain a great deal of information about Hilbert in personal interviews with people who, although they were not his students, were close to the Göttingen circle at various times. These included Hans Lewy, Alexander Ostrowski, George Pólya, Brigitte Rellich, Carl Ludwig Siegel, Gabor Szegö, Olga Taussky-Todd, Jan van der Corput, B. L. van der Waerden, Ellen Weyl-Bär. Letters from Kurt and Elizabeth Reidemeister and from Helmut Hasse described Hilbert's last years. Alfred Tarski and Kurt Gödel, as well as Professor Bernays, answered my questions about Hilbert's work in logic and foundations.

I am grateful to Lily Rüdenberg and Ruth Buschke for their kindness in allowing me to quote from the letters which their father, Hermann Minkowski, wrote to Hilbert during the many years of their close friendship. Unfortunately, the Hilbert half of the correspondence, which was returned to Mrs. Hilbert by Mrs. Minkowski in 1933, is — as far as I have been able to determine — no longer in existence. The few quotations from Hilbert's letters to Minkowski which do appear in this book are from Blumenthal, who had the opportunity to read Hilbert's letters before he wrote the biographical sketch for the collected works.

Horst Hilbert, the son of Hilbert's cousin, supplied many details about the family. J. K. von Schroeder of the Geheimes Staatsarchiv der Stiftung Preußischer Kulturbesitz searched out vital statistics. Kin-ya Honda translated his biographical sketch of Hilbert into English for me. H. Vogt, director of the Niedersächsische Staats- und Universitätsbibliothek, made available the letters from Hilbert which are in the Klein and Hurwitz papers. Martin Kneser, the present director of the Mathematical Institute, provided me with office space and gave me access to the Hilbert papers. Ursula Drews, the secretary of the Institute, was most helpful. Irma Neumann, whose mother was the Hilbert's well-loved housekeeper for many years, shared with me the Hilbert family pictures.

Special thanks are due to my sister, Julia Robinson, who never faltered in providing assistance, advice and encouragement; to Volker Strassen, who introduced me to Göttingen and its mathematical tradition; to Ursula Lawrenz, Christa Strassen and Edith Fried, who supplemented my knowledge of German and of Germany.

It makes me very happy that the book is being published by Springer-Verlag, who had close ties with Hilbert and Göttingen and who, by taking the risks of publication, substantially contributed to the revival of German science after the first world war.

The manuscript has been read at various stages by Paul Bernays, Richard Courant, Paul Ewald, Lothar Nordheim, Julia Robinson, R. M. Robinson, Volker Strassen, Gabor Szegö, John Addison Jr., and Max Born.

After all this very generous assistance, any errors which remain are most certainly my own.

San Francisco, California
August 3, 1969

Constance Reid

Contents

I	Youth	1
II	Friends and Teachers	9
III	Doctor of Philosophy	15
IV	Paris	22
V	Gordan's Problem	28
VI	Changes	39
VII	Only Number Fields	47
VIII	Tables, Chairs, and Beer Mugs	57
IX	Problems	65
X	The Future of Mathematics	74
XI	The New Century	84
XII	Second Youth	91
XIII	The Passionate Scientific Life	102
XIV	Space, Time and Number	111
XV	Friends and Students	117
XVI	Physics	125
XVII	War	137
XVIII	The Foundations of Mathematics	148
XIX	The New Order	158
XX	The Infinite!	168
XXI	Borrowed Time	178
XXII	Logic and the Understanding of Nature	190
XXIII	Exodus	198
XXIV	Age	206
XXV	The Last Word	216

I

Youth

The fortuitous combination of genes that produces an unusually gifted individual was effected by Otto Hilbert and his wife Maria sometime in the spring of 1861; and on January 23, 1862, at one o'clock in the afternoon, their first child was born in Wehlau, near Königsberg, the capital of East Prussia. They named him David.

Thanks to an autobiography and family chronicle left by the founder of the Königsberg branch of the Hilbert family, something is known about David's background on the paternal side. During the seventeenth century there were Hilberts in Saxony. For the most part they were artisans and tradespeople, but frequently enough for the fact to be commented upon they took as their wives the daughters of teachers. They were Protestants, and their Biblical names seem to indicate that they were Pietists, members of a fundamentalist sect of the time which emphasized "repentance, faith as an attitude of heart, and regeneration and sanctification as experiential facts."

At the beginning of the eighteenth century, one Johann Christian Hilbert, although trained as a brass worker, became a successful lace merchant with more than a hundred people in his employ, "the most distinguished man" in the little town of Brand near Freiberg. Unfortunately, though, he died while his children were still young, and the fortune he had left was dissipated by unscrupulous guardians. Forced by necessity, his son Christian David Hilbert apprenticed himself to a barber, served as a military barber in the army of Frederick the Great, and came eventually to Königsberg. He seems to have been a man of exceptional energy and industry. He purchased a barbershop, then enrolled at the local university, studied medicine, and became the city's licensed surgeon and accoucheur. From that time on, the Hilberts were professional men, who chose as their wives the daughters of merchants. One of Christian David's many children was David Fürchtegott Leberecht (*Fear God Live Right*) Hilbert. This was David's grandfather. He was a judge and a Geheimrat, which was a title of some honor. His son Otto was David's father, a county judge at the time of David's birth. One uncle was a lawyer, another was a gymnasium director — a position equivalent to that of a high school principal but of considerably more prestige.

Not much is known about David's background on the maternal side. Karl Erdtmann was a Königsberg merchant, and his daughter Maria Therese was David's mother. She was an unusual woman — in the German way of expression "an original" — interested in philosophy and astronomy — and prime numbers. There was Huguenot blood in her, and some people later wondered if that might not explain "a certain strangeness" in her son.

David's birth coincided almost exactly with the birth of German nationalism. A few months before, the brother of the dead Prussian king had made the traditional pilgrimage to Königsberg and in the ancient castle church had placed on his own head the crown of the Prussians. A short time later he chose as his chief minister the Count Otto von Bismarck-Schönhausen. In the ensuing period of the wars for the unification of Germany under Prussia, David's father became a city judge and moved his family into Königsberg proper.

The East Prussian capital dated from the middle of the thirteenth century, when the knights of the teutonic order had built their castle upon the rising ground which lies behind the point where the two branches of the Pregel river meet and flow into the Baltic. In David's time the stout castle still stood — the heart of a recently modernized city of gas lights and horse-drawn trams. From the Hilbert house at No. 13 Kirchenstrasse it was only a few blocks to the river, "our gate to freedom," as the citizens of Königsberg liked to call it. Although the city was four and a half miles from the mouth of the Pregel, the sharp salt tang of the Baltic was everywhere. Gulls settled gently on the green lawns. Sea breezes fluttered the bright sails of the fishing boats. Odors of salt water and fish, of pitch and lumber, of smoke hung over the city. Boats and barges coming up the Pregel brought romantic cargoes, unloaded and reloaded in front of tall warehouses which stood at the river's edge, and went back down to the sea with *Bernstein* (amber) and a fine white claylike substance which was used in the manufacture of pipes and called *Meerschaum*. Seven great bridges, each with a distinct personality, spanned the Pregel, five of them joining its banks with an island, called the Kneiphof. Königsberg had entered the history of mathematics through these bridges. They had provided the problem, solved the century before by Euler, which lies at the foundation of what is now known as topology. The cathedral of Königsberg was on the Kneiphof, and next to it the old university and the grave of Königsberg's greatest son, Immanuel Kant.

Like all boys in Königsberg, David grew up with the words of Kant in his mind and ears. Every year on the twenty-second of April, the anniversary of the philosopher's birth, the crypt next to the cathedral was opened to the public. On these occasions David undoubtedly accompanied his philosophically inclined mother to pay his respects, saw the bust with the familiar features crowned for this special day with a fresh wreath of laurel, and spelled out the words on the wall of the crypt:

"The greatest wonders are the starry heavens above me and the moral law within me."

His mother must also have pointed out the constellations to him and introduced him to those interesting "first" numbers which, unlike the other numbers, are divisible only by themselves and 1.

Under his father, his early training stressed the Prussian virtues of punctuality, thrift and faithfulness to duty, diligence, discipline and respect for law. A judgeship in Prussia was obtained through civil service promotion. It was a comfortable, safe career for a conservative man. Judge Hilbert is reported to have been rather narrow in his point of view with strict ideas about proper behavior, a man so set in his ways that he walked the same path every day and so "rooted" in Königsberg that he left it only for his annual vacation on the Baltic.

David was to be the Hilberts' only son. When he was six, a sister was born and christened Elise.

The year that David was seven, the King himself, soon to be Kaiser, returned to Königsberg for the first time since his coronation. Here in person for the boy to see was the man "destined," in the words of the town chronicler, "to elevate his house to its greatest brilliance, his land to its greatest strength." A large crowd gathered on the wooden bridge over the castle lake to see the King, the bridge broke under the weight, and 67 persons were drowned.

The next year, Prussia challenged France. In a few months the triumphant news blazed through the East Prussian capital — the French emperor had been taken prisoner. While Bismarck and the generals prepared to lay siege to Paris, David, now eight, started to school. This was two years later than the usual starting age of six, and it indicates that he may have received his first instruction at home, probably from his mother. She was something of an invalid and is said to have spent much of her time in bed.

In the Vorschule of the royal Friedrichskolleg, David now received the preliminary instruction required for the humanistic gymnasium, which he would have to attend if he wished to become a professional man, a clergyman, or a university professor. This included reading and writing the German and Roman scripts, spelling, parts of speech and analysis of simple sentences, important Biblical stories, and simple arithmetic involving addition, subtraction, multiplication and division of small numbers.

In autumn of 1872, when he was ready for the gymnasium proper, the Prussian army paid a triumphal visit to Königsberg. What was to be eventually more important for David was that at the same time a Jewish family named Minkowski came to the city from Alexoten near Kovno. They had left their native Russia because of the persecution to which Jews were being subjected by the Czar's government. The father, who had been a successful merchant, had been forced to sell everything hurriedly without profit. Now, in Königsberg, he turned to a new trade – the export of white linen rags. When his children were disturbed by this change in the family fortunes, the mother explained to them that the father's new occupation was one of the noblest, since the paper for the fine books which they loved could be made only from such rags. Eventually the father was again successful, but at first things were hard. The family moved into a big old house by the railroad station, on the other side of the Pregel river from where Judge Hilbert and his family lived.

In Russia, Max, the oldest Minkowski boy, had not been permitted to enroll in the gymnasium because he was a Jew; and he was never to obtain a formal education, becoming instead a partner in his father's business and, after the latter's death, the "father" of the family. Oskar, the second son, was now one of the few Jews to attend the Altstadt Gymnasium in Königsberg. Later, as a doctor and medical researcher, he discovered the relationship between the pancreas and diabetes and became well known as the "grandfather of insulin." Eight and a half year old Hermann, the third son, entered the Vorschule of the same gymnasium. According to a loving biography compiled for the family by the sister, Fanny, and entitled "Three Universal Geniuses," the Minkowski boys were "a sensation" in Königsberg, "not only because of their great talents but also because of their charming personalities." Little Hermann's mathematical abilities were particularly impressive; and in one class, when the teacher failed to understand a mathematical problem on the board, the students chorused, "Minkowski, help!"

There is no sister's record that anyone was particularly impressed at this time by the abilities of the Hilbert boy. He later recalled himself as dull and silly in his youth — "dammelig" was the word he used. This was perhaps an exaggeration; but, as a friend later commented, "Behind everything Hilbert said, no matter how paradoxically it was phrased, one always felt his intense, and often quite touching, desire for the truth."

The gymnasium which David's parents had chosen for him was reputed to be the best in Königsberg, a venerable private institution of learning which dated back to the early seventeenth century and counted among its graduates Kant himself. But it was an unfortunate selection. There was at that time a rare concentration of youthful scientific talent in Königsberg. At one point Max and Willy Wien, Arnold Sommerfeld and Hermann Minkowski were all simultaneously in attendance at the Altstadt Gymnasium. David, because he attended Friedrichskolleg, did not have an opportunity to become acquainted with any of these boys during his school days.

Friedrichskolleg was also, unfortunately for David, very traditional and rigid in its curriculum. The name "Gymnasium" derived from the fact that such a school was conceived as offering a mental gymnastic which would develop a boy's mind as physical exercise develops his body. To this end the study of Latin and Greek was regarded of the utmost importance. It was believed that by cultivating these languages and their literatures, the student could acquire skill in all the mental operations. The grammar would assist him in formulating his ideas; the poetry would awaken his sense of the aesthetic and cultivate his taste; the study of historical and philosophical writers would broaden his horizons and furnish a basis for "the proper conception" of the present. After the ancient languages, mathematics was traditionally most valued for strengthening the mental muscles; but at Friedrichskolleg it ran a poor second to Latin and Greek. Science was not offered.

The language classes formed by far the largest part of the curriculum. Emphasis was on obtaining a firm foundation in grammar — the study of literature had to wait upon that. There was little opportunity for independent thinking or expression. Sometimes, however, David scribbled little verses in the margins of his notebooks.

During this same period the youngest Minkowski boy, in sheet and pillowcase, was playing Othello in a family production. Curled up in the window seat of a room where only Fanny came to practice the piano, he was reading Shakespeare, Schiller and Goethe, learning almost all of the latter by heart, "so that," according to Fanny, "for the rest of his life he needed no other books except his scientific works."

David found memorization exceedingly difficult and for the most part, at Friedrichskolleg, learning was equated with remembering. The language classes particularly, according to a friend, "caused him more sorrow than joy." He was also not particularly quick at comprehending new ideas. He seemed never really able to understand anything until he had worked it through his own mind. But in spite of his difficulties in school, he never fell behind his classmates. He was industrious, and he had a clear perception of the realities of the Prussian educational system. There were no foolish gestures. Unlike Einstein, later, he did not leave the gymnasium before taking the Abitur (the examination by means of which a student qualified to enter the university).

A member of the Hilbert family said years afterwards when she was an old lady:

"All I know of Uncle David is that his whole family considered him a bit off his head. His mother wrote his school essays for him. On the other hand, he could explain mathematics problems to his teachers. Nobody really understood him at home."

Already he had found the school subject which was perfectly suited to his mind and a source of inexhaustible delight. He said later that mathematics first appealed to him because it was "bequem" — *easy, effortless*. It required no memorization. He could always figure it out again for himself. But he recognized that according to the regulations he could not go on to the university and study mathematics, become a mathematician, unless he first obtained a diploma from a gymnasium. So now he slighted this favorite subject and concentrated on "getting by" in Latin and Greek.

The days at Friedrichskolleg were always to be remembered as unhappy ones.

There was a bright time in the year, however. That was the summer vacation. Then he accompanied his family to Rauschen, a tiny fishing village a short distance from the sea. Although Rauschen was to become a very popular summer resort, it was frequented at that time by only a few people. Among these was the large family of Karl Schmidt. Like Otto Hilbert, Schmidt had been trained as a lawyer; but he was a radical Social Democrat and had chosen eventually to be a master mason and housebuilder instead. The Schmidts' fifth child, whose name was Käthe, showed already an exceptional gift in sketching the workers and sailors of Königsberg. Many years later, as the famous artist Käthe Kollwitz, she was to recall the annual trip to Rauschen much as it must have been experienced by David:

"The trip to Rauschen took some five hours, since there was no railroad at that time. We rode in a *journalière*, which was a large covered wagon with four or five rows of seats. The back seats were taken out and filled with things needed for the stay, bedding, clothing, baskets, boxes of books and cases of wine. Three and sometimes four horses were needed. The driver sat on a high seat in front. Then off we would go through the narrow streets of Königsberg, through the clanging Tragheim Gate, and then out across the whole of Samland. Shortly before we reached Sassau we would catch sight of the sea for the first time. Then we would all stand on tiptoe and shout: *The sea, the sea!* Never again would the sea ... be to me what the Baltic Sea at Samland was. The inexpressible splendor of the sunsets seen from the high coastline; the emotion when we saw it again for the first time, ran down the sea-slope, tore off our shoes and stockings and felt once again the cool sand underfoot, the metallic slapping of the waves...."

Summers at Rauschen were idyllic. It was "a children's paradise" to those who came. In September school took up again. In November the Pregel froze and did not thaw until March.

Once, recalling his boyhood, Hilbert explained, "I did not do much mathematics in school, because I knew I would do that later on." But at the time he must have felt less philosophical about the postponement. In September 1879, at the beginning of his last gymnasium year, he transferred from Friedrichskolleg to the Wilhelm Gymnasium, a state school which placed considerably more emphasis on mathematics, treating even some of the new developments in geometry.

By this time the precocious Hermann Minkowski, although two years younger, was passing David by. That spring, "by virtue of his splendid memory and rapid comprehension" (as Hilbert later reported), Minkowski completed the eight year course at the Altstadt Gymnasium in five and a half years and went on to the local university.

At the Wilhelm Gymnasium, David was much happier than he had been at Friedrichskolleg. The teachers seemed to recognize and encourage his originality, and in later years he was often to recall them with affection. Grades improved — "good's" in almost everything (German, Latin, Greek, theology and physics) and in mathematics "vorzüglich," the highest possible mark given at the time. He did so well on his written examinations that he was excused from taking the final oral examination for the leaving certificate. The evaluation which appeared on the back of the certificate rated

Hilbert

his deportment as "exemplary," commented on his industry and "serious scientific interest," and then concluded:

"For mathematics he always showed a very lively interest and a penetrating understanding: he mastered all the material taught in the school in a very pleasing manner and was able to apply it with sureness and ingenuity."

This is the earliest recorded glimpse of the mathematician Hilbert.

II

Friends and Teachers

It was Hilbert's good fortune that the university in his native city, although far from the center of things in Berlin, was in its scientific tradition one of the most distinguished universities in Germany.

Jacobi had taught at Königsberg when, in the time of Gauss, he was considered the "second best" mathematician in Europe. Richelot, who had been Jacobi's successor, had had the distinction of discovering in the work of an unknown gymnasium teacher the genius of Karl Weierstrass. He had then persuaded the University to award Weierstrass an honorary degree and had travelled to the little town where Weierstrass taught to make the presentation in person — "We have all found our master in Herr Weierstrass!" The versatile Franz Neumann had established at Königsberg the first institute of theoretical physics at a German university and had originated the format of the seminar.

By the time that Hilbert enrolled in the autumn of 1880, Weierstrass was the most distinguished mathematician in Germany; Jacobi and the magnanimous Richelot were dead; but Franz Neumann, who was to live to be almost a hundred, was still to be seen at university gatherings and sometimes still lectured. Every student quickly learned the story of how when a great academy had attempted to set up regulations for the apportionment of scientific credit, Neumann — many of whose discoveries were never published — had said simply, "The greatest luck is the discovery of a new truth; to that, recognition can add little or nothing."

Hilbert now found the university as free as the gymnasium had been confined. Faculty members chose the subjects they wished to teach; and students, the subjects they wished to learn. There were no specified requirements, no minimum number of units, no roll call, no examinations until the taking of the degree. Many quite naturally responded to this sudden freedom by spending their first university years in the traditional occupations of the student fraternal organizations — drinking and duelling. But for the 18-year-old Hilbert the university offered something more alluring — the freedom to concentrate, at last, upon mathematics.

There was never any doubt in Hilbert's own mind about his future vocation. Although his father disapproved, he enrolled, not in law, but in mathematics, which was at that time part of the Philosophical Faculty.

He was beginning his study at a time when much of the exuberant development of mathematics in the first part of the century had been firmly pruned into shape by Weierstrass and others. The general atmosphere was self-congratulatory. It was felt that mathematics had at last reached a level of logical strictness, or rigor, which would never need to be, and indeed could not be, surpassed. At the same time, however, a professor at Halle named Georg Cantor was developing an original theory of sets in which he treated the infinite in a new and disturbing way. According to the traditional conception, the infinite was something "unlimitedly increasing." But in this work of Cantor's it was something entirely different — not increasing but being "fixed mathematically by numbers in the definite form of a completed infinite." This conception of a "completed infinite" was one to which (Cantor later wrote) he had been "logically forced, almost against my will, because in opposition to traditions which had become valued by me." It was to be the subject of the most violent and bitter controversy among mathematicians during the coming decade.

His first semester at the university, Hilbert heard lectures on integral calculus, determinant theory, and the curvature of surfaces. The second semester, following the popular custom of moving from university to university, he set out for Heidelberg, the most delightful and most romantic of all the German universities.

At Heidelberg, Hilbert attended lectures by Lazarus Fuchs, whose name was already synonymous with linear differential equations. Fuchs's lectures were very impressive but in a rather unusual way. Rarely prepared, he customarily produced on the spot what he wished to say. Thus his students, as one of them later wrote, "had the opportunity of seeing a mathematical mind of the highest order actually in operation."

The following semester Hilbert could have moved on to Berlin, where there was a constellation of scientists that included Weierstrass, Kummer, Kronecker and Helmholtz. But he was deeply attached to the city of his birth, resembling in this at least his father; and so he returned to the University of Königsberg.

During this time there was but one full professor of mathematics at Königsberg. This was Heinrich Weber, an extremely gifted and versatile man and a worthy successor to Jacobi and Richelot. He made important contributions to fields as diverse as number theory and mathematical physics. He also wrote many important books. He co-authored with Richard Dedekind a famous book on the arithmetic theory of algebraic functions of one variable, and his book on algebra and the Riemann-Weber book on methods of mathematical physics were both classics in their field.

From Weber, Hilbert heard lectures on number theory and function theory and made his first acquaintance with the theory of invariants, the most fashionable mathematical theory of the day. He carefully saved the notes of these first lectures and also all the others he heard during his university days. The hand in which they are written is boyish, there are youthful misspellings, but no doodles. Only one set of notes appears to have been extensively worked over at a later date. These are the notes he took on the number theory lectures of Weber.

The following semester — the spring of 1882 — Hilbert chose to remain again at his home university. That same spring, Hermann Minkowski returned to Königsberg from Berlin, where he had studied for the past three semesters.

Minkowski was a chubby-faced boy with a scholar's pince-nez perched rather incongruously on his still unformed nose. While he was in Berlin, he had won a monetary prize for his mathematical work and then given it up in favor of a needy classmate. But this was not known in Königsberg. (Even his family learned of the incident only much later when the brother of the classmate told them about it.) Although Minkowski was still only 17 years old, he was involved in a deep work with which he hoped to win the Grand Prix des Sciences Mathématiques of the Paris Academy.

The Academy had proposed the problem of the representation of a number as the sum of five squares. Minkowski's investigations, however, had led him far beyond the stated problem. By the time the deadline of June 1, 1882, arrived, he still had not had his work translated into French as the rules of the competition required. Nevertheless, he decided to submit it. At the last minute, at the suggestion of his oldest brother Max, he wrote a short prefatory note in which he explained that his neglect had been due to the attractions of his subject and expressed the hope that the Academy would not think "I would have given more if I had given less." As an epigraph he inscribed a line from Boileau: "Rien n'est beau que le vrai, le vrai seul est aimable."

During the year that this work was under consideration by the Academy, Minkowski attended lectures at Königsberg. In spite of his youth, he was very stimulating to the other mathematics students. Because of his experiences in Berlin he brought to the young men isolated in the East Prussian capital a sense of participation in the mathematics of the day. He was, however, extremely shy, stammered slightly, and turned a deep red whenever attention was directed at him. It does not seem likely that during this first year he developed a close relationship with any of the other mathematics students, most of whom, like Hilbert, were several years older than he was.

Then, in the spring of 1883, came the announcement that this boy, still only 18 years old, had been awarded jointly with the well-known English mathematician Henry Smith the Grand Prix des Sciences Mathématiques. The impression which the news made in Königsberg can be gauged by the fact that Judge Hilbert admonished David that presuming on acquaintance with "such a famous man" would be "impertinence."

For a while it seemed, though, that Minkowski might not actually receive his prize. The French newspapers pointed out that the rules of the competition had specifically stated that all entries must be in French. The English mathematicians let it be known that they considered it a reflection upon their distinguished countryman, who had since died, that he should be made to share a mathematical prize with a boy. ("It is curious to contemplate at a distance," an English mathematician remarked some forty years later, "the storm of indignation which convulsed the mathematical circles of England when Smith, bracketed after his death with the then unknown German mathematician, received a greater honor than any that had been paid to him in life.") In spite of the pressures upon them, the members of the prize committee never faltered. From Paris, Camille Jordan wrote to Minkowski: "Work, I pray you, to become a great mathematician."

Hilbert knew his luck when he saw it. In spite of his father's disapproval, he soon became friends with the shy, gifted Minkowski. He was shortly to comment of another shy young mathematician that "with skillful treatment I am sure he would open up," and now he apparently applied such skillful treatment to Minkowski.

Although the two young men came from different family backgrounds and had in many ways quite different personalities, they had — under the surface — many traits in common; and, years later, when Hilbert had occasion to write about Minkowski, he revealed more about himself than he ever did at any other time.

In addition to their enthusiastic love for mathematics, they shared a deep, fundamental optimism. As far as science in general was concerned, the period of their university days was one of triumphant pessimism, a reaction against the almost religious belief in the power of science which had flourished in the previous century. The works of Emil duBois-Reymond, a physiologist turned philosopher, were widely read and much quoted. DuBois-Reymond concerned himself with the limits of the knowledge of nature — this was, in fact, the title of his most famous lecture. He maintained that certain problems, which he called transcendental, or supersensible, were unsolvable even in principle. These included the nature of matter and force, the origin of motion, the origin of sensation and consciousness. His gloomy concession, "Ignoramus et ignorabimus" — *we are ignorant and we shall remain ignorant* —, was the catchword of many of the scientific-philosophical discussions at the university. But to both Hilbert and Minkowski such a concession was thoroughly abhorrent. Already they shared the conviction (as Hilbert later put it) "that every definite mathematical problem must necessarily be susceptible of an exact settlement, either in the form of an actual answer to the question asked, or by the proof of the impossibility of its solution and therefore the necessary failure of all attempts."

This belief in the solvability of every mathematical problem had recently received spectacular support. The German mathematician Ferdinand Lindemann had proved the long suspected transcendence of the number π and had thus established the impossibility of the ancient dream of "squaring the circle." Up to the time of this achievement, Lindemann's career had not been a great success. An ambitious work which he had published had been cruelly (and rather unfairly) criticized. But now, he had recouped everything by solving a famous problem. He was the man of the hour in mathematics. When Weber left Königsberg for Charlottenburg, Lindemann was invited to take his place.

In spite of his current fame, Lindemann was not a mathematician of the same caliber as Weber. He was to have little influence on Hilbert (and none at all on Minkowski); but he was responsible for bringing to Königsberg shortly the young man who was to be, rather than either Weber or Lindemann, Hilbert's real teacher. This was Adolf Hurwitz.

In the spring of 1884 when Hurwitz arrived from Göttingen to take over the duties of an Extraordinarius, or associate professor, he was still not 25 years old. Like Minkowski, he had behind him a record of mathematical precocity. His gymnasium teacher, Hannibal Schubert, had been so impressed with his mathematical abilities that he had devoted Sundays to instructing Hurwitz in his own specialty, which became known as the "Schubert calculus." He had also managed to convince Hurwitz's father, a Jewish manufacturer as doubtful of the rewards of an academic life as Judge Hilbert, that the gifted boy must be permitted to continue his study of mathematics. Encouraged by Schubert, the father borrowed the necessary money from a friend.

Hurwitz's first mathematical work was published in collaboration with Schubert while he was still in the gymnasium. His later studies gave him an exceptionally wide background in the mathematics of the day. He received his doctor's degree from Felix Klein, one of the most spectacular of the younger mathematicians in Germany at that time. He attended the lectures of the great men in Berlin and then moved on to Göttingen, where he did impressive work in function theory.

He was a sweet-tempered young man who loved music almost as much as mathematics and played the piano beautifully. But already, before he came to Königsberg, he had suffered a nearly fatal bout with typhoid fever. He had frequent, very severe migraine headaches, which may have been at least partially caused by the fact that he was a perfectionist in everything he did.

Hilbert found the new teacher "unpretentious in his outward appearance" but saw that "his wise and gay eyes testified as to his spirit." He and Minkowski soon established a close relationship with Hurwitz. Every afternoon, "precisely at five," the three met for a walk "to the apple tree." It was at this time that Hilbert found a way of learning infinitely preferable to poring over dusty books in some dark classroom or library.

"On unending walks we engrossed ourselves in the actual problems of the mathematics of the time; exchanged our newly acquired understandings, our thoughts and scientific plans; and formed a friendship for life."

Learning "in the most easy and interesting way," the three young men explored every kingdom of the mathematical world. Hurwitz with his vast "well founded and well ordered" knowledge was always the leader. He quite overwhelmed the other two.

"We did not believe," Hilbert recalled later, "that we would ever bring ourselves so far."

But there was no need for them to feel like Alexander, who complained to his schoolmates, "Father will conquer everything and there will be nothing left for us to conquer."

The world of mathematics is inexhaustible.

III

Doctor of Philosophy

Having completed the eight university semesters required for a doctor's degree, Hilbert began to consider possible subjects for his dissertation. In this work he would be expected to make some sort of original contribution to mathematics. At first he thought that he might like to investigate a generalization of continued fractions; and he went to Lindemann, who was his "Doctor-Father," with this proposal. Lindemann informed him that unfortunately such a generalization had already been given by Jacobi. Why not, Lindemann suggested, take instead a problem in the theory of algebraic invariants.

Although the theory of algebraic invariants was considered a very modern subject, its roots lay in the seventeenth century invention of analytic geometry by René Descartes. On Descartes's map of the plane, horizontal coordinates are real numbers which are designated by x; vertical coordinates, real numbers designated by y. Since any point on the plane is then equivalent to a pair of real numbers x, y, geometric figures can be formalized by algebraic equations and, conversely, algebraic equations can be graphed as geometric figures. Concepts and relations in both fields are clarified — geometric ideas becoming more abstract and easily handled; algebraic ideas, more vivid and more intuitively comprehensible.

There is also a great gain in generality. Just as the size and the shape of figures do not change when their position in relation to the axes is changed, so certain properties of their related algebraic forms remain unchanged. These "invariants" serve to characterize the given geometric figure. Thus, quite naturally, the development of projective geometry, which concerns itself with the often dramatic transformations effected by projection, led eventually to a parallel development in algebra which concentrated on the invariants of algebraic forms under various groups of transformations.

Because of its sheer sophisticated power, the algebraic approach soon outstripped the geometric one; and the theory of algebraic invariants became a subject of consuming interest for a number of mathematicians.

The pioneers in the new theory had been Englishmen — Arthur Cayley and his good friend, Joseph Sylvester, both men, as it happened, lawyers turned mathematicians. But the Germans had been quick to take up the theory; and now the great German mathematical journal, *Mathematische Annalen*, was almost exclusively an international forum for papers on algebraic invariants.

The problem which Lindemann suggested to Hilbert for his doctoral dissertation was the question of the invariant properties for certain algebraic forms. This was an appropriately difficult problem for a doctoral candidate, but not so difficult that he could not be expected to solve it. Hilbert showed his originality by following a different path from the one generally believed to lead to a solution. It was a very nice piece of work, and Lindemann was satisfied.

A copy of the dissertation was dispatched to Minkowski, who after his father's recent death had gone to Wiesbaden with his mother.

"I studied your work with great interest," Minkowski wrote to Hilbert, "and rejoiced over all the processes which the poor invariants had to pass through before they managed to disappear. I would not have supposed that such a good mathematical theorem could have been obtained in Königsberg!"

On December 11, 1884, Hilbert passed the oral examination. The next and final ordeal, on February 7, 1885, was the public promotion exercise in the Aula, the great hall of the University. At this time he had to defend two theses of his own choice against two fellow mathematics students officially appointed to be his "opponents." (One of these was Emil Wiechert, who later became a well-known seismologist.) The contest was generally no more than a mock battle, its main function being to establish that the candidate could perceive and frame important questions.

The two propositions which Hilbert chose to defend spanned the full breadth of mathematics. The first concerned the method of determining absolute electromagnetic resistance by experiment. The second pertained to philosophy and conjured up the great ghost of Immanuel Kant.

It had been the position of Kant, who had lectured on mathematics as well as philosophy when he taught at Königsberg, that man possesses certain notions which are not *a posteriori* (that is, obtained from experience) but *a priori*. As examples of *a priori* knowledge he had cited the fundamental concepts of logic, arithmetic and geometry — among these the axioms of Euclid.

The discovery of *non*-euclidean geometry in the first part of the nineteenth century had cast very serious doubt on this contention of Kant's; for it had shown that even with one of Euclid's axioms negated, it is still possible to derive a geometry as consistent as euclidean geometry. It thus became clear that the knowledge contained in Euclid's axioms was *a posteriori* — from experience — not *a priori*.

Could this also possibly be true of the fundamental concepts of arithmetic?

Gauss, who was apparently the first mathematician to be aware of the existence of non-euclidean geometries, wrote at one time:

"I am profoundly convinced that the theory of space occupies an entirely different position with regard to our knowledge *a priori* from that of [arithmetic]; that perfect conviction of the necessity and therefore the absolute truth which is characteristic of the latter is totally wanting in our knowledge of the former. We must confess, in all humility, that number is solely a product of our mind. Space, on the other hand, possesses also a reality outside our mind, the laws of which we cannot fully prescribe *a priori*."

Hilbert appears to have felt this way too; for in his second proposition he maintained:

That the objections to Kant's theory of the a priori *nature of arithmetical judgments are unfounded.*

There is no record of his defense of this proposition. Apparently, his arguments were convincing; for at the conclusion of the disputation he was awarded the degree of Doctor of Philosophy.

The Dean administered the oath:

"I ask you solemnly whether by the given oath you undertake to promise and confirm most conscientiously that you will defend in a manly way true science, extend and embellish it, not for gain's sake or for attaining a vain shine of glory, but in order that the light of God's truth shine bright and expand."

That night the new Doctor of Philosophy and his celebrating friends wired the news to Minkowski.

Hilbert was now set upon the first step of an academic career. If he were fortunate and qualified, he would arrive ultimately at the goal, the full professorship — a position of such eminence in the Germany of the day that professors were often buried with their title and the subject of their specialty on their gravestones. As a mere doctor of philosophy, however, he was not yet eligible even to lecture to students. First he had to turn out still another piece of original mathematical research for what was known as "Habilitation." If this was acceptable to the faculty, he would then be awarded the *venia legendi*, which carried with it the title of Privatdozent and the privilege of delivering lectures without pay under the sponsorship of the university. As such a docent he would have to live on fees paid by students who chose to come to hear his lectures. Since the courses which all the students took, such as calculus, were always taught by a member of the official faculty, he would be fortunate to draw a class of five or six. It was bound to be a meager time. Eventually, however, if he attracted attention by his work and abilities (or better yet, it was rumored, if he married a professor's daughter), he would become an Extraordinarius, or associate professor, and receive a salary from the university. The next step would be an offer of an Ordinariat, or full professorship. But this final step was by no means automatic, since the system provided an almost unlimited supply of docents from which to draw a very limited number of professors. Even in Berlin, there were only three mathematics professors; at most Prussian universities, two; at Königsberg, only one.

As a hedge against the vicissitudes of such a career, a young doctor could take the state examination and qualify himself for teaching at the gymnasium level. This was not a prize to be scorned. Although many, their eyes on the prestige-ladened professorship, didn't consider the alternative, one needed only to match the number of docents with the number of professorial chairs which might reasonably become vacant in the next decade to see its advantages. Hilbert now began to prepare himself for the state examination, which he passed in May 1885.

That same summer Minkowski returned to Königsberg, received the degree of doctor of philosophy, and then left almost immediately for his year in the army. (Hilbert was one of his official opponents for the promotion exercises.)

Hilbert had not been called up for military service. He considered a study trip, and Hurwitz urged Leipzig — and Felix Klein.

Although Klein was only 36 years old, he was already a legendary figure in mathematical circles. When he had been 23 (Hilbert's present age) he had been a full professor at Erlangen. His inaugural lecture there had made mathematical history as the Erlangen Program — a bold proposal to use the group concept to classify and unify the many diverse and seemingly unrelated geometries which had developed since the beginning of the century. Early in his career he had shown an unusual combination of creative and organizational abilties and a strong drive to break down barriers between pure and applied science. His mathematical interest was all-inclusive. Geometry, number theory, group theory, invariant theory, algebra — all had been combined for his master work, the development and completion of the great Riemannian ideas on geometric function theory. The crown of this work had been his theory of automorphic functions.

But the Klein whom Hilbert now met in Leipzig in 1885 was not this same dazzling prodigy. Two years before, in the midst of the work on automorphic functions, a young mathematician from a provincial French university had begun to publish papers which showed that he was concentrating his efforts in the same direction. Klein immediately recognized the caliber of his competitor and began a nervous correspondence with him. With almost super-human effort he drove himself to reach the goal before Henri Poincaré. The final result was essentially a draw. But Klein collapsed. When Hilbert came to him, he had only recently recovered from the long year of deep mental depression and physical lassitude which had followed his breakdown. He had passed the time writing a little book on the icosahedron which was to become a classic; but the future direction of his career was still not determined.

Hilbert attended Klein's lectures and took part in a seminar. He could not have avoided being impressed. Klein was a tall, handsome, dark haired and dark bearded man with shining eyes, whose mathematical lectures were universally admired and circulated even as far as America. As for Klein's reaction to the young doctor from Königsberg — he carefully preserved the Vortrag, or lecture, which Hilbert presented to the seminar and he later said: "When I heard his Vortrag, I knew immediately that he was the coming man in mathematics."

Since his breakdown, Klein had received two offers of positions — the first, to Johns Hopkins University in America, he had refused; the second, to Göttingen, he had just recently accepted. Apparently Hilbert had already absorbed a feeling for Göttingen, the university of Gauss, Dirichlet and Riemann. Inspired by Klein's appointment, he now scribbled one of his little verses on the inside cover of a small notebook purchased in Leipzig. The writing is so illegible that the German cannot be made out exactly, but the sense of the verse is essentially "Over this gloomy November day/ Lies a glow, all shimmery/ Which Göttingen casts over us/ Like a youthful memory."

In Leipzig, Hilbert soon became acquainted with several other young mathematicians. One of these was Georg Pick, whose knowledge of pollination and breeding as well as his admiration for Hurwitz's work recommended him to Hilbert. Another was Eduard Study, whose main interest, like Hilbert's own, was invariant theory. The two should have had a lot in common, but this was not to be the case. Study was a "strange person," Hilbert wrote to Hurwitz, "and almost completely at opposite poles from my nature and, as I think I can judge, from yours too. Dr. Study approves, or rather he knows, only one field of mathematics and that's the theory of invariants, very exclusively the symbolic theory of invariants. Everything else is unmethodical 'fooling around' .... He condemns for this reason all other mathematicians; even in his own field he considers himself to be the only authority, at times attacking all the other mathematicians of the symbolic theory of invariants in the most aggressive fashion. He is one who condemns everything he doesn't know whereas, for example, my nature is such I am most impressed by just that which I don't yet know." (Hurwitz wrote back, "This personality is more repugnant to me than I can tell you; still, in the interest of the young man, I hope that you see it a little too darkly.")

There were a considerable number of people at Leipzig who were interested in invariant theory; but Klein went out of his way to urge both Study and Hilbert to go south to Erlangen to pay a visit to his friend Paul Gordan, who was universally known at that time as "the king of the invariants." For some reason the expedition was not made. Perhaps because Hilbert did not care to make it with Study.

Hilbert was soon a member of the inner mathematical circle in Leipzig. At the beginning of December 1885, a paper of his on invariants was presented by Klein to the scientific society. On New Year's Eve he was invited to a "small but very select" party at Klein's — "Professor Klein, his honored spouse, Dr. Pick and myself." That same evening Minkowski, stranded and cold at Fort Friedrichsburg in the middle of the Pregel, was sending off New Year's greetings to his friend with the plaintive question, "Oh, where are the times when this poor soldier was wont to busy himself over the beloved mathematics?" But at the Kleins' the conversation was lively — "on all possible and impossible things." Klein tried to convince Hilbert that he should go to Paris for a semester of study before he returned to Königsberg. "He said," Hilbert wrote to Hurwitz, "that Paris is at this time a beehive of scientific activity, particularly among the young mathematicians, and a period of study there would be most stimulating and profitable for me, especially if I could manage to get on the good side of Poincaré."

Klein himself in his youth had made the trip to Paris in the company of his friend Sophus Lie, and both he and Lie had brought away their knowledge of group theory, which had played an important part in their careers. Now, according to Hurwitz, Klein always tried to send promising young German mathematicians to Paris.

Hurwitz himself seconded Klein's recommendation: "I fear the young talents of the French are more intensive than ours, so we must master all their results in order to go beyond them."

By the end of March 1886, Hilbert was on his way.

IV

Paris

On the train to Paris, Hilbert had the good luck to be in the same compartment with a student from the École Polytechnique who knew all the French mathematicians "at least from having looked at them." But in Paris, of necessity, he had to join forces with the disagreeable Study, who was already established there, also on Klein's recommendation.

Together, Hilbert and Study paid the mathematical visits which Klein had recommended. When they wrote to Klein, they read their letters aloud to one another so that they would not repeat information.

As soon as Hilbert was settled, he wrote to Klein. The letter shows how important he considered the professor. It was carefully drafted out with great attention to the proper, elegant wording, then copied over in a large, careful Roman script rather than the Gothic which he continued to use in his letters to Hurwitz.

"The fact that I haven't allowed myself at an earlier point in time to entrust the international post with a letter to you is due to the various impediments and the unforeseen cares which are always necessary on the first stay in a strange country. Fortunately, I have now adjusted to the climate and accommodated myself to the new environment well enough that I can start to spend my time in the way that I wish . . . ."

He tried hard to follow Klein's instructions about becoming friendly with Poincaré. The Frenchman was eight years older than he. Already he had published more than a hundred papers and would shortly be proposed for the Academy with the simple statement that his work was "above ordinary praise." In his first letter, Hilbert reported to Klein that Poincaré had not yet returned the visit which he and Study had paid on him; however, he added, he had heard him lecture at the Sorbonne on potential theory and the mechanics of fluids and had later been introduced to him.

"He lectures very clearly and to my way of thinking very understandably although, as a French student here remarks, a little too fast. He gives the impression of being very youthful and a bit nervous. Even after our introduction, he does not seem to be very friendly; but I am inclined to attribute this to his apparent shyness, which we have not yet been in a position to overcome because of our lack of linguistic ability."

By the time that Hilbert wrote to Klein again, Poincaré had returned the visit of the young Germans. "But about Poincaré I can only say the same — that he seems reserved because of shyness, but that with skillful treatment he would open up."

In replying to the letters from Paris, Klein (who was now established in Göttingen) played no favorites between his two young mathematicians. "It is thoroughly necessary that you and Hilbert have personal contact with Gordan and Noether," he wrote to Study. "Next time," he concluded his letter, "I shall write to Dr. Hilbert." But Hilbert seems to have placed a higher value on Klein's letters, for he preserved those written to Study as well as those written to himself.

The French mathematicians — Hilbert wrote to Klein — welcomed him and Study with great warmth. Jordan was most kind "and he is the one who presents the most devoted greetings to you." He gave a dinner for Hilbert and Study "to which only Halphen, Mannheim and Darboux were invited." Since, however, everyone spoke German in deference to the visitors, the conversation on mathematics was "very superficial."

Hilbert was not impressed with the mathematical lectures he heard. "French students do not have much that would interest us." Picard's lectures seemed "least elementary." Although Hilbert found Picard's pronunciation hard to understand, he attended his lectures regularly. "He gives the impression of being very energetic and positive in his conversation as well as in his teaching."

Some of the well-known mathematicians were a disappointment. "Concerning Bonnet — the trouble we went through to find him — a cruel fate sent us to three different houses first — was scarcely in proportion to the advantage we expected from such an old mathematician. He is obviously no longer responsive to mathematical things."

Hilbert and Study attended the meeting of the Société Mathématique in the hope of becoming acquainted with some young — or, at least, younger — mathematicians: "one reason is to be not always towered over by men who are so much greater than we are." Among those they met, Hilbert found Maurice d'Ocagne especially outstanding "because of his pleasant manners and approachability." In the course of the Société meeting he saw how he could sketch out a more direct proof than the one given by d'Ocagne of a theorem in his communication. "So I took courage, supported by Halphen, to point out this way of proving it." D'Ocagne asked Hilbert to write down his proof and offered to correct the French in case he wanted to publish it in the *Comptes Rendus*. "But I do not want to go into this thing because I think neither the theorem itself nor the proof is important enough to be put in the *Comptes Rendus*."

"As to the publications of Poincaré," Klein commented in this connection, "I have always the impression that there is the intention to publish something even if none or few new results are present. Do you approve of this? Have you happened to hear in Paris that people there have the same opinion?"

Among the French mathematicians, it was Hermite who seemed the most attractive to Hilbert:

"He not only showed us all his well-known politeness by returning our visit promptly, but also showed himself very kind ... by offering to spend the morning with me when he doesn't have a lecture."

The young Germans went back for a second visit. Hermite seemed very old to them — he was 64 — "but extraordinarily friendly and hospitable." He talked of his law of reciprocity in binary forms and encouraged them to extend it to ternary forms. In fact, much of the conversation was on invariants, since that was the subject in which Hermite knew his young visitors were most interested. He directed their attention to the most famous unsolved problem in the theory — what was known as "Gordan's Problem" after Klein's friend Gordan at Erlangen — and told them at length about his correspondence with Sylvester concerning the latter's efforts to solve it.

"The way Hermite talked about other, non-scientific topics proves that he has kept his youthful attitude into old age," Hilbert wrote admiringly to Klein.

While Hilbert was having these stimulating contacts in Paris, Minkowski was still soldiering in Königsberg. "I have stood guard duty at 20°, and been forgotten to be relieved on Christmas Eve . . . ." But he was hoping soon "to renew an old acquaintance with Frau Mathematika," and begged for news in the smallest detail of all that had happened to his friend "in enemy territory."

"And if one of the great gentlemen, Jordan or Hermite, still remembers me, please give him my best regards and make it clear that I less through nature than through circumstance am such a lazybones."

1886

In Paris, Hilbert was concentrating singlemindedly on mathematics. The letters to Klein record no sightseeing expeditions but mention only his desire to visit the observatory. In addition to the mathematical calls and lectures, he was attempting to edit and copy "in pretty writing" the paper for his habilitation. The work was progressing well.

At the end of April 1886, Study went home to Germany and reported to Klein in person on the activities in Paris.

"Not as much about mathematics as I expected," Klein commented disapprovingly to Hilbert. He then proceeded to fire off half a dozen questions and comments which had occurred to him while glancing through the most recent number of the *Comptes Rendus:* "Who is Sparre? The so-called Theorem of M. Sparre is already in a Munich dissertation (1878, I think). Who is Stieltjes? I have an interest in this man. I have come across an earlier paper by Humbert — it would be very interesting if you could check on the originality of his work (perhaps via Halphen?) and find out for me a little more about his personality. It is strange that the geometry in the style of Veronese-Segré happens to be coming back in fashion again..." The tone was more intimate than when Klein had been writing to the two young men together. "Hold it always before your eyes," he admonished Hilbert, "that the opportunity you have now will never come again."

This letter of Klein's found poor Hilbert spending a miserable month. The doctor diagnosed an illness of acclimatization "while I think it is a terrible poisoning of the stomach from H_2SO_4, which one has to drink here in a thin and pallid form under the name of wine." There were no more calls and the copying of the habilitation paper had to be postponed. He managed only to drag himself to lectures and meetings. "Everything stops when the inadequacy of the human organism shows itself...."

He may also have been just a little homesick.

By the end of June, on his way back to Königsberg, he was happy and full of enthusiasm. He stopped in Göttingen and reported to Klein on the Paris experiences. It was his first visit to the University, and he found himself charmed by the little town and the pretty, hilly countryside, so different from the bustling city of Königsberg and the flat meadows beyond it. He also stopped in Berlin, where he "paid a visit to everything that has anything to do with mathematics." This included even the formidable Leopold Kronecker.

Kronecker was a tiny man, scarcely five feet tall, who had so successfully managed his family's business and agricultural affairs that he had been able to retire at the age of 30 and devote the rest of his life to his hobby, which was mathematics. As a member of the Berlin Academy, he had regularly taken advantage of his prerogative to deliver lectures at the University. He was now 63 and only recently, since the retirement of Kummer, had he become an official professor.

Kronecker had made very important contributions, especially to the higher algebra; but he once remarked that he had spent more time thinking about philosophy than about mathematics. He was now disturbing his fellow mathematicians, particularly in Germany, by his loudly voiced doubts about the soundness of the foundations of much of the contemporary mathematics. His principal concern was the concept of the arithmetic continuum, which lies at the foundation of analysis. The continuum is the totality of real numbers — positive and negative — integers, fractions or rationals, and irrationals — which provides mathematicians with a unique number for every point on a line. Although the real numbers had been used for a long time in mathematics, it was only during the current century that their nature had been clarified in a precise and rigorous manner in the work of Cauchy and Bolzano and, more recently, of Cantor and Dedekind.

The new formulation did not satisfy Kronecker. It was his contention that nothing could be said to have mathematical existence unless it could actually be constructed with a finite number of positive integers. In his view, therefore, common fractions exist, since they can be represented as a ratio of two positive integers, but irrational numbers like π do not exist — since they can be represented only by an infinite series of fractions. Once, discussing with Lindemann the proof that π is transcendental, Kronecker objected: "Of what use is your beautiful investigation regarding π? Why study such problems when irrational numbers do not exist?" He had not yet made his remark that "God made the natural numbers, all else is the work of man," but already he was talking confidently of a new program which would "arithmetize" mathematics and eliminate from it all "nonconstructive" concepts. "And if I can't do this," he said, "it will be done by those who come after me!"

Although a man of many admirable qualities, Kronecker had been virulent and very personal in his attacks on the men whose mathematics he disapproved. ("In fact," recalled Minkowski in a letter to Hilbert, "I did not hear much good about Kronecker even when I was in Berlin.") The distinguished old Weierstrass had been reduced almost to tears by Kronecker's remarks about "the incorrectness of *all* those conclusions with which *so-called* analysis works at present." The high-strung, sensitive

Cantor, as a result of Kronecker's attacks on the theory of sets, had broken down completely and had had to seek asylum in a mental institution.

Hilbert had been warned not to expect a welcome from Kronecker, but surprisingly he was received — he wrote to Klein — "in a very friendly way."

Back home in Königsberg, he settled down to the serious business of habilitation. The work he had prepared was a much more ambitious paper than the doctoral dissertation had been, but still on the subject of invariant theory. To a later mathematician, who studied "every line" of Hilbert's work during his own student days, the habilitation paper was to seem a curiously false start: "He begins with the claim that it is a most important point of view, then it just goes out like a burnt match. Nothing came of it . . . . I was always surprised that for several years Hilbert went around in a direction that didn't lead anywhere, perhaps because of the too formal point of view which he took; and this may have been partly due to his contact with Study."

In addition to his paper, the candidate for habilitation also had to deliver a lecture on a topic selected by the faculty from a choice which he offered. Hilbert proposed "The Most General Periodic Functions" and "The Concept of the Group." The faculty selected the first topic, which was also the one which he himself preferred. The lecture was presented to the satisfaction of everyone concerned; the colloquium examination, passed successfully. On July 8, 1886, Hilbert was able to write to Klein: "The title with which you undeservedly addressed me in your last letter is now in actuality mine."

Earlier, there had been some discussion between Hilbert and Klein over the advisability of Hilbert's habilitating at Königsberg. The East Prussian capital was very much on the outskirts of mathematical activity. Few mathematics students were willing to come that far, so few, in fact, that Lindemann had to refuse Minkowski's request to habilitate at Königsberg when he got out of the army.

"But, after all, I am content and full of joy to have decided myself for Königsberg," Hilbert wrote to Klein. "The constant association with Professor Lindemann and, above all, with Hurwitz is not less interesting than it is advantageous to myself and stimulating. The bad part about Königsberg being so far away from things I hope I will be able to overcome by making some trips again next year, and perhaps then I will get to meet Herr Gordan . . . ."

Almost half of the great creative years between twenty and thirty were gone.

V

Gordan's Problem

Hilbert was resolved that as a docent he would educate himself as well as his students through his choice of subjects and that he would not repeat lectures, as many docents did. At the same time, on the daily walk to the apple tree, he and Hurwitz set for themselves the goal of "a systematic exploration" of mathematics.

The first semester he prepared lectures on invariant theory, determinants and hydrodynamics, the last at the suggestion of Minkowski, who was habilitating at Bonn and showing an interest in mathematical physics. There were not many who took advantage of this earliest opportunity to hear David Hilbert. Only in the lectures on invariant theory was he able to draw the number of students required by the University for the holding of a class. "Eleven docents depending on about the same number of students," he complained disgustedly to Minkowski. In honor of his new status he had a formal picture taken. It showed a young man with glasses, a somewhat straggly moustache and already thinning hair, who looked as if he might be expected to go after what he wanted.

In Bonn, Minkowski was having his troubles. He did not find the other docents congenial, and the mathematics professor had been taken ill. "I feel his absence especially. He was the only one here to whom I could put a mathematical question, or with whom I could speak at all on a mathematical subject." Whenever he had the opportunity, he returned to Königsberg and joined Hilbert and Hurwitz on their daily walks.

During these years the friendship between Hilbert and Minkowski deepened. Minkowski was a frequent vacation guest at Rauschen. Receiving the photograph of Hilbert after one of the Rauschen visits, Minkowski wrote, "If I had not seen you in it so stately and dignified, I would otherwise have had to think of the outlandish impression which you made on me in your Rauschen outfit and hairstyle at our brief meeting this summer." He added, musingly: "That we, although so close, could not at all open up to one another was for me more than a little surprising."

In their correspondence they continued to address each other by the formal pronoun "Sie"; but Hilbert, sending Minkowski a reprint of his first published work — the paper which Klein had presented to the Leipzig Academy the previous year — inscribed it: "To his friend and colleague in the closest sense ... from the author."

That first year as a docent, Hilbert made none of the trips which he had so optimistically planned in order to compensate himself for the isolation of Königsberg. Later he was to recall the years in the "security" of his native city as a time of "slow ripening." The second semester he gave the lectures on determinants and hydrodynamics which he had originally hoped to give the first semester. He began to plan lectures on spherical harmonics and numerical equations. In spite of the variety of his lectures, his own published work continued to be entirely in the field of algebraic invariants; but he also interested himself in questions in other fields.

Finally, at the beginning of 1888, he felt that he was at last ready to take the trip which he had so long promised himself. He drew up an itinerary which would allow him to call on 21 prominent mathematicians, and in March he set out. In his letters to Minkowski he jokingly referred to himself as "an expert invariant-theory man." Now he went first to Erlangen, where the "king of the invariants" held his court.

Paul Gordan was an impressive personality among the mathematicians of the day. Twenty-five years older than Hilbert, he had come to science rather late. His merchant father, while recognizing the son's unusual computational ability, had refused for a long time to concede his mathematical ability. A one-sided, impulsive man, Gordan was to leave a curiously negative mark upon the history of mathematics; but he had a sharp wit, a deep capacity for friendship, and a kinship with youth. Walks were a necessity of life to him. When he walked by himself, he did long computations in his head, muttering aloud. In company he talked all the time. He liked to "turn in" frequently. Then, sitting in some cafe in front of a foaming stein of the famous Erlangen beer, surrounded by young people, a cigar always in his hand, he talked on, loudly, with violent gestures, completely oblivious of his surroundings. Almost all of the time he talked about the theory of algebraic invariants.

It had been Gordan's good fortune to enter this theory just as it moved onto a new level. The first years of development had been devoted to determining the laws which govern the structure of invariants; the next concern had been the orderly production and enumeration of the invariants, and this was Gordan's meat. Sometimes a piece of his work would contain nothing but formulas for 20 pages. "Formulas were the indispensable supports for the formation of his thoughts, his conclusions and his mode of expression," a friend later wrote of him. Gordan's strength, however, in the invention and execution of the formal algebraic processes was considerable. At the beginning of his career, he had made the first break-through in a famous invariant problem. For this he had been awarded his title as king of the invariants. The general problem, which was still unsolved and now the most famous problem in the theory, was called in his honor "Gordan's Problem." This was the problem which Hermite had discussed with Hilbert and Study in Paris.

"Gordan's Problem" was far removed from the "solving for x" with which algebra had begun so many centuries before. It was a sophisticated "pure mathematical" question posed, not by the physical world, but by mathematics itself. The internal structure of all invariant forms was by this time known. Although there would be certain ambiguities and repetitions, different invariant forms of specified order and degree could be written down and counted, at least in principle. The next question was of a quite different nature, for it concerned the totality of invariants. Was there a *basis*, a finite system of invariants in terms of which all other invariants, although infinite in number, could be expressed rationally and integrally?

Gordan's great achievement, exactly 20 years before the meeting with Hilbert, had been to prove the existence of a finite basis for the binary forms, the simplest of all algebraic forms. Characteristically, his proof had been a computational one, based on the nature of certain elementary operations which generate invariants. Today it is dismissed as "crude computation"; but that it was, in its day, a high point in the history of invariant theory is apparent from the fact that in 20 years of effort by English, German, French and Italian mathematicians, no one had been able to extend Gordan's proof beyond binary forms, although in certain specific cases the theorem was known to be true. The title won in 1868 remained unchallenged. Just before Hilbert's arrival in Erlangen, Gordan had published the second part of his "Lectures on Invariant Theory," the plan of this work being primarily "to expound and exemplify worthily" (as a writer of the day explained) the theorem which he had proved at that time.

Hilbert had been familiar with Gordan's Problem for some time; but now, listening to Gordan himself, he seems to have experienced a phenomenon which he had not experienced before. The problem captured his imagination with a completeness that was almost supernatural.

Here was a problem which had every one of the characteristics of a great fruitful mathematical problem as he himself was later to list them:

Clear and easy to comprehend ("for what is clear and easily comprehended attracts, the complicated repels").

Difficult ("in order to entice us") *yet not completely inaccessible* ("lest it mock our efforts").

Significant ("a guidepost on the tortuous paths to hidden truths").

The problem would not let him go. He left Gordan, but Gordan's Problem accompanied him on the train up to Göttingen, where he went to visit Klein and H. A. Schwarz. Before he left Göttingen, he had produced a shorter, more simple, more direct version of Gordan's famous proof of the theorem for binary forms. It was, according to an American mathematician of the period, "an agreeable surprise to learn that the elaborate proofs of Gordan's theorem formerly current could be replaced by one occupying not more than four quarto pages."

From Göttingen, Hilbert went on to Berlin and visited Lazarus Fuchs, who was now a professor at the university there; also Helmholtz; and Weierstrass, who had recently retired. He then paid another call on Kronecker. He had a great deal of admiration for Kronecker's mathematical work, but still he found the older man's authoritarian attitude toward the nature of mathematical existence extremely distasteful. Now he discussed with Kronecker some plans for future investigations in invariant theory. Kronecker does not seem to have been much impressed. He cited a work of his own and said, Hilbert noted, "that my investigation on the subject is contained therein." They had a long talk, however, about Kronecker's ideas on what constitutes mathematical existence and his objections to Weierstrass's use of irrational numbers. "Equal is only $2 = 2$.... Only the discreet and singular have significance," Hilbert wrote in the little booklet in which he kept notes on the conversations with the mathematicians he visited. The importance the conversation with Kronecker had in Hilbert's mind at this time is indicated by the fact that he devoted four pages of his notebook to it — the other mathematicians visited, including Gordan, never received more than a page.

He left Kronecker, still thinking about Gordan's Problem.

Back home in Königsberg, the problem was with him in the midst of pleasure and work, even at dances, which he loved to attend. In August he went up to Rauschen, as was still his custom; and from Rauschen, on September 6, 1888, he sent a short note to the *Nachrichten* of the Göttingen Scientific Society. In this note he showed in a totally unexpected and original way how Gordan's Theorem could be established, by a uniform method, for forms in any desired number of variables.

No one was prepared for the announcement of the solution of the famous old problem, and the first reaction was almost sheer disbelief.

Since Gordan's own solution of the simplest case, the solution of the general problem had been sought in essentially the same manner, by means of the same kind of elaborate algorithmic apparatus which had been used so successfully by Gordan. With many variables and a complicated transformation group, this approach became fantastically difficult. It was not unusual for a single formula to run from page to page in the *Annalen*. "Comparable only to the formulas which describe the motion of the moon!" a later mathematician complained. In this atmosphere of absolute formalism it had occurred to Hilbert that the only way to achieve the desired proof would be to approach it from a path entirely different from the formalistic one which all investigators to date had taken and found impenetrable. He had set aside the whole elaborate apparatus and rephrased the question essentially as follows:

"If an infinite system of forms be given, containing a finite number of variables, under what conditions does a finite set of forms exist, in terms of which all the others are expressible as linear combinations with rational integral functions of the same variables for coefficients?"

The answer he came to was that such a set of forms *always* exists.

The foundation on which this sensational proof of the existence of a finite basis of the invariant system rested was a lemma, or auxiliary theorem, about the existence of a finite basis of a module, a mathematical idea he had obtained from the study of Kronecker's work. The lemma was so simple that it seemed almost trivial. Yet the proof of Gordan's general theorem followed directly from it. The work was the first example of the characteristic quality of Hilbert's mind — what one of his pupils was to describe as "a natural naiveté of thought, not coming from authority or past experience."

When the proof of Gordan's Theorem appeared in print in December, Hilbert promptly fired off a copy to Arthur Cayley, who half a century before had laid the foundation of the theory. ("The theory of algebraic invariants," a later mathematician once wrote, "came into existence somewhat like Minerva: a grown-up virgin, mailed in the shining armor of algebra, she sprang forth from Cayley's jovian head. Her Athens, over which she ruled and which she served as a tutelary and beneficent goddess, was projective geometry. From the beginning she was dedicated to the proposition that all projective coordinate systems are created equal ....")

"Dear Sir," Cayley replied politely from Cambridge on January 15, 1889, "I have to thank you very much for the copy of your note.... It [seems] to me that the idea is a most important valuable one, and that it ought to lead to a demonstration of the theorem as to invariants, but I am unable to satisfy myself as yet that you have obtained such a demonstration."

By January 30, however, having received two explanatory letters from Hilbert in the intervening time, Cayley was congratulating the young German: "My difficulty was an *a priori* one, I thought that the like process should be applicable to semi-invariants, which it seems it is not; and now I quite see.... I think you have found the solution of a great problem."

Hilbert had solved Gordan's Problem very much as Alexander had untied the Gordian Knot.

At Gordium [Plutarch tells us] he saw the famous chariot fastened with cords made of the rind of the cornel-tree, which whosoever should untie, the inhabitants had a tradition, that for him was reserved the empire of the world. Most authors tell the story that Alexander, finding himself unable to untie the knot, the ends of which were secretly twisted round and folded up within it, cut it asunder with his sword. But Aristobulus tells us it was easy for him to undo it, by only pulling the pin out of the pole, to which the yoke was tied, and afterwards drawing of the yoke itself from below.

To prove the finiteness of the basis of the invariant system, one did not actually have to construct it, as Gordan and all the others had been trying to do. One did not even have to show how it could be constructed. All one had to do was to prove that a finite basis, of logical necessity, *must exist*, because any other conclusion would result in a contradiction — and this was what Hilbert had done.

The reaction of some mathematicians was similar to what must have been the reaction of the Phrygians to Alexander's "untying" of the knot. They were not at all sure that he had untied it. Hilbert had not produced the basis itself, nor had he given a method of producing it. His proof of Gordan's Theorem could not be utilized to produce in actuality a finite basis of the invariant system of even a single algebraic form.

Lindemann found his young colleague's methods "unheimlich" — *uncomfortable, sinister, weird*. Only Klein seemed to recognize the power of the work — "wholly simple and, therefore, logically compelling" — and it was at this time that he decided he must get Hilbert to Göttingen at the first

Hilbert

opportunity. Gordan himself announced in a loud voice that has echoed in mathematics long after his own mathematical work has fallen silent:

"Das ist nicht Mathematik. Das ist Theologie."

Hilbert had now publicly taken a position in the current controversy provoked by Kronecker over the nature of mathematical existence. Kronecker insisted that there could be no existence without construction. For him, as for Gordan, Hilbert's proof of the finiteness of the basis of the invariant system was simply not mathematics. Hilbert, on the other hand, throughout his life was to insist that if one can prove that the attributes assigned to a concept will never lead to a contradiction, the mathematical existence of the concept is thereby established.

In spite of the philosophical difference, Hilbert was at this time greatly under the influence of the mathematical ideas of Kronecker — in fact, the fundamental significance of his work in invariants was later to be seen as the application of arithmetical methods to algebraic problems. He sent a copy of every paper he published to Kronecker. Nevertheless, Kronecker remarked petulantly to Minkowski that he was going to stop sending papers to Hilbert if Hilbert did not send papers to him. Hilbert promptly composed a letter which managed to be formal and respectful but firm:

"I remember exactly, and my list of mailed papers also shows it clearly, that I have taken the liberty of sending you a copy of each paper without exception immediately after its publication; and you have had the kindness to send your thanks on postcards for some of the last mailings. On the other hand, most honorable professor, it has never happened that a reprint of one of your papers has arrived as a gift from you to me. When I had the honor of calling on you about a year ago, however, you mentioned that you would choose something from your papers and send it to me. Under the circumstances I believe that there must be some misunderstanding, and I write these lines to remove it as fast and as surely as possible."

Then, with many crossings-out, he struggled to express the idea that what he had written should not be construed as expressing any other meaning than the stated one: not reproaches, but just explanations. He finally gave up, and simply signed himself, "Most respectfully, David Hilbert."

During the next two years, Hilbert, still a docent, sent two more notes to the *Nachrichten* and then in 1890 brought all his papers on algebraic forms together into a unified whole for the *Annalen*. By this time the revolutionary effect of Hilbert's work was being generally recognized and accepted. Gordan, offering another proof of one of Hilbert's theorems, was deferential to the young man — Herr Hilbert's proof was "completely correct," he wrote, and his own proof would not even have been possible "if Herr Hilbert had not utilized in invariant theory concepts which had been developed by Dedekind, Kronecker and Weber in another part of mathematics."

While Hilbert was thus involved in the purest of pure mathematics, Minkowski was moving increasingly away from it. Heinrich Hertz, two years after his discovery of the electromagnetic waves predicted by Maxwell, and still only 31 years old, had recently become professor of physics at Bonn. Minkowski, complaining of "a complete lack of half-way normal mathematicians" among his colleagues, found himself attracted more and more by Hertz and by physics. At Christmas he wrote that, contrary to his custom, he would not be spending the vacation at Königsberg:

"I do not know if I need console you though, since this time you would have found me thoroughly infected with physics. Perhaps I even would have had to pass through a 10-day quarantine period before you and Hurwitz would have admitted me again, mathematically pure and unapplied, to your joint walks."

At another time he wrote:

"The reason that I am now almost completely swimming in physical waters is because here at the moment as a pure mathematician I am the only feeling heart among wraiths. So for now," he explained, "in order to have points in common with other mortals, I have surrendered myself to magic — that is to say, physics. I have my laboratory periods at the Physics Institute; at home I study Thomson, Helmholtz and consorts. And from the end of next week on, I will even work several days a week in a blue smock in an institute for the production of physical instruments, a technician, therefore, and as practical as you can imagine!"

But the diverging of scientific interests did not affect the friendship; and, in fact, it was at this time that the two young men made the significant transition in their correspondence from the formal pronoun "Sie" to the intimate "du."

The Privatdozent years seemed to stretch out interminably. The letters were much concerned with the possibility of promotion. In 1891 Minkowski wrote that he had been told that he might be proposed for a position in Darmstadt. "But this ray of hope could easily shine so long that it shines upon mostly grey hair." That same year — apparently with special permission from the University — Hilbert was delivering his lectures on analytic functions to only one student — an American from Baltimore — a man somewhat older than the young lecturer but, in his opinion, "very sharp and extraordinarily interested." This was Fabian Franklin, an important man in invariant theory and the successor of Sylvester at Johns Hopkins.

Because there were few mathematics students at Königsberg, Hilbert attended the meetings of the natural scientists as well as those of the mathematicians. But Königsberg was surprisingly full of congenial young people. Wiechert was a docent too; and he had recently been joined by a student named Arnold Sommerfeld, with whom he was devising a harmonic analyzer. Both Wiechert and Sommerfeld were to become masters of electrodynamic theory, but when "Little Sommerfeld" heard Hilbert lecture on ideal theory, he became convinced that his interest lay entirely with the most pure and abstract mathematics. "Already," he later commented, "it was clear that a spirit of a special sort was at work."

There was lots of happy social life. Hilbert was a gay young man with a reputation as a "snappy dancer" and a "charmeur," according to a relative. He flirted, outrageously, with a great number of girls. His favorite partner for all activities, however, was Käthe Jerosch, the daughter of a Königsberg merchant, an outspoken young lady with an independence of mind that almost matched his own.

Even after the work of 1890, Gordan's Problem still would not let Hilbert go. As a mathematician he preferred an actual construction to a proof of existence. "There is," as one mathematician has said, "an essential difference between proving the existence of an object of a certain type by constructing a tangible example of such an object, and showing that if none existed one could deduce contrary results. In the first case one has a tangible object, while in the second one has only the contradiction." He would very much have liked to produce for old Kronecker, Gordan and the rest a constructive proof of the finiteness of the basis of the invariant system. At the moment there was simply no method at hand.

In the course of the next two years, however, the nature of his work began to change. It became infused with the ideas of algebraic number fields. Again, Kronecker's ideas were important. And it was here that Hilbert found at last the powerful new tools he had been seeking. In a key work, in 1892, he took up the question of exactly what was needed to produce in actuality a full system of invariants in terms of which all the other invariants could be represented. Using as a foundation the theorem which he had earlier proved, he was able to produce what was in essence a finite means of executing the long sought construction.

Although Hilbert was not the first to make use of indirect, non-constructive proofs, he was the first to recognize their deep significance and value and to utilize them in dramatic and extremely beautiful ways. Kronecker had recently died; but to those who like Kronecker still declared that existence statements are meaningless unless they actually specify the object asserted to exist, Hilbert was always to reply:

"The value of pure existence proofs consists precisely in that the individual construction is eliminated by them, and that many different constructions are subsumed under one fundamental idea so that only what is essential to the proof stands out clearly; brevity and economy of thought are the *raison d'être* of existence proofs.... To prohibit existence statements... is tantamount to relinquishing the science of mathematics altogether."

Now, through a proof of existence, Hilbert had been able to obtain a construction. The impetus which his achievement gave to the use of existential methods can hardly be overestimated.

Minkowski was utterly delighted:

"For a long while it has been clear to me that it could be only a question of time until the old invariant question was settled by you — only the dot was lacking on the 'i'; but that it all turned out to be so surprisingly simple has made me very happy, and I congratulate you."

He was inspired to literary flight and an assortment of metaphors. The first existence proof might have got smoke in Gordan's eyes, but now Hilbert had found a smokeless gunpowder. The castle of the robber barons — Gordan and the rest — had been razed to the ground with the danger that it might never rise again. Hilbert would be doing a service to his fellow mathematicians if he would bring together the materials in this area on which one could rebuild. But he probably would not want to spend his time doing that. There were still too many other things that he was capable of doing!

Gordan himself conceded gracefully.

"I have convinced myself that theology also has its merits."

When Klein went to Chicago for what was billed as an "International Congress of Mathematicians" to celebrate the founding of the University of Chicago, he took with him a paper by Hilbert in which that young man matter-of-factly summarized the history of invariant theory and his own part in it:

"In the history of a mathematical theory the developmental stages are easily distinguished: the naive, the formal, and the critical. As for the theory of algebraic invariants, the first founders of it, Cayley and Sylvester, are together to be regarded as the representatives of the naive period: in the drawing up of the simplest invariant concepts and in the elegant applications to the solution of equations of the first degrees, they experienced the immediate joy of first discovery. The inventors and perfecters of the symbolic calculation, Clebsch and Gordan, are the champions of the second period. The critical period finds its expressions in the theorems I have listed above...."

The theorems he referred to were his own.

It was a rather brash statement for a young mathematician who was still not even an Extraordinarius, but it had considerable truth in it. Cayley and Sylvester were both alive, one at Cambridge and the other at Oxford. Clebsch was dead, but Gordan was one of the most prominent mathematicians of the day. Now suddenly, in 1892, as a result of Hilbert's work, invariant theory, as it had been treated since the time of Cayley, was finished. "From the whole theory," a later mathematician wrote, "the breath went out."

With the solution of Gordan's Problem, Hilbert had found himself and his method — an attack on a great individual problem, the solution of which would turn out to extend in significance far beyond the problem itself. Now something totally unexpected occurred. The problem which had originally aroused his interest had been solved. The solution released him.

At the conclusion of his latest paper on invariants he had written: "Thus I believe the most important goals of the theory of function fields generated by invariants have been obtained." In a letter to Minkowski, he announced with even more finality: "I shall definitely quit the field of invariants."

VI

Changes

During the next three years Hilbert rose in the academic ranks, did all the things that most young men do at this time of their lives, married, fathered a child, received an important assignment, and made a decision which changed the course of his life.

This sudden series of events was set into motion by the death of Kronecker and the game of "mathematical chairs" which ensued in the German universities. Suddenly it seemed that the meager docent years might be coming to an end. Minkowski calling in Berlin on Friedrich Althoff, who was in charge of all matters pertaining to the universities, heralded the news:

"A. says . . . the following are supposed to receive paid Extraordinariats: you, I, Eberhard, and Study. I have not neglected to represent you to A. as the coming man in mathematics. . . . As to Study, in conscience I could only praise his good intentions and his diligence. A. is very devoted to you and Eberhard."

At almost the same time Hurwitz, who had been an associate professor (Extraordinarius) at Königsberg for eight years, received an offer of a full professorship from the Swiss Federal Institute of Technology in Zürich. This meant an end to the daily mathematical walks, but opened up the prospect of Hilbert's being appointed to Hurwitz's place.

"Through this circumstance," Minkowski wrote affectionately, "your frightful pessimism will have been allayed so that one dares again to venture a friendly word to you. In some weeks now, hopefully, the Privatdozent-sickness will be definitely over. You see — at last comes a spring and a summer."

In June, Hurwitz married Ida Samuels, the daughter of the professor of medicine. Hilbert had recently become engaged to Käthe Jerosch, and after Hurwitz's wedding he was increasingly impatient with the slow pace of promotion. At last, in August, the faculty unanimously voted him to succeed to Hurwitz's place. He announced the setting of the date of his wedding at the same time he communicated the news of the promotion to Minkowski.

Minkowski replied happily with his congratulations: "You will now have finally been converted to the idea that those in the decisive positions are sincerely well disposed toward you. Your prospects for the future, therefore, are excellent."

The Hilbert and Jerosch families had long been friends. From the outset it was generally agreed that Hilbert had found the perfect mate for himself. "She was a full human being in her own right, strong and clear," one of Hilbert's earliest pupils wrote of Käthe, "and always stood on the same footing with her husband, kindly and forthright, always original."

A photograph, taken about this time, shows the young couple. He is 30; she is 28. Already they look rather like one another. They are almost the same height, mouths wide and firm, strong noses, a level clear-eyed look. Hilbert's head seems relatively small. He has grown a beard. Already his hair has receded until the high scholar's forehead stands out impressively. Neither pretty nor homely, Käthe has good features, but she seems more interested in things other than her own appearance. Her dark hair is parted in the middle, drawn back rather severely, and coiled on the top of her head toward the back.

On October 12, 1892, Hilbert and Käthe Jerosch were married.

("The pleasant frame of mind in which you find yourself cannot help but have repercussions in your scientific work," Minkowski wrote. "I expect another great discovery.")

At almost the same time that Hilbert succeeded Hurwitz in Königsberg, Minkowski received his promised associate professorship in Bonn. He had hoped to go somewhere else, but "it will be better for you to remain in Bonn," Althoff told him. By now Heinrich Hertz had been struck down with the illness which was soon to take his life at the age of 37; Minkowski's interest in physics had abated; and he had returned to his first love, the theory of numbers. But later he once said to Hilbert that if "Papa" Hertz had lived he might have become a physicist instead of a mathematician.

Minkowski's approach to number theory was geometrical, it being his aim to express algebraic conjectures about the rational numbers in terms of geometric figures, an approach which frequently made the proofs more obvious. He was deeply absorbed in a book on this new subject, and his letters to Hilbert were filled with his concern about the presentation of his material. All must be "klipp und klar" before it went to the publisher. Although he called Poincaré "the greatest mathematician in the world," he told Hilbert, "I could not bring myself to publish things in the form in which Poincaré publishes them."

The book frequently kept Minkowski from Königsberg at vacation time. Hilbert complained about a lack of mathematical conversation now that Hurwitz was gone. "I am in a much more unhappy situation than you," Minkowski reminded him. "Just as closed off as Königsberg is from the rest of the world, just so closed off is Bonn from all other mathematicians. One is here a pure mathematics Eskimo!"

By the beginning of the new year (1893) Minkowski was happier. The book was half finished, accompanied by praise from Hermite which Hilbert found very touching.

"You are so kind as to call my old research works a point of departure for your magnificent contribution," the old Frenchman wrote to Minkowski, "but you have left them so far behind that they cannot claim now any other merit than to have suggested to you the direction in which you have chosen to proceed."

Hilbert began the year with a new proof of the transcendence of e (first proved by Hermite) and of π (proved by Lindemann). His proof was a considerable improvement over these earlier ones, astonishingly simple and direct. Here was the great work which Minkowski had been anticipating since the previous fall. Receiving it, he sat down and wrote immediately.

"An hour ago I received your note on e and π ... and I cannot do other than to express to you right away my sincere heartfelt astonishment.... I can picture the exhilaration of Hermite upon reading your paper and, as I know the old gentleman, it won't surprise me if he should shortly inform you of his joy that he is still permitted to experience this."

Along with the professional and personal changes in his life, Hilbert was beginning to show a new mathematical interest. "I shall devote myself to number theory from now on," he had told Minkowski after the completion of the last work on invariants. Now he turned to this new subject.

Gauss, as is well known, placed the theory of numbers at the pinnacle of science. He described it as "an inexhaustible storehouse of interesting truths." Hilbert saw it as "a building of rare beauty and harmony." He was as charmed as Gauss had been by "the simplicity of its fundamental laws, the economy of its concepts, and the purity of its truth"; and both men were equally fascinated by the contrast between the obviousness of the many numerical relationships involved and the "monstrous" difficulty of demonstrating them. Yet, in spite of the similarity of their comments, they were talking about two different versions of number theory.

Gauss was praising the classical theory of numbers, which goes back to the Greeks and deals with the relationships which exist among the ordinary whole, or natural, numbers. Most important are those between the prime numbers, called the "building blocks" of the number system, and the other numbers which, unlike the primes, can be divided by some number other than themselves and 1. By Gauss's time the concept of number had been extended far beyond the natural numbers. But Gauss himself had become the first mathematician to extend the notions of number theory itself beyond the rational "field" in which every sum, difference, product and (unlike among the natural numbers) quotient of two numbers is another number in the field. He did this for those numbers of the form $a + b\sqrt{-1}$ where a and b are rational numbers. These numbers also form a field, an algebraic number field, as do the numbers of the form $a + b\sqrt{2}$, and so on; and they are among the fields which are the subject of what is called algebraic number theory. It was this development, the number theory creation of Gauss, which Hilbert praised.

The greatest obstacle to the extension of number theory to algebraic number fields had been the fact that in most algebraic number fields the fundamental theorem of arithmetic, which states that the representation of any number as the product of primes is unique, does not hold. This obstacle had been eventually overcome by Kummer with the invention of "ideal numbers." Since Kummer, two mathematicians with very different mathematical approaches had been at work in algebraic number fields. Even before Hurwitz had left for Zürich, he and Hilbert had been devoting their daily walks to discussions of the modern number theory works of these two. "One of us took the Kronecker demonstration for the complete factorization in prime ideals and the other took Dedekind's," Hilbert later recalled, "and we found them both abominable." Now he began his work in algebraic number fields in much the same way that he had opened his attack on Gordan's Problem. He went back and thought through the basic idea. His first paper in the new subject was another proof for the unique decomposition of the integers of a field into prime ideals.

Hilbert had scarcely settled down into his new position as an assistant professor with a salary and a wife when there was welcome news. Lindemann had received an offer from Munich and would be leaving Königsberg.

"I take it for granted — and with any sense of justice Lindemann cannot think otherwise — that you should be his successor," Minkowski wrote to Hilbert. "If he succeeds in putting it through, he will at least leave with honor the place which he has occupied for 10 years."

Hilbert of course agreed. The final decision in the matter was not Lindemann's, however, but Althoff's. The faculty nominated Hilbert and three other more established mathematicians for the vacant professorship and sent the list to Berlin.

Althoff was no bureaucrat, but an administrator who had been academically trained. His great goal was to build up mathematics in Germany. He was a good friend of Klein's — the two had served in the army together during the Franco-Prussian War — and he thought very highly of Klein's opinion. Now, from the faculty's impressive list of names, he selected that of the 31-year-old Hilbert. He then proceeded to consult him about the appointment of a successor to his post as Extraordinarius — something almost unheard of.

Here was an opportunity to bring Minkowski back to Königsberg. In spite of the difficult situation which existed at Bonn because of the long illness of the professor of mathematics, Hilbert embarked enthusiastically upon the unfamiliar course of academic diplomacy. He wrote to Minkowski of the possibility that they might soon be together again.

"I would consider it special luck to step into your place at Königsberg," Minkowski replied. "The association with my mathematical colleagues here is really deplorable. One complains of migraine; as for the other, his wife trots in every five minutes in order to give another, non-mathematical direction to the conversation. If I could exchange this association with yours, it would be the difference between day and night for my scientific development."

But the sick professor at Bonn wanted to hang on to Minkowski, to whom he had become accustomed. Althoff liked to keep his professors happy. The negotiations dragged on.

In the meantime, in the new household, things were going along well and according to form. On August 11, 1893, at the seaside resort of Cranz, a first child, a son, was born to the Hilberts and named Franz.

A few weeks after Franz's birth, Hilbert went south to Munich for the annual meeting of the German Mathematical Society, which had recently been organized by a group of mathematicians — Hilbert among them — for the purpose of providing more contact among the different branches of mathematics. At the meeting Hilbert presented two new proofs of the decomposition of the numbers of a field into prime ideals. Although he had only begun to publish in the area of algebraic number theory, his competence apparently impressed the other members. One of the Society's projects was the yearly publication of comprehensive surveys of different fields of mathematics (the first had been on the theory of invariants); and now it was voted that Hilbert and Minkowski, who was of course already well known as a number theorist, be asked to prepare such a report on the current state of affairs in the theory of numbers "in two years." The note of urgency in the assignment was occasioned by the fact that the revolutionary work of Kummer, Kronecker and Dedekind was so extremely complicated or so far in advance of its time that it was still incomprehensible to most mathematicians. That Hilbert and Minkowski could be expected to rectify this situation was a tribute, not only to their mathematical ability, but also to the simplicity and clarity of their mathematical presentation.

That fall the letters that went between Königsberg and Bonn were devoted almost equally to three topics: the organization of the report for the Mathematical Society, the progress of the negotiations to bring Minkowski to Königsberg again, and the fact that baby Franz could already "outshriek" all other babies in his father's opinion.

The situation in Bonn did not improve; by New Year's Day 1894, Minkowski wrote that he had given up almost all hope of obtaining the appointment in Königsberg. Then three days later, following an interview with Althoff, he sent a joyful letter to Hilbert.

"End good, all good.... Hearty thanks for all your kind efforts which have led to this happy result; and may we have a pleasant and profitable collaboration which will make the prime numbers and the reciprocity laws *wiggeln und waggeln*."

On his way up to Königsberg in March, Minkowski stopped in Göttingen. H. A. Schwarz had by this time moved on to Berlin — his place being taken by Heinrich Weber — and Klein had a free hand to put into practice his ideas. Minkowski seems to have been tremendously impressed by the stimulating situation which Klein had already created at the University. "Who knows when I shall have another opportunity to inspire the mathematical workshop which is now of the highest repute?"

With Minkowski's arrival in the spring of 1894, the daily walks to the apple tree and the number theory discussions were happily resumed. It was Hilbert's feeling that he could not have had a better collaborator on the *Zahlbericht*, as the number theory report was called. In spite of Minkowski's mild disposition, he was fundamentally critical, insisted on literary as well as intellectual clarity, "and even to the work of others applied a strict standard."

The *Zahlbericht* now began to take shape in Hilbert's mind. Such an assignment as the one made by the Mathematical Society might be expected to be an unwelcome chore to a young mathematician, but this was not to be the case with Hilbert. Already his own work showed that his particular interest was the extension of the reciprocity laws to algebraic number fields. Now he willingly set aside these plans, seeing in the assigned report an opportunity to lay the foundation needed for deeper investigations. Although he still had no fondness for learning from books, he read everything that had been published on number theory since the time of Gauss. The proofs of all known theorems would have to be weighed carefully. Then he would have to decide in favor of those "the principles of which are capable of generalization and most useful for further research." But before such a selection could be made, the "further research" itself would have to be carried out. The difficulties of thought and style which had barred the way to general appreciation and understanding of his predecessors' work would have to be eliminated. It had been decided that the report should be divided into two parts. Minkowski would treat rational number theory; Hilbert, algebraic number theory. During the year 1894 Hilbert laid the foundations of his share of the *Zahlbericht*.

But, again, the two friends were not to be together for long. Early in December a letter labeled "Very Confidential" arrived from Göttingen.

"Probably you do not yet know," Klein wrote to Hilbert, "that Weber is going to Strassburg. This very evening we will have a meeting of the faculty to choose a committee to set up a list; and as little as I can predict the results, I want to inform you that I shall make every effort to see that no one other than you is called here.

"You are the man whom I need as my scientific complement because of the direction of your work and the power of your mathematical thinking and the fact that you are still in the middle of your productive years. I am counting on it that you will give a new inner strength to the mathematical school here, which has grown continuously and, as it seems, will grow even more — and that perhaps you will even exercise a rejuvenating effect upon me . . . .

"I can't know whether I will prevail in the faculty. I know even less whether the offer will follow from Berlin as we propose it. But this one thing you must promise me, even today: that you will not decline the call if you receive it!"

There is no record that Hilbert ever considered declining. In fact, he wrote to Klein, "Without any doubt I would accept a call to Göttingen with great joy and without hesitation." But he may have had some doubts. Klein was the acknowledged leader of mathematics in Germany. He was a regal man, the word "kingly" being now used most frequently to describe him. Sometimes even "kingly" wasn't strong enough, and one former student referred to him as "the divine Felix." A man who knew him well and was proud of the fact that Klein once took his advice in a personal matter, later confessed that he felt even to the end a distance between himself and Klein "as between a mortal and a god."

As for Klein's feelings — already it was clear that Hilbert questioned any authority, personal or mathematical, and went his own way. Klein was not unaware of the reasons against his choice. When in the faculty meeting his colleagues accused him of wanting merely a comfortable younger man, he replied, "I have asked the most difficult person of all."

Hilbert worked very hard on his reply to Klein's letter, crossing out and rewriting extensively to get exactly the effect he wanted. When he was satisfied, he had Käthe copy his letter in her best handwriting. It was a custom he was to follow often throughout his career.

"Your letter has surprised me in the happiest way," he began. "It has opened up a possibility for the realization of which I might have hoped at best in the distant future and as the final goal of all my efforts...

"Decisive for me above all would be the scientific stimulation which would come from you and the greater sphere of influence and the glory of your university. Besides, it would be the fulfillment of mine and my wife's dearest wish to live in a smaller university town, particularly one which is so beautifully situated as Göttingen."

Upon receiving this letter from Hilbert, Klein proceeded to plan out a campaign.

"I have already told Hurwitz that we will not propose him this time so that we will be more successful in proposing you. We will call Minkowski in second place. I have discussed this with Althoff and he thinks that will make it easier then for Minkowski to get your place in Königsberg."

Within a week he was writing triumphantly to Hilbert:

"This has been just marvellous, much faster than I ever dared to hope it could be. Please accept my heartiest welcome!"

VII

Only Number Fields

The red-tiled roofs of Göttingen are ringed by gentle hills which are broken here and there by the rugged silhouette of an ancient watch tower. Much of the old wall still surrounds the inner town, and on Sunday afternoons the townspeople "walk the wall" — it is an hour's walk. Outside the wall lie the yellow-brick buildings of the Georg August Universität, founded by the Elector of Hannover who was also George II of England. Inside, handsome half-timbered houses line crooked, narrow streets. Two thoroughfares, Prinzenstrasse and Weender Strasse, intersect at a point which the mathematicians call the origin of the coordinates in Göttingen. The center of the town, however, is the Rathaus, or town hall. On the wall of its Ratskeller there is a motto which states unequivocally: *Away from Göttingen there is no life.*

The great scientific tradition of Göttingen goes back to Carl Friedrich Gauss, the son of a man who was at different times a gardener, a canal-tender and a brick-layer. Gauss enrolled at the University in the autumn of 1795 as the protégé of the Duke of Brunswick. During the next three years he had so many great mathematical ideas that he could often do no more than record them in his journal. Before he left the University, at the age of 21, he had virtually completed one of the masterpieces of number theory and of mathematics, the *Disquisitiones Arithmeticae*. Later he returned to Göttingen as director of the observatory with incidental duties of instruction. He spent the rest of his life there, leaving his mark on every part of pure and applied mathematics. But when he was an old man and had won a place with Archimedes and Newton in the pantheon of his science, he always spoke of the first years he had spent at Göttingen as "the fortunate years."

Hilbert arrived in Göttingen in March 1895, almost exactly one hundred years after Gauss. It was not immediately apparent to the students that another great mathematician had joined the tradition. Hilbert was too different from the bent, dignified Heinrich Weber whom he replaced and the tall, commanding Klein. "I still remember vividly," wrote Otto Blumenthal, then a student in his second semester, "the strange impression I received of the medium-sized, quick, unpretentiously dressed man with a reddish beard, who did not look at all like a professor."

Klein's reputation drew students to Göttingen from all over the world, but particularly from the United States. The *Bulletin* of the newly founded American Mathematical Society regularly listed the courses of lectures to be given in Göttingen, and at one time the Americans at the University were sufficient in number and wealth to have their own letterhead: *The American Colony of Göttingen*. "There are about a dozen . . . in our lectures," a young Englishwoman named Grace Chisholm (later Mrs. W. H. Young) wrote to her former classmates at Cambridge. "We are a motley crew: five are Americans, one a Swiss-French, one a Hungarian, and one an Italian. This leaves a very small residuum of German blood."

The center of mathematical life was the third floor of the Auditorienhaus. Here Klein had established a reading room, the Lesezimmer, which was entirely different from any other mathematical library in existence at that time. Books were on open shelves and the students could go directly to them. Klein had also established on the third floor what was to become almost his signature: a tremendous collection of mathematical models housed in a corridor where the students gathered before lectures. Although not in actuality a room, it was always referred to as the Room of the Mathematical Models.

Klein's lectures were deservedly recognized as classics. It was his custom often to arrive as much as an hour before the students in order to check the encyclopedic list of references which he had had his assistant prepare. At the same time he smoothed out any roughness of expression or thought which might still remain in his manuscript. Before he began his lecture, he had mapped out in his mind an arrangement of formulas, diagrams and citations. Nothing put on the blackboard during the lecture ever had to be erased. At the conclusion the board contained a perfect summary of the presentation, every square inch being appropriately filled and logically ordered.

It was Klein's theory that students should work out proofs for themselves. He gave them only a general sketch of the method. The result was that a student had to spend at least four hours outside class for every hour spent in class if he wished to master the material. Klein's forte was the comprehensive view. "He possessed the ability to see the unifying idea in far apart problems and knew the art of explaining this insight by amassing the necessary details," a student has said. In the selection of his lecture subjects, Klein pursued a characteristically noble plan: "to gain in the course of time a complete view of the whole field of modern mathematics."

In contrast, Hilbert delivered his lectures slowly and "without frills," according to Blumenthal, and with many repetitions "to make sure that everyone understood him." It was his custom to review the material which he had covered in the previous lecture, a gymnasium-like technique disdained by the other professors. Yet his lectures, so different from Klein's, were shortly to seem to many of the students more impressive because they were so full of "the most beautiful insights."

In a well-prepared lecture by Hilbert the sentences followed one another "simply, naturally, logically." But it was his custom to prepare a lecture in general, and often he was tripped up by details. Sometimes, without especially mentioning the fact, he would develop one of his own ideas spontaneously in front of the class. Then his lectures would be even farther from the perfection of Klein's and exhibit the rough edges, the false starts, the sometimes misdirected intensity of discovery itself.

In the eight and a half years of teaching at Königsberg, Hilbert had not repeated a single subject "with the one small exception" of a one-hour course on determinants. In Göttingen now he was easily able to choose his subjects to adjust to Klein's wishes. The first semester he lectured on determinants and elliptic functions and conducted a seminar with Klein every Wednesday morning on real functions.

Although Hilbert had accepted the professorship in Göttingen with alacrity, there were two aspects of the new situation that bothered him. Käthe was not happy. The society in Göttingen, while more scientifically stimulating for him, lacked the warmth to which she had been accustomed in Königsberg. Carefully observed distinctions of rank cut the professors off from the docents and advanced students. In spite of Klein's kindness, he maintained with the Hilberts, as he did with everyone else, a certain distance. Mrs. Klein (granddaughter of the philosopher Hegel) was a very quiet woman, not the kind who likes to gather people around her. The Klein house at 3 Wilhelm Weber Strasse, big, square and impressive with a bust of Jupiter on the stairs that led to Klein's study, looked already like the institute it was eventually to become. For Hilbert "comradeship" and "human solidarity" were essential to scientific production. Like Käthe, he found the atmosphere at Göttingen distinctly cool.

Hilbert was also concerned, in the beginning, that he might not prove worthy of the confidence which Klein had shown in him. He recognized that he had been taken on faith. Before he had left Königsberg, he had written to Klein, "My positive achievements — which I indeed know best myself — are still very modest." In the draft of a later letter he had returned to this same subject, adding hopefully, "As to my scientific program, I think that I will eventually succeed in shaping the theory of ideals into a general and usable tool (applicable also to analytic functions and differential equations) which will complement the great and promising concept of the group." Then he had carefully crossed out this sentence and noted in the margin: *I have not written this.*

Now, in Göttingen, Hilbert concentrated all his powers on his share of the number theory report for the German Mathematical Society, which he saw as the necessary foundation for his future hopes.

In Königsberg, Minkowski almost immediately received the appointment as his friend's successor. "The whole thing has taken place so quickly that I still have not come to complete consciousness of my astounding luck. In any case, I know I have you alone to thank for everything. I shall see I break out of my cocoon so that no one will hold it against you for proposing me." Minkowski was happy in his new position — professors now went out of their way to describe to him the virtues of their daughters — but since Hilbert's departure, he wrote, he had walked "not once" to the apple tree.

With encouragement from Hilbert, Minkowski now took advantage of the fact that he was a full professor to deliver a course of lectures on Cantor's theory of the infinite. It was at a time when, according to Hilbert, the work of Cantor was still actually "taboo" in German mathematical circles, partly because of the strangeness of his ideas and partly because of the earlier attacks by Kronecker. Although Minkowski admired Kronecker's mathematical work, he deplored as much as Hilbert the way in which the older man had tried to impose his restrictive personal prejudices upon mathematics as a whole.

"Later histories will call Cantor one of the deepest mathematicians of this time," Minkowski said. "It is most regrettable that opposition based not alone on technical grounds and coming from one of the most highly regarded mathematicians could cast a gloom over his joy in his scientific work."

As the year 1895 progressed, the letters between Göttingen and Königsberg became less frequent.

"We both try in silence to crack the difficult and not really very tasty nut of our common report," Minkowski wrote, taking up the correspondence again, "you perhaps with sharper teeth and more exertion of energy."

The idea of the joint report did not really appeal to Minkowski. "I started somewhat too late with my share," he wrote unhappily. "Now I find many little problems it would have been nice to dispose of." He was more interested in his book on the geometry of numbers. "The complete presentation of my investigations on continued fractions has reached almost a hundred printed pages but the all-satisfying conclusion is still missing: the vaguely conceived characteristic criterion for cubic irrational numbers .... But I haven't been able to work on this problem because I have really been working on our report."

Hilbert, on the other hand, was devoting himself wholeheartedly to the report. He was fascinated by the deep connections which had recently been revealed between the theory of numbers and other branches of mathematics. Number theory seemed to him to have taken over the leading role in algebra and function theory. The fact that this had not occurred earlier and more extensively was, in his opinion, due to the disconnected way in which number theory had developed and the fact that its treatment had always been chronological rather than conceptual. Now, he believed, a certain and continuous development could be effected by the systematic building up of the theory of algebraic number fields.

After the Wednesday morning seminars he walked with the students up to a popular restaurant on the Hainberg for lunch and more mathematics. On these excursions he talked freely to them "as equals," according to Blumenthal, but always the subject of conversation at this time was "only algebraic number fields."

By the beginning of 1896, Hilbert's share of the *Zahlbericht* was almost finished, but Minkowski's was not. In February Hilbert proposed that either Minkowski's share should be published with his as it stood, or else it should be published separately the following year.

"I accept your second plan," Minkowski wrote gratefully. "The decision ... is hard on me only insofar as I'll have the guilty feeling for a whole year that I didn't meet the expectations of the Society and, in some degree, your expectations. You, it is true, haven't made any remark of this kind, but .... The reproaches may lose some of their force if now the biggest part of my book is appearing and the rest is following soon. Finally, I can imagine that I am doing what I think is in the interest of the project. I beg you not to think I left you in the lurch."

Within a month after receiving this letter, Hilbert had completed his report on algebraic number fields. It was exactly a year since his arrival in Göttingen. The manuscript, which was to run to almost 400 pages in print, was carefully copied out by Käthe Hilbert in her clear round hand and sent to the printer. The proof-sheets were mailed to Minkowski in Königsberg as they arrived. Minkowski's letters during this period show the affectionate and yet sharp and unrelenting care with which he read them.

"One more remark seems to be necessary on page 204." "I have read till where the long calculations start. They still seem pretty tangled." "This thought is not so simple that it can be silently omitted."

Minkowski had recently received an offer of a position in Zürich. Such an offer, known as "a call," was customarily the subject of complicated ritual and negotiation, since it was the only means by which a man who had become a full professor could further improve his situation. Minkowski had no gift for such parrying. Althoff, he wrote to Hilbert, did not seem eager to keep him at Königsberg. Rather regretfully, he finally accepted the position in Zürich for the fall of 1896.

In Zürich, however, he was again in the company of Hurwitz ("just the same except for a few white hairs"), and the two friends read the remaining proof-sheets of Hilbert's report together. Corrections and suggestions kept coming to Göttingen.

Hilbert began to grown impatient.

Minkowski soothed him: "I understand that you want to be through with the report as soon as possible . . . but as long as there are so many remarks to be made, I can't promise you any great speed . . . ." "A certain care is advisable . . . ." "Comfort yourself with the thought that the report will be finished soon and will gain high approval."

The careful proofreading continued.

By this time Hilbert was beginning to feel more at home in Göttingen. He had found a congenial colleague in Walther Nernst, a professor of physics and chemistry who, like himself, was the son of a Prussian judge. But Hilbert also liked to be with younger people, and now he cheerfully ignored convention in choosing his friends. These included Sommerfeld, who had come to Göttingen to continue his studies and had become Klein's first assistant. He also selected the brightest, most interesting students in his seminar for longer walks. His "Wunderkinder," he called them.

Although even advanced students and docents stood in awe of Klein, they easily fell into a comradely relationship with Hilbert. His Königsberg accent with its distinctive rhythm and inflection seemed to them to give a unique flavor to everything he said. They delighted in mimicking his manner and opinions, were quick to pick up the "Aber nein!" — *But no!* —with which he announced his fundamental disagreement with an idea, whether in mathematics, economics, philosophy, human relations, or simply the management of the University. ("It was very characteristic the way he said it, but very difficult to catch in English, even in twenty words.")

In the seminar they found him surprisingly attentive to what they had to say. As a rule he corrected them mildly and praised good efforts. But if something seemed too obvious to him he cut it short with "Aber das ist doch ganz einfach!" — *But that is completely simple!* — and when a student made an inadequate presentation he would chastise him or her in a manner that soon became legendary. "Ja, Fräulein S-----, you have given us a very interesting report on a beautiful piece of work, but when I ask myself what have you really said, it is chalk, chalk, nothing but chalk!" And he could also be brutal. "You had better think twice before you uttered a lie or an empty phrase to him," a later student recalled. "His directness could be something to be afraid of."

After a year in Göttingen, the Hilberts decided to build a house on Wilhelm Weber Strasse, the broad linden-lined avenue favored by professors. ("Very likely," wrote Minkowski, "Fate will feel challenged now and try to seduce you from Göttingen with many spectacular offers.") The house was a forthright yellow-brick structure with none of the "new style" ornateness favored by its neighbors. It was large enough that the activities of 4-year-old Franz would not disturb his father as they had in the apartment. The yard in back was large too. They got a dog, the first of a long line of terriers, all to be named Peter. Hilbert, who worked best "under the free sky," hung an 18-foot blackboard from his neighbor's wall and built a covered walk-way so that he could be outdoors even in bad weather.

The house was almost finished when Hilbert wrote the introduction to the *Zahlbericht*. To a later student with a love of language not characteristic of most mathematicians, the introduction was to seem one of the most beautiful parts of German prose, "the style in the literary sense being the accurate image of the way of thinking." In it Hilbert emphasized the esteem in which number theory had always been held by the greatest mathematicians. Even Kronecker was quoted approvingly as "giving expression to the sentiment of his mathematical heart" when he made his famous pronouncement that God made the natural numbers . . . .

"I still find many things to criticize," Minkowski wrote patiently. "...Will you not in your foreword perhaps mention the fact that I read the last three sections in manuscript?"

Thus instructed, Hilbert wrote an acknowledgment of what he owed to his friend. Minkowski was still not satisfied.

"That you omitted the thanks to Mrs. Hilbert both Hurwitz and I find scandalous and this simply can't be allowed to remain so."

This last addition was made in the study of the new house at 29 Wilhelm Weber Strasse. The final date on the introduction to the *Zahlbericht* was April 10, 1897.

"I wish you luck that finally after the long years of work the time has arrived when your report will become the common property of all mathematicians," Minkowski wrote upon receiving his specially bound copy, "and I do not doubt in the near future you yourself will be counted among the great classicists of number theory.... Also I congratulate your wife on the good example which she has set for all mathematicians' wives, which now for all time will remain preserved in memory."

The report on algebraic number fields exceeded in every way the expectations of the members of the Mathematical Society. They had asked for a summary of the current state of affairs in the theory. They received a masterpiece, which simply and clearly fitted all the difficult developments of recent times into an elegantly integrated theory. A contemporary reviewer found the *Zahlbericht* an inspired work of art; a later writer called it a veritable jewel of mathematical literature.

The quality of Hilbert's creative contribution in the report is exemplified by that theorem which is still known today simply as "Satz 90." The development of the ideas contained in it were to lead to homological algebra, which plays an important role in algebraic geometry and topology. As another mathematician has remarked, "Hilbert was not only very thorough, but also very fertile for other mathematicians."

For Hilbert, the spring of 1897 was a memorable one — the new house completed, the *Zahlbericht* at last in print. Then came sad news. His only sister, Elise Frenzel, wife of an East Prussian judge, had died in childbirth.

According to a cousin, the relationship between brother and sister was reputed in the family to have been "cool." But for Minkowski, writing to Hilbert at the time, it seemed impossible to find comforting words:

"Whoever knew your sister must have admired her for her always cheerful and pleasant disposition and must have been carried along by her happy approach to life. I still remember... how gay she was in Munich, and how she was in Rauschen. It is really unbelievable that she should have left you so young. How close she must have been to your heart, since you have no other brother or sister and you grew up together for so many years! It seems sometimes that through a preoccupation with science, we acquire a firmer hold over the vicissitudes of life and meet them with greater calm, but in reality we have done no more than to find a way to escape from our sorrows."

Minkowski's next letter, however, contained happy personal news. He had become engaged to Auguste Adler, the daughter of the owner of a leather factory near Strassburg. "My choice is, I am convinced, a happy one and I certainly hope . . . it will be good for my scientific work." In a postscript he added a little information about his fiancée for the Hilberts. "She is 21 years old, she looks very *sympathisch*, not only in my judgment, but also in the judgment of all those who know her. She has grown up with six brothers and sisters, is very domestic, and possesses an unusual degree of intelligence."

Minkowski planned to be married in September, but first there was an important event. An International Congress of Mathematicians was going to take place in August in Zürich, which being Swiss was considered appropriately neutral soil. Klein was asked to head the German delegation. "Which will have the consequence," Minkowski noted, "that nobody will come from Berlin."

Although for some reason Hilbert did not attend this first congress, he read the papers which were presented and was most impressed by two of the featured addresses. One of these was a lecture on the modern history of the general theory of functions by Hurwitz. The other was an informal talk by Poincaré on the way in which pure analysis and mathematical physics serve each other.

Shortly after the Congress, Minkowski was married in Strassburg.

He did not write to Hilbert again until the end of November:

"After my long silence, you must think that my marriage has changed me completely. But I stay the same for my friends and for my science. Only I could not show any interest for some time in the usual manner."

With the *Zahlbericht* completed, Hilbert was now involved in investigations of his own which he had long wished to pursue. The focal point of his interest was the generalization of the Law of Reciprocity to algebraic number fields. In classical number theory, the Law of Quadratic Reciprocity, known to Euler, had been rediscovered by Gauss at the age of 18 and given its first complete proof. Gauss always regarded it as the "gem" of number theory and returned to it five more times during his life to prove it in a different way each time. It describes a beautiful relationship which exists between pairs of primes and the remainders of squares when divided by these.

For treating the Law of Reciprocity in the generality which he had in mind, Hilbert needed a broad foundation; and this he had achieved in the *Zahlbericht*. In its introduction he had noted that "the most richly equipped part of the theory of algebraic number fields appears to me the theory of abelian and relative abelian fields which has been opened up by Kummer for us through his work on the higher reciprocity law and by Kronecker through his investigation of the complex multiplication of elliptic functions. The deep insights into this theory which the works of these two mathematicians give us show at the same time that ... an abundance of the most precious treasures still lies concealed, beckoning as rich reward to the investigator who knows their value and lovingly practices the art to win them."

Hilbert now proceeded to go after these treasures. As a result of his work on the *Zahlbericht* he had a knowledge of the terrain that was both "intimate and comprehensive." He moved cautiously but with confidence.

"It is a great pleasure," a later mathematician noted, "to watch how, step by step, in a succession of papers ascending from the particular to the general, the adequate concepts and methods are evolved and the essential connections come to light."

By studying the classical Law of Quadratic Reciprocity of Gauss, Hilbert was able to restate it in a simple, elegant way which also applied to algebraic number fields. From this he was then able to guess with brilliant clarity what the reciprocity law must be for degrees higher than 2, although he did not prove his conjectures in all cases. The crown of his work was the paper published the year after the *Zahlbericht* and entitled "On the theory of relative abelian fields." In this paper, which was basically programmatic in character, he sketched out a vast theory of what were to become known as "class-fields," and developed the methods and concepts needed to carry out the necessary investigations. To later mathematicians it was to seem that he had "conceived by divination" — nowhere else in his work is the accuracy of his mathematical intuition so apparent. Unlike the work in invariants, which had marked the end of a development, the work in algebraic number fields was destined to be a beginning. But for other mathematicians.

Hilbert himself now turned abruptly away.

VIII

Tables, Chairs, and Beer Mugs

The announcement that Hilbert would lecture during the winter of 1898–99 on the elements of geometry astonished the students to whom he had talked "only number fields" since his arrival in Göttingen three years before. Yet the new interest was not entirely without antecedent.

In his docent days Hilbert had attended a lecture in Halle by Hermann Wiener on the foundations and structure of geometry. In the station in Berlin on his way back to Königsberg, under the influence of Wiener's abstract point of view in dealing with geometric entities, he had remarked thoughtfully to his companions: "One must be able to say at all times — instead of points, straight lines, and planes — tables, chairs, and beer mugs." In this homely statement lay the essence of the course of lectures which he now planned to present.

To understand Hilbert's approach to geometry, we must remember that in the beginning mathematics was a more or less orderless collection of statements which either seemed self-evident or were obtained in a clear, logical manner from other seemingly self-evident statements. This criterion of evidence was applied without reservation in extending mathematical knowledge. Then, in the third century B. C., a teacher named Euclid organized some of the knowledge of his day in a form that was commonly followed. First he defined the terms he would use — *points, lines, planes*, and so on. Then he reduced the application of the criterion of evidence to a dozen or so statements the truth of which seemed in general so clear that one could accept them as true without proof. Using only these definitions and axioms (as the statements were later called), he proceeded to derive almost five hundred geometric statements, or theorems. The truth of these was in many cases not at all self-evident, but it was guaranteed by the fact that all the theorems had been derived strictly according to the accepted laws of logic from the definitions and the axioms already accepted as true.

Although Euclid was not the most imaginative of the Greek geometers and the axiomatic method was not original with him, his treatment of geometry was greatly admired. Soon, however, mathematicians began to recognize that in spite of its beauty and perfection there were certain flaws in Euclid's work; particularly, that the axioms were not really sufficient for the derivation of all the theorems. Sometimes other, unstated assumptions crept in — especially assumptions based on visual recognition that in a particular construction certain lines were bound to intersect. It also seemed that one of Euclid's axioms — the Parallel Postulate — went so far beyond the immediate evidence of the senses that it could not really be accepted as true without proof. In its various forms the Parallel Postulate makes a statement essentially equivalent to the statement that through any point not on a given line in a plane, at most one line can be drawn which will not intersect the given line. Generally, however, this flaw and the others in Euclid were dismissed as things which could be easily removed, first by enlarging the original number of axioms to include the unstated assumptions and then by proving the particularly questionable axiom as a theorem, or by replacing it with another more intuitively evident axiom, or — finally — by demonstrating that its negation led to a contradiction. This last and most sophisticated method of dealing with the problem of the Parallel Postulate represents the first appearance in mathematics of the concept of *consistency*, or freedom from contradiction.

Gauss was apparently the first mathematician to whom it occurred, perhaps as early as 1800, that the negation of Euclid's parallel postulate might not lead to a contradiction and that geometries other than Euclid's might be possible. But this idea smacked so of metaphysical speculation that he never published his investigations on the subject and only communicated his thoughts to his closest friends under pledges of secrecy.

During the 1830's, however, two highly individualistic mathematicians tried independently but almost simultaneously to derive from a changed parallel axiom and the other, unchanged traditional axioms of euclidean geometry what theorems they could. Their new axiom stated in essence that through any point not on a given line, infinitely many lines can be drawn which will never meet the given line. Since this was contrary to what they thought they knew as true, the Russian Lobatchewsky and the Hungarian J. Bolyai expected that the application of the axiomatic method would lead to contradictory theorems. Instead, they found that although the theorems established from the new set of axioms were at odds with the results of everyday experience (the angles of a triangle, for instance, did not add up to two right angles as in Euclid's geometry), none of the expected contradictions appeared in the new geometry thus established. It was possible, they had discovered, to build a consistent geometry upon axioms which (unlike Euclid's) did not seem self-evidently true, or which even appeared false.

Surprisingly enough, this discovery of *non*-euclidean geometries did not arouse the "clamors of the Boeotians" which, according to Gauss (in a letter to Bessel on January 27, 1829), had deterred him from publishing his own investigations on the subject. In fact, there was not very much interest in the discovery among mathematicians. For the majority it seems to have been *too* abstract.

It was not until 1870 that the idea was generally accepted. At that time the 21-year-old Felix Klein discovered a "model" in the work of Cayley by means of which he was able to identify the primitive objects and relations of non-euclidean geometry with certain objects and relations of euclidean geometry. In this way he established that non-euclidean geometry is every bit as consistent as euclidean geometry itself; for a contradiction existing in the one will have of necessity to appear in the other.

Thus the impossibility of demonstrating the Parallel Postulate was at last shown to be "as absolutely certain as any mathematical truth whatsoever." But, again, the full impact of the discovery was not immediately and generally felt. The majority of mathematicians, although they now recognized the several non-euclidean geometries resulting from various changes in the Parallel Postulate, held back from recognizing the fact, which automatically followed, that Euclid's other axioms were equally arbitrary hypotheses for which other hypotheses could be substituted and that still other non-euclidean geometries were possible.

A few mathematicians did try to achieve treatments of geometry which would throw into relief the full implication of the discovery of the non-euclidean geometries, and would at the same time eliminate all the hidden assumptions which had marred the logical beauty of Euclid's work. Such a treatment had been first achieved by Moritz Pasch, who had avoided inadvertently depending on assumptions based on visual evidence by reducing geometry to a pure exercise in logical syntax. Giuseppe Peano had gone even farther. In essence he had translated Pasch's work into the notation of a symbolic logic which he himself had invented. Peano's version of geometry was completely abstract — a calculus of relations between variables.

It was difficult to see how Hilbert could hope to go beyond what had already been done in this area of mathematical thought. But now in his lectures he proceeded to reverse the trend toward absolutely abstract symbolization of geometry in order to reveal its essential nature. He returned to Euclid's points, straight lines and planes and to the old relations of incidence, order and congruence of segments and angles, the familiar figures. But his return did not signify a return to the old deception of euclidean geometry as a statement of truths about the physical universe. Instead — within the classical framework — he attempted to present the modern point of view with even greater clarity than either Pasch or Peano.

With the sure economy of the straight line on the plane, he followed to its logical conclusion the remark which he had made half a dozen years before in the Berlin station. He began by explaining to his audience that Euclid's definitions of point, straight line and plane were really mathematically insignificant. They would come into focus only by their connection with whatever axioms were chosen. In other words, whether they were *called* points, straight lines, planes or were *called* tables, chairs, beer mugs, they would *be* those objects for which the relationships expressed by the axioms were true. In a way this was rather like saying that the meaning of an unknown word becomes increasingly clear as it appears in various contexts. Each additional statement in which it is used eliminates certain of the meanings which would have been true, or meaningful, for the previous statements.

In his lectures Hilbert simply *chose* to use the traditional language of Euclid:

"Let us conceive three distinct systems of things," he said. "The things composing the first system we will call *points* and designate them by the letters $A, B, C, \ldots$."

The "things" of the other two systems he called *straight lines* and *planes*. These "things" could have among themselves certain mutual relations which, again, he chose to indicate by such familiar terms as *are situated, between, parallel, congruent, continuous*, and so on. But, as with the "things" of the three systems, the meaning of these expressions was not to be determined by one's ordinary experience of them. For example, the primitive terms could denote any objects whatsoever provided that to every pair of objects called *points* there would correspond one and only one of the objects called *straight lines*, and similarly for the other axioms.

The result of this kind of treatment is that the theorems hold true for any interpretation of the primitive notions and fundamental relationships for which the axioms are satisfied. (Many years later Hilbert was absolutely delighted to discover that from the application of a certain set of axioms the laws governing the inheritance of characteristics in the fruit fly can be derived: "So simple and precise and at the same time so miraculous that no daring fantasy could have imagined it!")

In his lectures Hilbert now proposed to set up on this foundation a simple and complete set of independent axioms by means of which it would be possible to prove all the long familiar theorems of Euclid's traditional geometry. His approach — the original combination of the abstract point of view and the concrete traditional language — was peculiarly effective. "It was as if over a landscape wherein but a few men with a superb sense of orientation had found their way in murky twilight, the sun had risen all at once," one of his later students wrote. By developing a set of axioms for euclidean geometry which did not depart too greatly from the spirit of Euclid's own axioms, and by employing a minimum of symbolism, Hilbert was able to present more clearly and more convincingly than either Pasch or Peano the new conception of the nature of the axiomatic method. His approach could be followed by the students in his class who knew only the original *Elements* of Euclid. For established mathematicians, whose first introduction to real mathematics had invariably been the *Elements*, it was particularly attractive, "as if one looked into a face thoroughly familiar and yet sublimely transfigured."

At the time of these lectures on geometry arrangements were being made in Göttingen for the dedication of a monument to Gauss and Wilhelm Weber, the two men — one in mathematics, the other in physics — from whom the University's twofold scientific tradition stemmed. To Klein the dedication ceremony seemed to offer an opportunity to emphasize once again the organic unity of mathematics and the physical sciences. Gauss's observatory had been no ivory tower. In addition to his mathematical discoveries, he had made almost equally important contributions to physics, astronomy, geodesy, electromagnetism, and mechanics. The broadness of his interest had been reinforced by a collaboration with Wilhelm Weber. The two men had invented an electromagnetic telegraph which transmitted over a distance of more than 9,000 feet; the monument was to show the two of them examining this invention. Carrying on and extending the tradition of mathematical abstraction combined with deep interest in physical problems was central to Klein's dream for Göttingen. So now he asked Emil Wiechert to edit his recent lectures on the foundations of electrodynamics for a celebratory volume, and asked Hilbert to do the same for his lectures on the foundations of geometry. (This was the same Wiechert who had been Hilbert's official opponent for his promotion exercises at Königsberg, now also a professor at Göttingen.)

For the published work, as a graceful tribute to Kant, whose *a priori* view of the nature of the geometrical axioms had been discredited by the new view of the axiomatic method, Hilbert chose as his epigraph a quotation from his fellow townsman:

"All human knowledge begins with intuitions, then passes to concepts, and ends with ideas."

Time was short, but he took time to send the proof-sheets of the work to Zürich so that Minkowski could go over them. As always, Minkowski was appreciative and prophetic. The work was, in his opinion, a classic and would have much influence on the thinking of present and future mathematicians.

"It is really not noticeable that you had to work so fast at the end," he assured Hilbert. "Perhaps if you had had more time, it would have lost the quality of freshness."

Minkowski was not too happy in Switzerland. "An open word — take the surprise easy — I would love to go back to Germany." His style of thinking and lecturing was not popular in Zürich "where the students, even the most capable among them, ... are accustomed to get everything spoon-fed." But he hesitated to let his availability be known in Germany. "I feel that even if I had some hope of getting a position, I would still make myself ridiculous in the eyes of many."

Hilbert tried to cheer him up by inviting him to Göttingen for the dedication ceremonies of the Gauss-Weber monument. The days spent there seemed "like a dream" to Minkowski when at the end of a week he had to return to the "hard reality" of Zürich. "But their existence cannot be denied any more than your $18 = 17 + 1$ axiom of arithmetic.... No one who has been in Göttingen recently can fail to be impressed by the stimulating society there."

As soon as Hilbert's lectures, entitled in English *The Foundations of Geometry*, appeared in print, they attracted attention all over the mathematical world.

A German reviewer found the book so beautifully simple that he rashly predicted it would soon be used as a text in elementary instruction.

Poincaré gave his opinion that the work was a classic: "[The contemporary geometers who feel that they have gone to the extreme limit of possible concessions with the non-euclidean geometries based on the negation of the Parallel Postulate] will lose this illusion if they read the work of Professor

Hilbert. In it they will find the barriers behind which they have wished to confine us broken down at every point."

In Poincaré's opinion, the work had but one flaw.

"The logical point of view alone appears to interest Professor Hilbert," he observed. "Being given a sequence of propositions, he finds that all follow logically from the first. With the foundation of this first proposition, with its psychological origin, he does not concern himself.... The axioms are postulated; we do not know from whence they come; it is then as easy to postulate A as C .... His work is thus incomplete, but this is not a criticism I make against him. Incomplete one must indeed resign oneself to be. It is enough that he has made the philosophy of mathematics take a long step forward...."

The American reviewer wrote prophetically, "A widely diffused knowledge of the principles involved will do much for the logical treatment of all science and for clear thinking and writing in general."

The decisive factor in the impact of Hilbert's work, according to Max Dehn, who as a student attended the original lectures, was "the characteristic Hilbertian spirit... combining logical power with intense vitality, disdaining convention and tradition, shaping that which is essential into antitheses with almost Kantian pleasure, taking advantage to the fullest of the freedom of mathematical thought!"

To a large extent, Hilbert, like Euclid himself, had achieved success because of the style and logical perfection of his presentation rather than its originality. But in addition to formulating the modern viewpoint in a way that was attractive and easily grasped, he had done something else which was to be of considerable importance. Having set up in a thoroughly rigorous modern manner the traditional ladder of thought — primitive notions, axioms, theorems — he had proceeded to move on to an entirely new level. In after years, when the approach would have become common, it would be known as metamathematics — literally, "beyond mathematics." For, unlike Euclid, Hilbert required that his axioms satisfy certain logical demands:

That they were *complete*, so that all the theorems could be derived from them.

That they were *independent*, so that the removal of any one axiom from the set would make it impossible to prove at least some of the theorems.

That they were *consistent*, so that no contradictory theorems could be established by reasoning with them.

The most significant aspect of this part of Hilbert's work was the attempted proof of the last requirement — that the axioms be shown to be consistent. This is the equivalent of establishing that reasoning with them will never lead to a contradiction: in short, that they can never be used to prove a statement and at the same time to prove its negation. Under the new conception of a mathematical theory as a system of theorems derived in a deductive way from a set of hypotheses arbitrarily chosen without any restriction as to their truth or meaning, this notion of the consistency of the theory was the only substitute for intuitive truth.

As we have seen, a method of establishing such consistency was already in use. By this method it had been established that any inconsistency which exists in non-euclidean geometry must also exist in euclidean geometry. Thus non-euclidean geometry had been shown to be at least as consistent as euclidean geometry.

Hilbert now took the next step, a step which had apparently occurred to no one else, although it was quite obvious. By the use of analytic geometry, he showed that any contradiction which exists in euclidean geometry must also appear as a contradiction in the arithmetic of real numbers. Both non-euclidean geometry and euclidean geometry were thus shown to be at least as consistent as the arithmetic of real numbers, which was accepted as consistent by all mathematicians.

Within a few months after its publication, Hilbert's little book on the foundations of geometry was a mathematical best seller. Plans were being made to translate it into French and English; it was later translated into other languages. Hilbert's students, who the year before had heard him talk "only algebraic number fields," watched the success of the book in amazement. How had Hilbert been able, once again, to enter a new area of mathematics and produce in it, immediately, great mature work? But even as they asked the question, Hilbert was beginning to publish in still another, entirely new area of mathematics.

IX

Problems

"Pure mathematics grows when old problems are worked out by means of new methods," Klein liked to tell his students. "As a better understanding is thus gained of the older questions, new problems naturally arise."

There is perhaps no better illustration of this statement of Klein's than the project which Hilbert now undertook. In the summer of 1899, immediately after the publication of *The Foundations of Geometry*, he turned to an old and celebrated problem known as the Dirichlet Principle, which involved all the greatest names in the mathematical tradition of Göttingen.

At the heart of this problem was a logical point which had been generally ignored up to the time of Weierstrass. Gauss, Dirichlet, Riemann and others had assumed that, in the case of what is known as the boundary value problem of the Laplace equation, a solution always exists. This assumption was intuitively plausible because in the corresponding physical situation described by the mathematical problem there has to be a definite physical result, or solution. Furthermore, from the mathematical side, Gauss had noted that the boundary value problem of this same equation can be reduced to the problem of minimizing a certain double integral for functions with continuous partial derivatives having the prescribed boundary values. Because of the positive character of this double integral, there is clearly a greatest lower bound for the value of the integral; from this it was further assumed that for one of the functions under consideration the integral would actually have the value of the greatest lower bound.

This mode of reasoning became known as the Dirichlet Principle when Bernhard Riemann used it very freely in his doctoral dissertation in 1851 as the foundation of his geometric function theory and named it in honor of his teacher, Lejeune Dirichlet, who had lectured on the principle in a less general form.

In retrospect, Riemann's dissertation is seen as one of the most important events of modern mathematical history. In its own day, however, it fell into disgrace when Weierstrass objected to the Dirichlet Principle. It was not legitimate mathematically, Weierstrass pointed out, to assume without proof that among the admissible functions there would be one for which the integral would actually have the minimum value.

To someone who is not a mathematician it may seem that this demand of Weierstrass's for a mathematical proof of a principle which obviously works in the physical situations to which it is applied is unreasonable. But this is not so, as Riemann himself recognized when Weierstrass made his criticism. Only a rigorous mathematical proof can establish the ultimate trustworthiness of a mathematical structure and insure that the mathematical description of the physical phenomena to which it is applied is always meaningful.

Riemann was not seriously disturbed by Weierstrass's criticism, however. He had made many of his function-theoretical discoveries on the basis of analogous physical situations, particularly in connection with the behavior of electric currents, and he believed that a problem which made "sensible physics" would make "sensible mathematics." He was confident that the existence of the desired minimum could be established by mathematical proof when and if necessary. But he died young, not yet forty; and a few years after his death, Weierstrass was able to show with finality that the Dirichlet Principle does not in fact invariably hold. He did this by constructing an example for which there is no suitable function that minimizes the integral under the prescribed boundary conditions.

This should have been the end of the Dirichlet Principle, but it was not. Although Riemann's theory was neglected for a while, it was simply too useful in mathematical physics to be discarded. Since the Principle itself was not in general valid, mathematicians invented various ingenious *ad hoc* methods of proving the existence theorems which Riemann had based on the foundation of the Dirichlet Principle. They thus managed to achieve essentially the same end results that he had achieved, but not with the same generality and elegance.

By the time that Hilbert turned his attention to the Dirichlet Principle, mathematicians had given up all hope of salvaging the Principle itself. Only recently Carl Neumann (the son of Franz Neumann), who had done some of the most important work on the subject, had mourned that the Dirichlet Principle, "which is so beautiful and could be utilized so much in the future, has forever sunk from sight."

Unlike many of his contemporaries, who found the demands of rigor a burden, Hilbert firmly believed that rigor contributes to simplicity. He had profound admiration for the way in which Weierstrass had transformed the intuitive analysis of continuity into a strict and logical system. But he refused to let himself be put off by Weierstrass's critique of the Dirichlet Principle. For him, as he said, its "alluring simplicity and indisputable abundance of possible applications" were combined with "a conviction of truth inherent in it."

It was characteristic of Hilbert's mathematical approach to go back to questions in their original conceptual simplicity, and this is what he now did — with, as one of his later pupils commented, "all the naiveté and the freedom from bias and tradition which is characteristic only of truly great investigators." In September 1899, almost fifty years after Riemann's dissertation, he was able to present to the German Mathematical Society a first attempt at what he called, in reference to Neumann's remark, the "resuscitation" of the Dirichlet Principle.

In a few minutes — the whole paper, including the introduction, was scarcely five pages in length — he showed how by placing certain limitations on the nature of the curves and boundary values he could remove Weierstrass's objections and so return Riemann's theory virtually to its original beauty and simplicity. This treatment of the famous problem excited "universal surprise and admiration," according to an American who was present at the meeting. The thought process was simple, but no longer intuitive in any way. Klein commented admiringly: "Hilbert has clipped the hair of the surfaces."

(Half a dozen years later, on the occasion of the 150th anniversary of the Göttingen Scientific Society, he was to return to the Dirichlet Principle and produce a second proof.)

"Hilbert's works on this subject belong to his deepest and most powerful achievements. They are more than the conclusion of a development," a later pupil who also did important work in this same field has written. "Not only was Hilbert's existence proof essentially simplified and generalized through the efforts of many mathematicians; but it was also given an important constructive turn. The physicist Walther Ritz, stimulated by Hilbert, invented from the rehabilitated Dirichlet Principle a powerful method for solving boundary value problems numerically by means of partial differential equations, a method which, just in our time, has shaped the computer into an increasingly succesful tool of numerical mathematics...."

After his success with the Dirichlet Principle, Hilbert decided that during the winter semester of 1899–1900 he would lecture — for the first time in his career — on the calculus of variations. This is the branch of analysis which deals with the type of extremum problems in which (as in the case of Dirichlet's problem) the variable for which a minimizing or maximizing value is sought is not a single numerical variable or a finite number of such variables, but a whole variable curve or function, or even a system of variable functions.

It was a matter of happy experience with Hilbert that great individual problems are the life blood of mathematics. For this reason the calculus of variations had a special charm for him. It was a mathematical theory which had developed from the solution of a single problem.

The problem of "the line of quickest descent" was proposed by Johann Bernoulli at the end of the seventeenth century as a challenge to the mathematicians of his time — and especially to his older brother Jakob, whom he had publicly derided as incompetent. Several people (including Newton) produced solutions; but the "rather inelegant" solution of the scorned older brother surpassed them all. For in it, he recognized what the others had not — that the problem of selecting from an infinity of possible curves the one having a given maximum or minimum property was essentially a new type of problem, demanding for its solution the invention of new methods.

A student who attended Hilbert's lectures on the calculus of variations at this time was Max von Laue.

"...the decisive impression," von Laue wrote of his student days, "was my astonishment at seeing how much information about nature can be obtained by the mathematical method. Profoundest reverence for theory would overcome me when it cast unexpected light on previously obscure facts. Pure mathematics, too, did not fail to impress me, especially in the brilliant courses of David Hilbert."

"This man," the future Nobel Prize winner added, "lives in my memory as perhaps the greatest genius I ever laid eyes on."

The calculus of variations had been made much more rigorous by Weierstrass, but it was still a relatively neglected branch of mathematics. During the winter of his lectures, Hilbert made several important contributions. These included a theorem in which he stated and proved the differentiability conditions of a minimizing arc which assures in many cases the existence of a minimum.

Essentially, however, Hilbert's mathematical interests at this time were more varied than they had been since his docent days in Königsberg. The investigations in geometry continued with a number of papers being published. There was also a paper entitled "The number concept," in which, stimulated by his new-found enthusiasm for the axiomatic method, he proposed that an axiomatic treatment be substituted for the usual "genetic" (as he called it) treatment of the real numbers and introduced the conception of a maximal (or non-extensable) model with his completeness axiom. It was in the midst of this uncharacteristically diversified activity that an invitation arrived for him to make one of the major addresses at the second International Congress of Mathematicians in Paris in the summer of 1900.

The new century seemed to stretch out before him as invitingly as a blank sheet of paper and a freshly sharpened pencil. He would like to make a speech which would be appropriate to the significance of the occasion. In his New Year's letter to Minkowski, he mentioned receiving the invitation and recalled the two speeches from the first International Congress which had so impressed him — the scintillating but technical lecture by Hurwitz on the history of the modern theory of functions and the popular discourse of Poincaré on the reciprocal relationship existing between analysis and physics. He had always wanted to reply to Poincaré with a defense of mathematics for its own sake, but he also had another idea. He had frequently reflected upon the importance of individual problems in the development of mathematics. Perhaps he could discuss the direction of mathematics in the coming century in terms of certain important problems on which mathematicians should concentrate their efforts. What was Minkowski's opinion?

Minkowski wrote he would have to give some thought to the matter.

On January 5, 1900, Minkowski wrote again.

"I have re-read Poincaré's lecture . . . and I find that all his statements are expressed in such a mild form that one cannot take exception to them . . . . Since you will be speaking before specialists, I find a lecture like the one by Hurwitz better than a mere chat like that of Poincaré . . . . Actually it depends not so much on the subject as on the presentation. Still, through the framing of the subject you can make twice as many listeners appear . . . .

"Most alluring would be the attempt at a look into the future and a listing of the problems on which mathematicians should try themselves during the coming century. With such a subject you could have people talking about your lecture decades later."

Minkowski did not fail to point out, however, that there were objections to this subject. Hilbert would probably not want to give away his own ideas for solving certain problems. An international audience would not be so interested in a philosophical discussion as a German audience. Prophecy would not come easy.

Hilbert

There was no reply from Hilbert.

On February 25 Minkowski wrote plaintively to Göttingen.

"How does it happen that one hears nothing from you? My last letter contained only the opinion that if you would give a beautiful lecture, then it would be very beautiful. But it is not easy to give good advice."

But Hilbert had not yet made up his mind about the subject of his speech to the Congress.

On March 29 he consulted Hurwitz.

"I must start preparing for a major talk at Paris, and I am hesitating about a subject.... The best would be a view into the future. What do you think about the likely direction in which mathematics will develop during the next century? It would be very interesting and instructive to hear your opinion about that."

There is no record of Hurwitz's reply.

Hilbert continued to mull over the future of mathematics in the twentieth century. By June he still had not produced a lecture, and the program for the Congress was mailed without its being listed.

Minkowski was greatly disappointed: "The desire on my part to travel to the Congress is now almost gone."

Then, in the middle of July, came a package of proof-sheets from Hilbert. Here at last was the text of the lecture. Entitled simply "Mathematical Problems," it was to be delivered in Paris and, at almost the same time, published in the *Nachrichten* of the Göttingen Scientific Society.

There was no more talk on Minkowski's part about not going to Paris.

He read the proof-sheets carefully and with interest. In them Hilbert set forth the importance of problems in determining the lines of development in a science, examined the characteristics of great fruitful problems, and listed the requirements for "solution." Then he presented and discussed 23 individual problems, the solution of which, he was confident, would contribute greatly to the advance of mathematics in the coming century.

The first half dozen problems pertained to the foundations of mathematics and had been suggested by what he considered the great achievements of the century just past: the discovery of the non-euclidean geometries and the clarification of the concept of the arithmetic continuum, or real number system. These problems showed strongly the influence of the recent work on the foundations of geometry and his enthusiasm for the power of the axiomatic method. The other problems were special and individual, some old and well known, some new, all chosen, however, from fields of Hilbert's own past, present, and future interest. The final, twenty-third problem was actually a suggestion for the future rather than a problem — that in the coming century mathematicians should pay more attention to what he considered an unjustly neglected subject — the calculus of variations.

Minkowski commented with special enthusiasm about the second problem on Hilbert's list. This was the first statement of what was to become known in twentieth century mathematics as the "consistency problem."

It will be remembered that in Hilbert's work on the foundations of geometry, he had established the consistency of the geometric axioms by showing that geometry is at least as consistent as the arithmetic of real numbers, which was accepted as consistent by all mathematicians. But what about arithmetic? Is it actually consistent? If arithmetic were to be set up as an axiomatic theory, as Hilbert had suggested in his recent paper on "The number concept," then this question had to be answered.

A lawyer might be satisfied that the "preponderance of evidence" indicated "beyond a reasonable doubt" that arithmetic is indeed free from contradiction. But Hilbert had not chosen to be a lawyer. For him as a mathematician the consistency of arithmetic would have to be established with a degree of certainty which is inconceivable in the law or any other human endeavor except mathematics. In his second problem he had asked for *a mathematical proof* of the consistency of the axioms of the arithmetic of real numbers.

To show the significance of this problem, he had added the following observation:

"If contradictory attributes be assigned to a concept, I say that *mathematically the concept does not exist* . . . . In the case before us, where we are concerned with the axioms of the real numbers in arithmetic, the proof of the consistency of the axioms is at the same time the proof of the mathematical existence of the complete system of real numbers or of the continuum. Indeed, when the proof for the consistency of the axioms shall be fully accomplished, the doubts which have been expressed occasionally as to the existence of the complete system of real numbers will become totally groundless."

This, Hilbert felt, would provide — at last — the answer to Kronecker.

"Highly original it is," Minkowski observed, "to set out as a problem for the future what mathematicians have for the longest time believed themselves already completely to possess — the arithmetic axioms. What will the large number of laymen in the audience say to that? Will their respect for us increase? You will have a fight on your hands with the philosophers too!"

During the next few weeks, Minkowski and Hurwitz studied the proof-sheets of Hilbert's lecture and made suggestions about its presentation to the Congress. They were both concerned that it was overly long. Hilbert had concluded the extensive introduction to his problems with a stirring reiteration of his conviction ("which every mathematician shares, but which no one has as yet supported by a proof") that every definite mathematical problem must necessarily be susceptible of an exact settlement, either in the form of an actual answer to the question asked, or by the proof of the impossibility of its solution and therewith the necessary failure of all attempts. He had then taken the opportunity to deny publicly and emphatically the "Ignoramus et ignorabimus" — *we are ignorant and we shall remain ignorant* — which the writings of Emil duBois-Reymond had made popular during the century which was passing:

"We hear within us the perpetual call. There is the problem. Seek its solution. You can find it by pure reason, for in mathematics there is no *ignorabimus*."

Both Minkowski and Hurwitz thought that this statement would be an effective conclusion to the speech. Then, perhaps, the list of problems itself could be distributed separately to the delegates.

"It is better," Minkowski admonished, "if you don't need the given time completely."

On July 28 Minkowski mailed the corrected proof-sheets back to Hilbert:

"Actually I believe that through this lecture, which indeed every mathematician in the world without exception will be sure to read, your attractiveness for young mathematicians will increase — if that is possible!"

On Sunday, August 5, the two friends met in Paris.

A thousand mathematicians had earlier signified their intention of coming to the Congress and bringing with them almost seven hundred members of their families to visit the Centennial Exhibition; but apparently rumors of crowds, high prices and hot weather had deterred them. On the morning of August 6, when the opening session of the Congress was called to order by Poincaré, the total attendance scarcely exceeded 250.

After the opening day, the mathematicians left the alien grounds of the Exhibition and retired to the hill on the left bank where the École Polytechnique and the École Normale Supérieure flank Napoleon's Panthéon and a narrow street leads down to the dingy buildings of the Sorbonne.

Although the past century had seen the development of many new branches of mathematics, the classification for the meetings of the Congress remained traditional. Pure mathematics was represented by sections on

Arithmetic, Algebra, Geometry and Analysis; applied mathematics, by a section on Mechanics. The general sections, which dealt with Bibliography and History in one and Teaching and Methods in the other, were considered of inferior rank to the mathematical sections, the lectures being of more general interest but "not necessarily the most valuable mathematically," according to an American who was there. Originally, Hilbert's talk had been planned for the opening session, but because of his tardiness it had to be delivered at the joint session of the two general sections on the morning of Wednesday, August 8.

The man who came to the rostrum that morning was not quite forty, of middle height and build, wiry, quick, with a noticeably high forehead, bald except for wisps of still reddish hair. Glasses were set firmly on a strong nose. There was a small beard, a still somewhat straggly moustache and under it a mouth surprisingly wide and generous for the delicate chin. Bright blue eyes looked innocently but firmly out from behind shining lenses. In spite of the generally unpretentious appearance of the speaker, there was about him a striking quality of intensity and intelligence.

He had already had an extract of his speech in French distributed. This was not at that time a common practice, and the members of his audience were surprised and grateful.

Slowly and carefully for those who did not understand German well, he began to speak.

X

The Future of Mathematics[1]

"Who of us would not be glad to lift the veil behind which the future lies hidden; to cast a glance at the next advances of our science and at the secrets of its development during future centuries? What particular goals will there be toward which the leading mathematical spirits of coming generations will strive? What new methods and new facts will the new centuries disclose in the wide and rich field of mathematical thought?

"History teaches the continuity of the development of science. We know that every age has its own problems, which the following age either solves or casts aside as profitless and replaces by new ones. If we would obtain an idea of the probable development of mathematical knowledge in the immediate future, we must let the unsettled questions pass before our minds and look over the problems which the science of today sets and whose solution we expect from the future. To such a review of problems the present day, lying at the meeting of the centuries, seems to me well adapted. For the close of a great epoch not only invites us to look back into the past but also directs our thoughts to the unknown future.

"The deep significance of certain problems for the advance of mathematical science in general and the important role which they play in the work of the individual investigator are not to be denied. As long as a branch of science offers an abundance of problems, so long is it alive: a lack of problems foreshadows extinction or the cessation of independent

---

[1] The general remarks from the talk on "Mathematical Problems," delivered by David Hilbert before the Second International Congress of Mathematicians at Paris in 1900, are reprinted with permission of the publisher, the American Mathematical Society, from the *Bulletin of the American Mathematical Society*, vol. 8, 1902, pp. 437–445, 478–479. Copyright 1902. The entire text of the talk appears in the *Bulletin* translated into English by Dr. Mary Winston Newson with the approval of Prof. Hilbert.

development. Just as every human undertaking pursues certain objectives, so also mathematical research requires its problems. It is by the solution of problems that the strength of the investigator is hardened; he finds new methods and new outlooks, and gains a wider and freer horizon.

"It is difficult and often impossible to judge the value of a problem correctly in advance; for the final award depends upon the gain which science obtains from the problem. Nevertheless, we can ask whether there are general criteria which mark a good mathematical problem. An old French mathematician said: 'A mathematical theory is not to be considered complete until you have made it so clear that you can explain it to the first man whom you meet on the street.' This clarity and ease of comprehension, here insisted on for a mathematical theory, I should still more demand for a mathematical problem if it is to be perfect; for what is clear and easily comprehended attracts, the complicated repels us.

"Moreover a mathematical problem should be difficult in order to entice us, yet not completely inaccessible, lest it mock our efforts. It should be to us a guidepost on the tortuous paths to hidden truths, ultimately rewarding us by the pleasure in the successful solution.

"The mathematicians of past centuries were accustomed to devote themselves to the solution of difficult individual problems with passionate zeal. They knew the value of difficult problems. I remind you only of the 'problem of the line of quickest descent,' proposed by Johann Bernoulli. Experience teaches, Bernoulli explained in the public announcement of this problem, that lofty minds are led to strive for the advance of science by nothing more than laying before them difficult and at the same time useful problems, and he therefore hoped to earn the thanks of the mathematical world by following the example of men like Mersenne, Pascal, Fermat, Viviani and others in laying before the distinguished analysts of his time a problem by which, as a touchstone, they might test the value of their methods and measure their strength. The calculus of variations owes its origin to this problem of Bernoulli's and to similar problems.

"Fermat has asserted, as is well known, that the diophantine equation $x^n + y^n = z^n$ (x, y and z integers) is unsolvable — except in certain self-evident cases. The attempt to prove this impossibility offers a striking example of the inspiring effect which such a very special and apparently unimportant problem may have upon science. For Kummer, spurred on by Fermat's problem, was led to the introduction of ideal numbers and to the discovery of the law of the unique decomposition of the numbers of a cyclotomic field into ideal prime factors — a law which today, in its generalization to any algebraic field by Dedekind and Kronecker, stands at the center of the modern theory of numbers and the significance of which extends far beyond the boundaries of number theory and into the realm of algebra and the theory of functions.

"To speak of a very different region of research, I remind you of the problem of the three bodies. The fruitful methods and the far-reaching principles which Poincaré has brought into celestial mechanics and which are today recognized and applied in practical astronomy are due to the circumstance that he undertook to treat anew that difficult problem and to come nearer a solution.

"The two last mentioned problems — that of Fermat and the problem of the three bodies — seem to us almost like opposite poles — the former a free invention of pure reason, belonging to the region of abstract number theory, the latter forced upon us by astronomy and necessary to an understanding of the simplest fundamental phenomena of nature.

"But it often happens also that the same special problem finds application in the most unlike branches of mathematical knowledge. So, for example, the problem of the shortest line plays a chief and historically important part in the foundations of geometry, in the theory of lines and surfaces, in mechanics and in the calculus of variations. And how convincingly has F. Klein, in his work on the icosahedron, pictured the significance of the problem of the regular polyhedra in elementary geometry, in group theory, in the theory of equations, and in that of linear differential equations.

"In order to throw light on the importance of certain problems, I may also refer to Weierstrass, who spoke of it as his happy fortune that he found at the outset of his scientific career a problem so important as Jacobi's problem of inversion on which to work.

"Having now recalled to mind the general importance of problems in mathematics, let us turn to the question of the sources from which this science derives its problems. Surely the first and oldest problems in every branch of mathematics stem from experience and are suggested by the world of external phenomena. Even the rules of calculation with integers must have been discovered in this fashion in a lower stage of human civilization, just as the child of today learns the application of these laws by empirical methods. The same is true of the first problems of geometry, the problems bequeathed to us by antiquity, such as the duplication of the cube, the squaring of the circle; also the oldest problems in the theory of the solution of numerical equations, in the theory of curves and the differential and integral calculus, in the calculus of variations, the theory of

Fourier series, and the theory of potential — to say nothing of the further abundance of problems properly belonging to mechanics, astronomy and physics.

"But, in the further development of a branch of mathematics, the human mind, encouraged by the success of its solutions, becomes conscious of its independence. By means of logical combination, generalization, specialization, by separating and collecting ideas in fortunate ways — often without appreciable influence from without — it evolves from itself alone new and fruitful problems, and appears then itself as the real questioner. Thus arose the problem of prime numers and the other problems of number theory, Galois's theory of equations, the theory of algebraic invariants, the theory of abelian and automorphic functions; indeed almost all the nicer questions of modern arithmetic and function theory arise in this way.

"In the meantime, while the creative power of pure reason is at work, the outer world again comes into play, forces upon us new questions from actual experience, opens up new branches of mathematics; and while we seek to conquer these new fields of knowledge for the realm of pure thought, we often find the answers to old unsolved problems and thus at the same time advance most successfully the old theories. And it seems to me that the numerous and surprising analogies and that apparently pre-established harmony which the mathematician so often perceives in the questions, methods and ideas of the various branches of his science, have their origin in this ever-recurring interplay between thought and experience.

"It remains to discuss briefly what general requirements may be justly laid down for the solution of a mathematical problem. I should say first of all, this: that it shall be possible to establish the correctness of the solution by means of a finite number of steps based upon a finite number of hypotheses which are implied in the statement of the problem and which must be exactly formulated. This requirement of logical deduction by means of a finite number of processes is simply the requirement of rigor in reasoning. Indeed the requirement of rigor, which has become a byword in mathematics, corresponds to a universal philosophical necessity of our understanding; on the other hand, only by satisfying this requirement do the thought content and the suggestiveness of the problem attain their full effect. A new problem, especially when it comes from the outer world of experience, is like a young twig, which thrives and bears fruit only when it is grafted carefully and in accordance with strict horticultural rules upon the old stem, the established achievements of our mathematical science.

"It is an error to believe that rigor in the proof is the enemy of simplicity. On the contrary, we find it confirmed by numerous examples that the rigorous method is at the same time the simpler and the more easily comprehended. The very effort for rigor forces us to discover simpler methods of proof. It also frequently leads the way to methods which are more capable of development than the old methods of less rigor. Thus the theory of algebraic curves experienced a considerable simplification and attained a greater unity by means of the more rigorous function-theoretical methods and the consistent introduction of transcendental devices. Further, the proof that the power series permits the application of the four elementary arithmetical operations as well as the term by term differentiation and integration, and the recognition of the utility of the power series depending upon this proof contributed materially to the simplification of all analysis, particularly of the theory of elimination and the theory of differential equations, and also of the existence proofs demanded in those theories. But the most striking example of my statement is the calculus of variations. The treatment of the first and second variations of definite integrals required in part extremely complicated calculations, and the processes applied by the old mathematicians had not the needful rigor. Weierstrass showed us the way to a new and sure foundation of the calculus of variations. By the examples of the simple and double integral I will show briefly, at the close of my lecture, how this way leads at once to a surprising simplification of the calculus of variations. For in the demonstration of the necessary and sufficient criteria for the occurrence of a maximum and minimum, the calculation of the second variation and in part, indeed, the tiresome reasoning connected with the first variation may be completely dispensed with — to say nothing of the advance which is involved in the removal of the restriction to variations for which the differential coefficients of the function vary but slightly.

"While insisting on rigor in the proof as a requirement for a perfect solution of a problem, I should like, on the other hand, to oppose the opinion that only the concepts of analysis, or even those of arithmetic alone, are susceptible of a fully rigorous treatment. This opinion, occasionally advocated by eminent men, I consider entirely erroneous. Such a one-sided interpretation of the requirement of rigor would soon lead to the ignoring of all concepts arising from geometry, mechanics and physics, to a stoppage of the flow of new material from the outside world, and finally, indeed, as a last consequence, to the rejection of the ideas of the continuum and of the irrational number. But what an important nerve, vital to mathematical science, would be cut by rooting out geometry and mathematical physics! On the contrary, I think that wherever mathematical ideas come up, whether from the side of the theory of knowledge or in geometry, or from the theories of natural or physical science, the problem arises for mathematics to investigate the principles underlying these ideas and so to establish them upon a simple and complete system of axioms, that the exactness of the new ideas and their applicability to deduction shall be in no respect inferior to those of the old arithmetical concepts.

"To new concepts correspond, necessarily, new symbols. These we choose in such a way that they remind us of the phenomena which were the occasion for the formation of the new concepts. So the geometrical figures are signs or mnemonic symbols of space intuition and are used as such by all mathematicians. Who does not always use along with the double inequality $a > b > c$ the picture of three points following one another on a straight line as the geometrical picture of the idea 'between'? Who does not make use of drawings of segments and rectangles enclosed in one another, when it is required to prove with perfect rigor a difficult theorem on the continuity of functions or the existence of points of condensation? Who could dispense with the figure of the triangle, the circle with its center, or with the cross of three perpendicular axes? Or who would give up the representation of the vector field, or the picture of a family of curves or surfaces with its envelope which plays so important a part in differential geometry, in the theory of differential equations, in the foundations of the calculus of variations, and in other purely mathematical sciences?

"The arithmetical symbols are written figures and the geometrical figures are drawn formulas; and no mathematician could spare these drawn formulas, any more than in calculation he could dispense with the insertion and removal of parentheses or the use of other analytical signs.

"The use of geometrical symbols as a means of strict proof presupposes the exact knowledge and complete mastery of the axioms which lie at the foundation of those figures; and in order that these geometrical figures may be incorporated in the general treasure of mathematical symbols, a rigorous axiomatic investigation of their conceptual content is necessary. Just as in adding two numbers, one must place the digits under each other in the right order so that only the rules of calculation, i.e., the axioms of arithmetic, determine the correct use of the digits, so the use of geometrical symbols is determined by the axioms of geometrical concepts and their combinations.

"The agreement between geometrical and arithmetical thought is shown also in that we do not habitually follow the chain of reasoning back to the axioms in arithmetical discussions, any more than in geometrical. On the contrary, especially in first attacking a problem, we apply a rapid, unconscious, not absolutely sure combination, trusting to a certain arithmetical feeling for the behavior of the arithmetical symbols, which we could dispense with as little in arithmetic as with the geometrical imagination in geometry. As an example of an arithmetical theory operating rigorously with geometrical ideas and symbols, I may mention Minkowski's work, *The Geometry of Numbers*.

"Some remarks upon the difficulties which mathematical problems may offer, and the means of overcoming them, may be in place here.

"If we do not succeed in solving a mathematical problem, the reason frequently consists in our failure to recognize the more general standpoint from which the problem before us appears only as a single link in a chain of related problems. After finding this standpoint, not only is this problem frequently more accessible to our investigation, but at the same time we come into possession of a method which is applicable also to related problems. The introduction of complex paths of integration by Cauchy and of the notion of the ideals in number theory by Kummer may serve as examples. This way for finding general methods is certainly the most practical and the most certain; for he who seeks for methods without having a definite problem in mind seeks for the most part in vain.

"In dealing with mathematical problems, specialization plays, as I believe, a still more important part than generalization. Perhaps in most cases where we unsuccessfully seek the answer to a question, the cause of the failure lies in the fact that problems simpler and easier than the one in hand have been either incompletely solved, or not solved at all. Everything depends, then, on finding those easier problems and on solving them by means of devices as perfect as possible and of concepts capable of generalization. This rule is one of the most important levers for overcoming mathematical difficulties; and it seems to me that it is used almost always, though perhaps unconsciously.

"Occasionally it happens that we seek the solution under insufficient hypotheses or in an incorrect sense, and for this reason do not succeed. The problem then arises: to show the impossibility of the solution under the given hypotheses, or in the sense contemplated. Such proofs of impossibility were effected by the ancients; for instance, when they showed that the ratio of the hypotenuse to the side of an isosceles right triangle is irrational. In later mathematics, the question as to the impossibility of certain solutions plays a prominent part; and we perceive in this way that old and difficult problems, such as the proof of the axiom of parallels, the squaring of the circle, or the solution of equations of the fifth degree by radicals, have finally found fully satisfactory and rigorous solutions, although in another sense from that originally intended. It is probably this remarkable fact along with other philosophical reasons that gives rise to the conviction (which every mathematician shares, but which no one has as yet supported by a proof) that every definite mathematical problem must necessarily be susceptible of an exact settlement, either in the form of an actual answer to the question asked, or by the proof of the impossibility of its solution and therewith the necessary failure of all attempts. Take any definite unsolved problem, such as the question as to the irrationality of the Euler-Mascheroni constant C or the existence of an infinite number of prime numbers of the form $2^n + 1$. However unapproachable these problems may seem to us and however helpless we stand before them, we have, nevertheless, the firm conviction that their solution must follow by a finite number of purely logical processes.

"Is this axiom of the solvability of every problem a peculiarity characteristic only of mathematical thought, or is it possibly a general law inherent in the nature of the mind, a belief that all questions which it asks must be answerable by it? For in other sciences also one meets old problems which have been settled in a manner most satisfactory and most useful to science by the proof of their impossibility. I cite the problem of perpetual motion. After seeking unsuccessfully for the construction of a perpetual motion machine, scientists investigated the relations which must subsist between the forces of nature if such a machine is to be impossible; and this inverted question led to the discovery of the law of the conservation of energy, which, again, explained the impossibility of perpetual motion in the sense originally intended.

"This conviction of the solvability of every mathematical problem is a powerful incentive to the worker. We hear within us the perpetual call: There is the problem. Seek its solution. You can find it by pure reason, for in mathematics there is no *ignorabimus*."

At this point, to shorten his talk, Hilbert presented only 10 of his 23 problems. However, all the problems are listed here essentially under the headings he used when he published the talk. The problems which he mentioned in Paris have been starred:

---

[2] A summary in French of Hilbert's talk and the list of the problems presented appears in *L'enseignement mathématique*, vol. 2, 1900, pp. 349–355.

Hilbert

*1. Cantor's problem of the cardinal number of the continuum.
*2. The compatibility of the arithmetical axioms.
3. The equality of the volumes of two tetrahedra of equal bases and equal altitudes.
4. The problem of the straight line as the shortest distance between two points.
5. Lie's concept of a continuous group of transformations without the assumption of the differentiability of the functions defining the group.
*6. The mathematical treatment of the axioms of physics.
*7. The irrationality and the transcendence of certain numbers.
*8. Problems of prime numbers (including the Riemann hypothesis).
9. The proof of the most general law of reciprocity in any number field.
10. The determination of the solvability of a Diophantine equation.
11. The problem of quadratic forms with any algebraic numerical coefficients.
12. The extension of Kronecker's theorem of Abelian fields to any algebraic realm of rationality.
*13. The proof of the impossibility of the solution of the general equation of the 7th degree by means of functions of only two arguments.
14. The proof of the finiteness of certain complete systems of functions.
15. A rigorous foundation of Schubert's enumerative calculus.
*16. The problem of the topology of algebraic curves and surfaces.
17. The expression of definite forms by squares.
18. The building up of space from congruent polyhedra.
*19. The determination of whether the solutions of "regular" problems in the calculus of variations are necessarily analytic.
20. The general problem of boundary values.
*21. The proof of the existence of linear differential equations having a prescribed monodromic group.
*22. Uniformization of analytic relations by means of automorphic functions.
23. The further development of the methods of the calculus of variations.

"The [ten] problems mentioned," Hilbert told his audience, "are merely samples of problems; yet they are sufficient to show how rich, how manifold and how extensive the mathematical science is today; and the question is urged upon us whether mathematics is doomed to the fate of those other sciences that have split up into separate branches, whose representatives scarcely understand one another and whose connection becomes ever more loose. I do not believe this nor wish it. Mathematical science is in my opinion an indivisible whole, an organism whose vitality is conditioned upon the connection of its parts. For with all the variety of mathematical knowledge, we are still clearly conscious of the similarity of the logical devices, the *relationship* of the *ideas* in mathematics as a whole and the numerous analogies in its different departments. We also notice that, the farther a mathematical theory is developed, the more harmoniously and uniformly does its construction proceed, and unexpected relations are disclosed between hitherto separate branches of the science. So it happens that, with the extension of mathematics, its organic character is not lost but only manifests itself the more clearly.

"But, we ask, with the extension of mathematical knowledge will it not finally become impossible for the single investigator to embrace all departments of this knowledge? In answer let me point out how thoroughly it is ingrained in mathematical science that every real advance goes hand in hand with the invention of sharper tools and simpler methods which at the same time assist in understanding earlier theories and cast aside older, more complicated developments. It is therefore possible for the individual investigator, when he masters these sharper tools and simpler methods, to find his way more easily in the various branches of mathematics than is possible in any other science.

"The organic unity of mathematics is inherent in the nature of this science, for mathematics is the foundation of all exact knowledge of natural phenomena. That it may completely fulfill this high destiny, may the new century bring it gifted prophets and many zealous and enthusiastic disciples!"

XI

The New Century

It was hot and sultry in the lecture hall at the Sorbonne by the time that Hilbert finished his lecture on mathematical problems, and the discussion which followed was "somewhat desultory," according to the reporter for the American Mathematical Society.

". . . the claim was made, although apparently without adequate grounds, that more had been done as regards the equation of the 7th degree (by some German writer) than the author of the paper was willing to allow. A more precise objection was taken to M. Hilbert's remarks on the axioms of arithmetic by M. Peano, who claimed that such a system as that specified as desirable had already been established by [some of] his compatriots . . . ."

The real news of the day, as reported in the special edition of The New York Times which was distributed on the grounds of the Exhibition, was that the United States, Great Britain, Germany and Japan were going to have to carry out their military program in China without any more troops from Russia and France, who were otherwise occupied on the Siberian frontier and in Indo-China. The King of Italy had recently been assassinated and the country was in turmoil. Queen Victoria planned to address Parliament. William Jennings Bryan was informed that he had again been chosen to head the Democratic ticket against President McKinley.

But up at the Sorbonne, in the remaining days of the Congress, it became quite clear that David Hilbert had captured the imagination of the mathematical world with his list of problems for the twentieth century. His practical experience seemed to guarantee that they met the criteria which he had set up in his lecture; his judgment, that they could actually be solved in the years to come. His rapidly growing fame — exceeded now only by that of Poincaré — promised that a mathematician could make a reputation for himself by solving one of the Paris problems.

Immediately after the Congress adjourned, Hilbert went for a short vacation at Rauschen. Receiving a note from him, Minkowski recalled the "beautiful times" they used to have upon the strand: "With pleasure I also see what I have of course known for a long time — that one can learn much from you, not only in mathematics, but also in the art of enjoying life sensibly like a philosopher."

The first important result in connection with one of Hilbert's problems came within the year 1900. His own student, 22-year-old Max Dehn, showed that (as Hilbert had conjectured) a regular tetrahedron cannot be cut up and reassembled into a cube of equal volume. This provided a partial solution for the third problem. The next year Dehn completed the solution. He thus became the first mathematician to pass into what was later to be known as "the honors class" of mathematicians who had solved or contributed to the solution of one of David Hilbert's 23 Paris problems.

After Paris, Hilbert himself continued to investigate geometric questions, but for the most part he devoted himself to questions of analysis. This is an area of mathematics that differs in an important respect from those in which he had previously worked. In arithmetic and algebra, calculations ordinarily involve only a finite number of quantities and end after a finite number of steps. Analysis operates in a continuum. Solutions are achieved by showing that infinite series of numbers converge to a limit. By now, Hilbert was the enthusiastic champion of the axiomatic method; and he thought he saw in analysis an opportunity to exercise the impressive powers of this method to unify, order, and clarify.

"It appears to me of outstanding interest," he was later to remark, "to undertake an investigation of the convergence conditions which serve for the erection of a given analytic discipline so that we can set up a system of the simplest fundamental facts which require for their proof a specific convergence condition. Then by the use of that one convergence condition alone — without the addition of any other convergence condition whatsoever — the totality of the theorems of the particular discipline can be established."

This was somewhat similar to what Riemann had attempted to do with the Dirichlet Principle, and Hilbert thought that he also could find "the one simple fundamental fact" which he needed in the calculus of variations. Then, one day in the winter of 1900–1901, a Swedish student brought to Hilbert's seminar a recently published paper on integral equations by a countryman named Ivar Fredholm.

Integral equations are functional equations, the history of which is closely tied up with the problems of mathematical physics, particularly the problem of the oscillations of a continuum. A theory of sorts for this type of equation had been very slowly evolving. But now Fredholm presented the solution of particular equations (which came to be named after him) in an original and elegant way that revealed a suggestive analogy between integral equations and the linear equations of algebra.

Hilbert promptly recognized that in this work Fredholm was coming closer to the desired goal of a unifying methodological approach to analysis than he himself was in his work in the calculus of variations. It was a matter of pride with him that he was not emotionally committed to any program — that he saw things as they were rather than as he would like them to be. So now, without regrets, he turned away from his own program and charged the subject of integral equations with impressive intensity. It would have to remain forever undecided (as Blumenthal observed) whether Hilbert would have been able to bring to the methods of the calculus of variations the flexibility and power necessary to penetrate and sustain the whole of analysis. For now Hilbert talked *only* about integral equations to his students.

It was at this same time that a young Japanese named Teiji Takagi came to Göttingen, his way paid by his country. He was to become one of the half dozen mathematicians who would develop the ideas of class-fields which Hilbert had sketched out in his last paper on algebraic number fields. He had already written a little book on *New Arithmetic*, very simple in comparison with the recent number theory work of Hilbert but much advanced for the mathematical level of his native land at that time; now he was looking forward to studying with the author of the *Zahlbericht*. But when Takagi arrived in Göttingen, Hilbert had nothing at all to say to him on the subject of number theory. Instead, he was already sketching out to his students in conversations and lectures some of the ideas which he would ultimately use in his general theory of integral equations.

Another student who arrived at this time was Erhard Schmidt. He had come down from Berlin to "scout" the mathematical education in Göttingen and compare it with that being offered in the capital by the formidable trio of Fuchs, H. A Schwarz, and Frobenius. Fuchs was the same Fuchs with whom Hilbert had studied at Heidelberg; Schwarz, who had been responsible for bringing Klein to Göttingen, conducted a twice-a-month colloquium that was internationally famous; Frobenius, it was said, delivered the most perfect mathematical lectures in Germany — "their only fault being," according to one student, "that because of their completeness they failed even to suggest the existence of unsolved problems." But young Schmidt was so impressed with the mathematics in Göttingen that he decided not to return to Berlin.

Although the subject of discussion had changed, the weekly seminar walks continued. Now, though, the calm of the countryside was frequently disturbed by the snorting of an engine-powered monster. Hilbert's friend Nernst had bought one of the new motor cars; and the hills that defeated other motorists were no problem for him. He merely turned the tap on the cylinder of N_2O he had installed on the dashboard, thus injecting laughing gas into the mixture, and roared triumphantly up any hill.

During the winter semester of 1901–1902, Hilbert was lecturing on potential theory and applying his first results on integral equations. Because of their newness, his ideas were sometimes almost impossible for his students to follow. Even the Lesezimmer notes, worked out by Albert Andrae, the assistant, were not too helpful. In fact, Andrae sometimes penciled warnings on the notes: "From page such-and-such to page so-and-so, no guarantee of correctness can be given." At the Christmas celebration of the Mathematics Club one befuddled student of potential theory read the following gently ironical lines: "Der eine bleibt erst unverständlich/ Der Andrae macht es klar." (The verse depended on the similarity between the name Andrae and the German word for "other" — *The one is at first obscure, the other makes it clear*.)

By spring Erhard Schmidt's enthusiasm for mathematics at Göttingen had brought some of his friends from Berlin. One of these was Constantin Carathéodory, the scion of a powerful Greek family, who at 26 had given up a promising career as an engineer to return to school and devote himself to the study of pure mathematics. His family considered his plan foolishly romantic: one does not usually begin a mathematical career at 26. "But I could not resist the obsession that through unrestrained preoccupation with mathematics my life would become worthwhile."

By now Hilbert was just about as famous as it is possible for a mathematician to be. He accepted his success, as Otto Blumenthal noted, "with a naive, mild pleasure, not letting himself be confused into false modesty." The string of victories which had been inaugurated by the solution of Gordan's Problem twelve short years ago reminded Blumenthal (now a Privatdozent himself) of Napoleon's Italian campaign: the climactic work in invariant theory — the *Zahlbericht* and the deep, fertile program for class-fields — the widely read and influential little book on the foundations of geometry — the salvaging of the Dirichlet Principle — the important theorems in the calculus of variations — the Paris Problems. Foreign academies elected Hilbert to membership. The German government awarded him the title of Geheimrat, roughly equivalent to an English knighthood.

Someone, trying to help Hilbert with something and addressing him over and over as "Herr Geheimrat," noted that he appeared irritated and asked worriedly, "Am I bothering you, Herr Geheimrat?"

"Nothing about you bothers me," Hilbert shot back, "except your obsequiousness!"

Klein, after becoming a Geheimrat, always insisted on being addressed by that title. How did Hilbert prefer to be addressed?

"Hilbert?" a former student says. "He didn't care. He was a king. He was Hilbert."

The parents were still alive at this time. Judge Hilbert had long remained "suspicious" of his son's profession and his success. Mathematics being what it is, he could never as an outsider really appreciate the quality of the achievement; but he must have been reassured at last by the honors.

Minkowski, on a visit to Göttingen, was much impressed with the mathematical atmosphere around his friend.

"Even through merely a sojourn in such air," he wrote after his return to the still unsatisfactory situation in Zürich, "a person receives an increased desire to do great things.... I am already at work on a paper for the *Annalen*."

And yet, as Hilbert passed his fortieth birthday on January 23, 1902, he was not entirely happy.

Although he and Klein shared "a perfect trust and common interest" (his own words), they were not intimate. More and more since Hilbert's coming to Göttingen, Klein had devoted himself to activities which were not in themselves mathematics. In addition to his teaching and administrative duties, he was the moving force behind a number of projects: the plan for the 30-volume mathematical encyclopedia; the newly organized International Schools Commission, which had been set up to study the development of teaching methods "in all civilized countries" from kindergarten to graduate school; the attempt to improve and enlarge the scientific education offered in the German middle schools and to bring together, at the university level, technical and mathematical training; the dream, cherished since his visit to America, of fertilizing technology itself with the methods of pure mathematics. Hilbert had scarcely any interest in these projects of Klein's.

Also, with age, Klein was becoming more olympian. A favorite joke among the students was the following: In Göttingen there are two kinds of mathematicians, those who do what they want and not what Klein wants —

and those who do what Klein wants and not what they want. Klein is not either kind. Therefore, Klein is not a mathematician.

Klein was interested in his students and spent long hours in conference with them, but he remained always a superior being. He dispensed ideas, one of the students later said, "with royal joyousness from his own wealth" and "directed each one with unfailing certainty to the point which best accorded with his individuality." Among themselves, they called him "the great Felix." It was told in Göttingen that at dinner in Klein's house a student was sometimes so awed by his host that he stood up when he was asked a question.

Hilbert did not feel personally threatened by Klein, however. A few years before, when he had been offered Sophus Lie's chair in Leipzig, he had consulted Minkowski about the advisability of his accepting the opportunity to leave Göttingen. Minkowski had pointed out that perhaps if he were "spatially separated" from Klein, "outsiders" might more easily recognize that it was Hilbert who was now the greatest German mathematician. But this argument had no effect. Hilbert had refused the Leipzig offer.

Now, however, he was becoming increasingly aware that in his relationship with Klein something was missing which was necessary to him — something which Klein because of his nature could not provide. Then, a few months after his fortieth birthday, he received another opportunity to leave Göttingen. Lazarus Fuchs died and the honored position — Fuchs's chair in Berlin — was offered to Hilbert.

When the news of Hilbert's "call" became known among the docents and advanced students, they were greatly upset. Many of them had come to Göttingen just because Hilbert was there. Some, like Erhard Schmidt and his friends, had come from Berlin itself. Yet it seemed to all only natural that Hilbert, the leading German mathematician, should want to take his place in the capital. Although they did not have much hope of affecting his decision, they appointed three students, headed by Walther Lietzmann, to go to his house and petition him to remain in Göttingen. Mrs. Hilbert served them punch in the garden. Hilbert listened without comment to what they had to say. They left discouraged. The length of time he was taking in coming to his decision, his frequent trips to the capital, his unusual nervousness in lectures, all led them to believe that he was planning to accept the Berlin offer.

Hilbert, however, was actually attempting to solve a personal problem in the way that he had grown accustomed to solving mathematical problems. He did not want to leave Göttingen. As he had explained to Minkowski at the time of the Leipzig offer, he felt that he had more vigor for his work in a little town, which made scientific interchange easy and provided many opportunities for contact with nature, than he would have in a big city. He was well aware of the benefits he reaped from the administrative genius of Klein. He also had a sense of the appropriateness of the leading German mathematician's being at the university of Gauss. But to stay happily in Göttingen, he knew he needed to have some colleague who would provide the close scientific and personal companionship which at Königsberg he had received from Minkowski. The solution to the problem was obvious. There was but one unwritten rule regarding an offer from another university. In addition to trying to improve one's own professonal situation, one was also expected to try to improve the situation of his department and his subject. Recently, Nernst, offered a position in Munich, had extracted as the price of his remaining in Göttingen the first physico-chemical laboratory in Germany. Now, with Klein's approval and support, Hilbert proposed to Althoff that a new professorship of mathematics be created at Göttingen and that it be offered to Minkowski.

He had not raised Minkowski's hopes of returning to Germany until the new position was definite. Then he saw to it that Minkowski was in Göttingen presenting a paper to the scientific society on the day the announcement was made.

The diplomatic skill with which he conducted this bold and ultimately successful maneuver is glimpsed in the fact that he always gave Althoff the full credit:

"It was again Althoff who transplanted Minkowski to soil better suited to him. With an intrepidity unprecedented in the history of the administration of the Prussian universities, Althoff created here in Göttingen a new professorship . . . ."

When the members of the Mathematics Club heard the news — "that Hilbert was staying and Minkowski was coming" — they delightedly organized a Festkommers. This was a formal drinking and smoking party. It was one of the two ways of expressing esteem for a professor. The other — the ultimate honor — was a torchlight procession, which was tendered to only a few professors, and then only at the end of a long and distinguished career.

The highlight of the celebration was a speech by Klein in which he developed a magnificent comprehensive picture of Hilbert's research and teaching and its influence on the future of mathematics. "Please give me that in writing," Hilbert was heard to say afterwards.

Minkowski, back in Zürich, wrote happily:

"I have the most beautiful hopes for my future life and work!"

XII

Second Youth

After Minkowski's arrival in Göttingen in the fall of 1902, Hilbert was no longer lonely: "A telephone call, or a few steps down the street, a pebble tossed up against the little corner window of his study, and there he was, always ready for any mathematical or non-mathematical undertaking."

Instead of conducting seminars with Klein, Hilbert now conducted them with Minkowski.

On Sunday mornings the two friends regularly set out with their wives on a picnic excursion.

The Hilberts had by this time left the Reformed Protestant Church in which they had been baptized and married. It was told in Göttingen that when Franz had started to school he could not answer the question, "What religion are you?" "If you do not know what you are," he was informed by the son of the philosopher Edmund Husserl, a Jew recently converted to Christianity, "then you are certainly a Jew."

The Sunday excursions were later enlarged to include the children of both families. Most frequently the destination was a resort called Mariaspring, where there was dancing outside under the trees. Here Hilbert would seek out his current "flame" — the pretty young wife of some colleague — and whirl her around the dance floor, much to the embarrassment of the little Minkowski girls, who found his energetic dancing very old fashioned. "It was a sport to him!" They were further embarrassed when, after the music stopped, he enveloped his partner in his great loden cape and made a show of hugging and kissing her.

The frequent parties at the Hilbert house were now more enjoyable for Hilbert because of the quiet presence of Minkowski. There was dancing at these affairs too — the rug rolled away, music furnished by the gramophone a manufacturer had presented to the famous mathematics professor, the commands given in French by Hilbert. The table was always ladened with a variety of food, but the staple was talk. A subject would come up, somebody would ask Hilbert what he thought about it. Astrology, for instance. What did he think about that? Without an instant's hesitation he would answer firmly in the still uncorrupted East Prussian accent which made everything he said sound amusing and memorable: "When you collect the 10 wisest men of the world and ask them to find the most stupid thing in existence, they will not be able to find anything stupider than astrology!" Perhaps the guests would be discussing Galileo's trial and someone would blame Galileo for failing to stand up for his convictions. "But he was not an idiot," Hilbert would object. "Only an idiot could believe that scientific truth needs martyrdom — that may be necessary in religion, but scientific results prove themselves in time." Minkowski did not offer his opinion so frequently as Hilbert. When he did speak, his observation — often in the form of an appropriate quotation from *Faust* — went to the heart of the matter, and Hilbert listened. But Hilbert was always the more intrepid in expressing opinions. What technological achievement would be the most important? "To catch a fly on the moon." Why? "Because the auxiliary technical problems which would have to be solved for such a result to be achieved imply the solution of almost all the material difficulties of mankind." What mathematical problem was the most important? "The problem of the zeros of the zeta function, not only in mathematics, but absolutely most important!"

Sometimes the little Hilbert boy Franz, who was not shown at parties, would stop at the door and listen.

In front of a group, Minkowski suffered from "Lampenfieber" — in English, *stagefright*. He was still embarrassed by attention, even of much younger people; and in Zürich his shy, stammering delivery had completely put off a student named Albert Einstein. But in Göttingen ("the shrine of pure thought," as it was called) the students recognized immediately that in Minkowski they had the privilege of hearing "a true mathematical poet." It seemed to them that every sentence he spoke came into being as he spoke it.

This was, at least once, quite literally true. Lecturing on topology, Minkowski brought up the Four Color Theorem — a famous unsolved problem in that field of mathematics. (The theorem states that four colors are always sufficient to color any map in such a way that no two adjoining regions will have the same color.)

"This theorem has not yet been proved, but that is because only mathematicians of the third rank have occupied themselves with it," Minkowski announced to the class in a rare burst of arrogance. "I believe I can prove it."

He began to work out his demonstration on the spot. By the end of the hour he had not finished. The project was carried over to the next meeting of the class. Several weeks passed in this way. Finally, one rainy morning, Minkowski entered the lecture hall, followed by a crash of thunder. At the rostrum, he turned toward the class, a deeply serious expression on his gentle round face.

"Heaven is angered by my arrogance," he announced. "My proof of the Four Color Theorem is also defective."

He then took up the lecture on topology at the point where he had dropped it several weeks before. (The Four Color Theorem remained unproved until 1976.)

Hilbert was now beginning to devote himself to integral equations with the same exclusiveness which he had earlier lavished on invariants and number fields. He had begun his investigations in a way reminiscent of his approach in the earlier subjects. In his first paper, sent as a communication to the Göttingen Scientific Society, he had presented a simple and original derivation of the Fredholm theory which exposed the fundamental idea more clearly than Fredholm's own work. There were also already glimpses of fresh, fruitful ideas to come. With an intuitive grasp of the underlying relationships existing among the different parts of mathematics and between mathematics and physics, Hilbert had recognized that Fredholm equations could open up a whole series of previously inaccessible questions in analysis and mathematical physics. It was now his goal to encompass within a uniform theoretical arrangement of equations the greatest possible domain of the linear problems of analysis.

Minkowski was occupying himself again with his beloved theory of numbers. It concerned him, according to Hilbert, that many mathematicians hardly get a breath of what he called the "special air" of number theory; and during the winter of 1903–04 he delivered a series of relatively untechnical lectures, later published as a book, in which he presented the methods he had created and some of his own most significant results in a way in which they could be easily grasped. Hilbert was just as interested as Minkowski in emphasizing "the insinuating melodies of this powerful music" — this metaphor was also Minkowski's — and when Legh Reid, one of his former American students, wrote a book on the subject, Hilbert endorsed it with enthusiasm. Number theory was "the pattern for the other sciences, ... the inexhaustible source of all mathematical knowledge, prodigal of incitement to investigations in all other domains ...." A number theory problem was never dated, it was "as timeless as a true work of art." Thanks to Minkowski, Germany had recently become, once again, the number theory center of the world. "But every devotee of the theory of numbers will desire that it shall be equally a possession of all nations and be cultivated and spread abroad, especially among the younger generation, to whom the future belongs."

During 1903, Hermann Weyl arrived in Göttingen. He was an 18 year old country boy, seemingly inarticulate, but with lively eyes and a great deal of confidence in his own abilities. He had chosen the University because the director of his gymnasium was the cousin of one of the mathematics professors "by the name of David Hilbert."

"In the fullness of my innocence and ignorance," Weyl wrote many years later from the Institute for Advanced Study in Princeton, New Jersey, "I made bold to take the course Hilbert had announced for that term, on the notion of number and the quadrature of the circle. Most of it went straight over my head. But the doors of a new world swung open for me, and I had not sat long at Hilbert's feet before the resolution formed itself in my young heart that I must by all means read and study whatever this man had written."

Hilbert's "optimism, his spiritual passion, his unshakable faith in the supreme value of science, and his firm confidence in the power of reason to find simple and clear answers to simple and clear questions" were irresistible. Weyl heard "the sweet flute of the Pied Piper ..., seducing so many rats to follow him into the deep river of mathematics." That summer he went home with a copy of the *Zahlbericht* under his arm and, although he did not have any previous knowledge of the mathematics involved, worked his way through it during the vacation.

He was a young man with a taste for words as well as for mathematics, and he found the peculiarly Hilbertian brand of thinking admirably reflected in the great *lucidity* of Hilbert's literary style:

"It is as if you are on a swift walk through a sunny open landscape; you look freely around, demarcation lines and connecting roads are pointed out to you before you must brace yourself to climb the hill; then the path goes straight up, no ambling around, no detours."

The summer months spent studying the *Zahlbericht* were, Weyl was always to say, the happiest months of his life.

It was also at this time, during Minkowski's first years, that Max Born, the son of a well-known medical researcher in Breslau, came to Göttingen on the advice of his friends, Ernst Hellinger and Otto Toeplitz. They had informed him that Göttingen was now "the mecca of German mathematics."

Born's stepmother had known Minkowski in Königsberg; and not long after his arrival, the new student was invited to lunch by the professor and presented to Guste Minkowski and the two little daughters. After lunch, Hilbert and Käthe came over and the whole party hiked to Die Plesse, a ruined castle overlooking the valley of the Leine and the red-tiled roofs of Göttingen.

Born was never to forget the afternoon.

"The conversation of the two friends was an intellectual fireworks display. Full of wit and humor and still also of deep seriousness. I myself had grown up in an atmosphere to which spirited discussion and criticism of traditional values was in no way foreign; my father's friends, most of them medical researchers like himself, loved lively free conversation; but doctors are closer to everyday life and as human beings simpler than mathematicians, whose brains work in the sphere of highest abstraction. In any case, I still had not heard such frank, independent, free-ranging criticism of all possibile proceedings of science, art, politics."

To Weyl and Born and the other students it seemed that Hilbert and Minkowski were "heroes," performing great deeds, while Klein, ruling above the clouds, was "a distant god." The older man now devoted more and more of his time and energy to the realization of his dream of Göttingen as the center of the scientific world. Before the turn of the century he had brought together economic leaders and scientific specialists in an organization called the Göttingen Association for Advancement of Applied Mathematics and Mechanics. As a result of the activities of this group (familiarly known as the Göttingen Association) the University was gradually being ringed by a series of scientific and technical institutes — the model for scientific-technological complexes which were later to grow up around various universities in America.

Sometimes Klein was now a little amusing for the seriousness with which he took himself and his many projects. It was said that he had but two jokes, one for the spring semester and one for the fall. He did not allow himself the pleasures of ordinary men. Every moment was budgeted. Even his daughter had to make an appointment to talk with her father.

Without ever making an issue of the matter, Hilbert and Minkowski saw to it that they themselves were never organized. Once, after Klein had completely filled a very large blackboard with figures on the German middle schools (the scientific education of which he was also trying to reform), he asked his colleagues if they had any comments. "Doesn't it seem to you, Herr Geheimrat," Minkowski asked softly, "that there is an unusually high proportion of primes among those figures?" On another occasion, when Klein, written agenda in hand, tried to turn the informal weekly walks of the mathematics professors into department meetings, Hilbert simply failed to turn up for the next walk. But for the most part the three men, so different in their personalities, worked together in rare harmony.

In 1904, when the Extraordinariat for applied mathematics became vacant, Klein proposed to Althoff that a full professorship in that subject be established, the first in Germany specifically designated for applied mathematics. For the new position he had in mind Carl Runge, then at Hannover. Runge was not only a distinguished experimental physicist, famous for his measurement of spectral lines, but also a first-rate mathematician, whose name is attached to the approximation of analytic functions by means of polynomials.

Runge had known and admired Klein for almost ten years and had recently become acquainted with Hilbert. "Hilbert is a charming human being," he had written his wife. "His idealism and his friendly disposition and unassuming honesty cause a person to like him very much." The possibility of associating with these two gifted mathematicians was so exciting to Runge, who had felt very much alone in Hannover, that he went to Berlin to talk to Althoff about the new position with the feeling that it was too good to be true. "However," as his daughter later wrote, "he had not reckoned with the fact that for Klein's most broad and comprehensive plans there was always more sympathy from Althoff than for the individual plan of another." The new position was his if he wanted it. The salary would be somewhat less than what he had been receiving in Hannover, however.

"But you must not let yourself be influenced by financial considerations," his wife wrote emphatically when she heard the news. "We will come through, even with a thousand Marks less, and it will not hurt either me or the children."

At the beginning of the winter semester 1904–05, Runge joined the faculty. The mathematics professors, a quartet now, took up the practice of a weekly walk every Thursday afternoon punctually at three o'clock. Klein gave up preparing agendas. The walks became pleasant informal rambles during the course of which anything, including department business, could be discussed; and, as Hilbert happily remarked, "science did not come out too short."

Runge had a gift for computation which impressed even his new colleagues. Once, when they were trying to schedule a conference several years in the future, it became necessary to know the date of Easter. Since the determination of Easter is no simple matter, involving as it does such things as the phases of the moon, the mathematicians began to search for a calendar. But Runge merely stood silent for a moment and then announced that Easter that year would fall on such and such a date.

It was equally amazing to the mathematicians how Runge was able to handle mechanics. When the Wright brothers made their first flight, he constructed a model of their plane from paper scraps which he weighted down with needles and then allowed to glide to the ground. In this way he estimated "rather correctly" the capacity of the motor, the details of which were still a secret.

When Runge arrived in Göttingen, the scientific faculty most closely connected with the mathematicians was also impressive. Physicists were Eduard Riecke and Woldemar Voigt. H. T. Simon was the head of the Institute for Applied Electricity; Ludwig Prandtl, of the Institute for Applied Mechanics; and Emil Wiechert, the Institute for Geophysics. Karl Schwarzschild was professor of astronomy.

But the stimulating society was not limited to these high-ranking men.

Otto Blumenthal, who was to be distinguished for the rest of his life as "Hilbert's oldest student," was very close to the professors although he was still merely a Privatdozent. He was a gentle, fun-loving, sociable young man who spoke and read a number of languages and was interested in literature, history and theology as well as mathematics and physics. Although born a Jew, he eventually became a Christian and frequently spoke of "we Protestants."

The unusually close relationship existing between the docents and the professors is evidenced by the fact that when Blumenthal and Ernst Zermelo, another docent, wished to give some experimental lectures on elementary arithmetic, Hilbert and Minkowski regularly attended to give more authority to their project.

Zermelo was somewhat older than Blumenthal, a nervous, solitary man who preferred whisky to company. He liked to prove at this time, which was before Peary's expedition, the impossibility of reaching the North Pole. The amount of whisky needed to reach a latitude, he maintained, is proportional to the tangent of the latitude, i.e., approaches infinity at the

Pole itself. When newcomers to Göttingen asked him about his curious name, he told them, "It used to be *Walzermelodie*, but then it became necessary to discard the first syllable and the last."

It was Zermelo who had recently pointed out to Hilbert a disturbing antinomy in set theory — the same antinomy which the young English logician Bertrand Russell pointed out to Gottlob Frege as Frege was ready to send to the printer his definitive work on the foundations of arithmetic. This antinomy — a contradiction reached by using methods of reasoning which had been accepted by mathematicians and everyone else since the time of Aristotle — had to do with the commonly recognized fact that some sets are members of themselves and others are not. For instance, the set of all sets having more than three members is a member of itself because it has more than three members. On the other hand, the set of all numbers is not a member of itself, since it is not a number. But now Zermelo and Russell, independently, had brought up the question of *the set of all sets that are not members of themselves*. Since the members of this set are the sets which are not members of themselves, the set is a member of itself if, and only if, it is not a member of itself.

By 1904, after its publication by Russell, the antinomy was having — in Hilbert's opinion — a "downright catastrophic effect" in mathematics. One after another, the great gifted workers in set theory — Frege himself as well as Dedekind — had all withdrawn from the field, conceding defeat. The simplest and most important deductive methods, the most ordinary and fruitful concepts seemed to be threatened; for this antinomy and others had appeared simply as a result of employing definitions and deductive methods which had been customary in mathematics. Even Hilbert had now to admit that perhaps Kronecker had been right — the ideas and methods of classical logic were in fact not equal to the strong demands of set theory.

In the past Hilbert had believed that Kronecker's doubts about the soundness of set theory and certain parts of analysis could be removed by substituting consistency, or freedom from contradiction, for construction by means of integers as the criterion of mathematical existence and then obtaining the necessary absolute proof of the consistency of the arithmetic of real numbers. Up until the discovery of the antinomies it had been his opinion that the desired consistency proof could be achieved relatively easily by a suitable modification of known methods of reasoning in the theory of irrational numbers. But with the discovery of the antinomies in set theory, upon which much of this reasoning was based, he saw that he was going to have to alter his view. In the late summer of 1904, when the third International Congress of Mathematicians met at Heidelberg, he departed from integral equations for the moment to take up the subject of the foundations of mathematics.

It had been Kronecker's contention that the integer was the foundation of arithmetic and that construction by means of a finite number of integers was therefore the only possible criterion of mathematical existence. Hilbert still, as always, passionately opposed such a restriction of mathematics and mathematical methods. Like Cantor, he firmly believed that the essence of mathematics is its freedom; and he saw in arbitrary restriction a real danger to the science. He was sure that there was a way to eliminate the antinomies without making the sacrifices which Kronecker's views demanded. His solution, however, involved going even farther than Kronecker had.

Hilbert now insisted that the integer itself "can and must" have a foundation.

"Arithmetic is often considered to be a part of logic, and the traditional fundamental logical notions are usually presupposed when it is a question of establishing a foundation for arithmetic," he told the mathematicians gathered at Heidelberg. "If we observe attentively, however, we realize that in the traditional exposition of the laws of logic certain fundamental arithmetical notions are already used; for example, the notion of set and, to some extent, also that of number. Thus we find ourselves turning in a circle, and that is why a partly simultaneous development of the laws of logic and of arithmetic is required if the antinomies are to be avoided."

He was convinced, he told them, that in this way "a rigorous and completely satisfying foundation" could be provided for the notion of number — "number" which would include, not only Kronecker's natural numbers and their ratios (the common fractions), but also the irrational numbers, to which Kronecker had always so violently objected, but without which "the whole of analysis," in Hilbert's opinion, "would be condemned to sterility."

It was Hilbert's proposal at Heidelberg that for the first time in the history of mathematics, proof itself should be made an object of mathematical investigation.

Poincaré commented several times, unfavorably, upon the idea. The Frenchman was convinced that the principle of complete, or mathematical, induction was a characteristic of the intellect ("in Kronecker's language," as Hilbert once explained in discussing Poincaré's position, "created by God") and that, therefore, the principle could not be established except by complete induction itself.

Hilbert did not follow up the Heidelberg proposal. Instead, he continued to work on his theory of integral equations and on the side, in the company of Minkowski and at Minkowski's suggestion, began a study of classical physics.

Minkowski already had considerable technical knowledge in the field of physics; Hilbert had almost none and was familiar only with the broad outlines of the subject. Nevertheless he took it up with enthusiasm. For the second time since leaving school — the first time had been in connection with the *Zahlbericht* — he embarked on a course of "book study." Blumenthal, who had already begun what was to be a life-long study of his teacher's character and personality, was most impressed. He remembered an occasion during his own student days when in the course of his reading he had discovered to his dismay that the most beautiful development in his dissertation had already appeared in another paper. Hilbert, he recalled, had merely shrugged and said, "Why do you also know so much literature?"

Klein followed the joint physics study with interest. When he was 17, he had been assistant to Julius Plücker in Bonn. At that time he had been determined that "after the attainment of the necessary mathematical knowledge" he would devote his life to physics. Then, two years later, Plücker had died. A transfer to Göttingen — where the mathematicians were a far livelier group than the physicists — had made Klein a mathematician instead of a physicist.

As the physics study progressed, Minkowski became increasingly fascinated by the riddles of electrodynamics as recently formulated in the works of H. A. Lorentz. But Hilbert did not waver in his personal concentration on integral equations. In 1904 he sent a second communication to the scientific society in which he developed a significant extension of Fredholm's idea. In his classic work, Fredholm had recognized the analogy between integral equations and the linear equations of algebra. Hilbert now went on to set up the analogue of the transformation of a quadratic form of n variables onto principal axes. Out of the resulting combination of analysis, algebra and geometry, he developed his theory of eigenfunctions and eigenvalues, a theory which, as it turned out, stood in direct relation to the physical theory of characteristic oscillations.

The spirit and significance of this work can best be glimpsed by a layman in the evaluation of it by a later student of Hilbert's:

"The importance of scientific achievement is often not alone in the new material which is added to material already on hand," Richard Courant has written. "Not less important for the progress of science can be an insight which brings order, simplicity and clarity into an existing but hard to reach area and thus facilitates or first makes possible the survey, comprehension and mastery of the science as a unified whole. We should not forget this point of view in connection with Hilbert's works in the field of analysis, ... for [all of these] exemplify his characteristic striving to find in the solution of new problems the methods which make the old difficulties easy, to establish new connections in existing materials and to bring the many branching streams of individual investigations back into a single bed."

It was during this happy, productive time that Hilbert received yet another offer to leave Göttingen. Leo Koenigsberger would give up his own chair in Heidelberg if Hilbert would accept it.

Although Käthe favored the change, Hilbert refused.

He did not neglect, however, to use the offer to negotiate for further advantages for mathematics as the price of his remaining in Göttingen. To one of his proposals Althoff objected: "But we do not have that even in Berlin!"

"Ja," Hilbert replied happily, "but Berlin is also not Göttingen!"

XIII

The Passionate Scientific Life

At the beginning of the twentieth century mathematics students all over the world were receiving the same advice:

"Pack your suitcase and take yourself to Göttingen!"

Sometimes it seemed that the little city was entirely populated with mathematicians. But it should be mentioned that there were other people, and for some *la grande affaire* was quite different. A French journalist, choosing Göttingen as the place where he would be best able to observe German students in their natural state, was most impressed by the prison on the third floor of the Great Hall of the University. On Weender Strasse he saw, not mathematicians, but young men who "walked like lords," their caps visored in the bright colors of the duelling student fraternities, their faces usually swathed in bandages. "They leave behind them," he reported, "a nauseating odor of iodoform, which penetrates everywhere in Göttingen." But the mathematicians preferred to recount how Minkowski, walking along Weender Strasse, saw a young man pondering an obviously grave problem and patted him on the shoulder, saying, "It is sure to converge" — and the young man smiled gratefully.

The days were long past when Hilbert had delivered his lectures on analytic functions for the sole benefit of Professor Franklin. Frequently now several hundred people jammed the hall to hear him, some perching even on the window sills. He remained unaffected by the size or the importance of the audience. "If the Emperor himself had come into the hall," says Hugo Steinhaus, who came to Göttingen at this time, "Hilbert would not have changed."

Was Hilbert so because of his position as the leading mathematician in Germany? "No, Hilbert would have been the same if he had had only one piece of bread."

Born was now Hilbert's "personal" assistant. At that time it was generally customary in German universities that only professors in the experimental sciences had assistants, who helped them with the laboratory work. But Klein, shortly after he took over mathematics in Göttingen, had contrived to obtain funds for a paid clerk for the Lesezimmer. The first person to hold this position had been Arnold Sommerfeld, and the Lesezimmer clerk had quite naturally become Klein's assistant. Hilbert's assistant was still unpaid at this time.

It was a "rather vague" job, according to Born, "but precious beyond description because it enabled me to see and to talk to him every day." In the morning Born came to Hilbert's house, where he usually found Minkowski already present. Together, the three discussed the subject matter of Hilbert's coming lecture, which was often taking place that same morning.

Hilbert had no patience with mathematical lectures which filled the students with facts but did not teach them how to frame a problem and solve it. He often used to tell them that "a perfect formulation of a problem is already half its solution."

"He would take a good part of the hour to explain a question," Steinhaus recalls. "Then the formal proof, when it followed, would appear to us so natural that we wondered why we had not thought of it ourselves."

In the preparation periods with Minkowski and Born, Hilbert was interested only in the general principles which he would present to the class. He refused to prepare to the point where, as he said contemptuously, "the students could easily fill up fine notebooks." Instead, his goal was to involve them in the scientific process itself, to illuminate difficulties and "to shape a bridge to the solution of actual problems." The details of the presentation would come to him later on the rostrum.

"It was a wonderful learning time for me," Born has written of these discussions with Hilbert and Minkowski, "not only in science, but also in the things of human life. I admired them and loved them both and they did not let me feel how great was the disparity in knowledge and experience between them and me, but treated me as a young colleague."

When it was time for Hilbert to go to his lecture, Minkowski, returning home, would frequently take Born with him. It was only two blocks from Hilbert's house to the Minkowski apartment on Planck Strasse; but often, deep in conversation, they "made a long walk" before they got there. The little girls would come running to meet their father, the one who arrived first being hoisted on his back and carried into the house, clutching his thick dark hair and shrieking happily. Unlike Hilbert, whose kinship with youth did not extend to the very young, Minkowski understood and enjoyed children. He had been the one whose encouragement and play had finally got the baby Franz to talk and his letters to Hilbert had always contained some message for Franz. His own children were to remember a father who saw each of them separately for a few minutes every day so that the younger would have a chance to talk to him too. "Uncle Hilbert" they were to remember as a man "not good with children."

Because of his very general method of preparation, Hilbert's lectures could turn into fiascos. Sometimes the details did not come to him, or came wrong. He got stuck. The assistant, if he were present, might be able to step in and rescue him. "The students are confused, Herr Professor, the sign is not right." But frequently both he and the class were beyond such help. He might shrug, "Well, I should have been better prepared," and dismiss the class. More often, he was inclined to push on.

And still, it was commonly agreed in Göttingen, there was no teacher who came close to Hilbert. In his classes mathematics seemed to the students to be still "in the making"; and most of them preferred his lectures to the more perfectly prepared, encyclopedic and "finished" lectures of Klein.

Rather unexpectedly, Hilbert had quite an interest in pedagogy. He did not have a very high opinion of the abilities of the ordinary student and believed that nothing much was absorbed until it was heard several times. "Five times, Hermann, *five* times!" was the memorable advice he gave to Weyl when that young man started his lecturing career. "Keep computations to the lowest level of the multiplication table" and "Begin with the simplest examples" — these were other favorite rules. He himself tried to present important ideas in especially vivid forms, looking always for contrasts to make them more striking and memorable.

The lectures on ordinary differential equations opened with two equations on the blackboard: $y'' = 0$ and $y'' + y = 0$. "Meine Herren," he would say, "from those you can learn the entire theory, even the difference in the significance of the initial value and the boundary value problem."

"The sentence, 'All girls who are named Käthe are beautiful,' is not a universal law," he told another class. "For it is dependent upon the naming and that is arbitrary."

The difference between a purely existential statement and a specified object was illustrated by the statement, which always brought a laugh from the students: "Among those who are in this lecture hall, there is one who has the least number of hairs."

In addition to his own lectures, Hilbert regularly conducted a joint seminar with Minkowski. In 1905, after a year of physics study, they decided to devote the seminar to a subject in physics — the electrodynamics of moving bodies. Although the stimulus to the undertaking came originally from Minkowski, Hilbert took an active part and was a real partner, according to Born, "frequently clarifying and always pressing toward clarity."

For Born and the other students the seminar meetings were exciting and stimulating hours. The Fitzgerald contraction, Lorentz local-time, the Michelson-Morley interference experiment, all were thoroughly discussed and "we heard completely fantastic statements about electrodynamics."

In one of those coincidences that occur not infrequently in the history of science, the year of the seminar saw similar ideas developed in papers on electrodynamics and special relativity by a patent clerk in Bern. "But of that," Born says, "nothing was as yet known in Göttingen and in the Hilbert-Minkowski seminar the name of Einstein was never mentioned."

Born, much impressed by the ideas discussed in the seminar, considered taking a topic in this field for his dissertation. But he had fallen into disgrace with Klein in another seminar, and it was axiomatic in Göttingen that those who were out of favor with the great Felix did not fare well. So, rather than risk being examined by Klein in geometry, Born switched to astronomy. He still would have to be examined by a mathematician, but now it would be Hilbert.

Before the examination, Born asked Hilbert's advice on how he should prepare for the mathematical questions.

"In what area do you feel yourself most poorly prepared?" Hilbert asked.

"Ideal theory."

Hilbert said nothing more, and Born assumed that he would be asked no questions in this area; but when the day of the examination arrived, all of Hilbert's questions were on ideal theory.

"Ja, ja," he said afterwards, "I was just interested to find out what you know about things about which you know nothing."

After 1905, Minkowski concentrated almost entirely upon electrodynamics. The work of the Bern patent clerk became known in Göttingen, and Minkowski recalled his former student. "Ach, der Einstein," he said ruefully, "der schwänzte immer die Vorlesungen — dem hätte ich das gar nicht zugetraut." (*Oh, that Einstein, always missing lectures — I really would not have believed him capable of it!*)

Hilbert continued with his investigations in the field of integral equations. To maintain a close bond between these and his teaching, he often brought his results into lectures and seminars before they were in finished form. Often it happened then that the investigations moved forward in a kind of collaboration with his students who, as he later remarked with pleasure, "repeatedly contributed to a more precise formulation and sometimes actually to an extension of the field." In 1904, for example, he had published his theory of eigenfunctions and eigenvalues. It was still "laborious" at a crucial point. Then in 1905, in his dissertation, Erhard Schmidt laid down a new foundation for the Hilbert theory which was destined, because of its clarity and brevity, to play an important role.

It was at this time, in the year 1905, that the Hungarian Academy of Science surprised the mathematical world with the announcement of an impressive prize of 10,000 gold crowns for that mathematician whose achievement during the past 25 years had most greatly contributed to the progress of mathematics. It was to be known as the Bolyai Prize in honor of Johann Bolyai, the Hungarian who had been one of the discoverers of non-euclidean geometry, and his father Wolfgang Bolyai, a fellow student and life-long friend of Gauss.

The Academy appointed a committee of Julius König, Gustav Rados, Gaston Darboux, and Felix Klein to decide the winner; but, even before the committee met, it was clear to everyone in the mathematical world that the choice would be between two men. The final vote was unanimous. The Bolyai Prize would go to Henri Poincaré, whose mathematical career had begun in 1879 while Hilbert was still a student in the gymnasium; but the committee also voted unanimously that, as a mark of their high respect for David Hilbert, the report which they made to the Academy on their choice would treat his mathematical work to the same extent that it treated Poincaré's.

"No cash, but honor," Klein wrote Hilbert regretfully from Budapest.

Later, back in Göttingen, Klein explained to Blumenthal that the decisive factor in the award's going to Poincaré was that the Frenchman had accomplished "a full orbit of the mathematical science."

"But," Klein prophesied, "Hilbert will yet encompass as comprehensive a field as Poincaré!"

It was an auspicious time for the prophecy. Hilbert was now creating what was to be the crowning achievement of his analytic work — the theory of infinitely many variables which was to become universally known as "Hilbert Space" theory.

The generalization of the algebraic theory of quadratic forms from 2 and 3 to any finite number of variables had been popular with algebraists during the preceding century; and since a pair of variables represents a point in the plane and a triple represents a point in 3-dimensional space, they had found it convenient to move to "spaces" of higher dimension as the number of variables increased. Such generalization is, as E. T. Bell once remarked, "an almost trivial project for a competent algebraist." But with the extension to an infinite number of variables, convergence has to be considered, and the resulting analytic problem is "trivial for nobody." A still further generalization is one in which, for example, the points in the space are represented by continuous functions.

Because of its extreme generality, the problem he was now attacking seemed almost inaccessible even to Hilbert. But he approached it boldly.

"If we do not let ourselves be confused by such considerations, it goes for us as for Siegfried, before whom the fire fell back, and as reward we are beckoned on — to the beautiful prize of a methodologically uniform framework for algebra and analysis!"

In the convenient and vivid imagery of space, many analytic relations can be expressed in the terms of familiar concepts, and much that is complicated and obscure when formulated analytically becomes almost intuitively obvious when formulated geometrically. For this reason Hilbert Space theory — which for technical reasons Hilbert himself originally called "Spectral Theory" — offered an immensely suggestive language for the easy and direct expression of very abstract results. Although it yielded up many of his own results and also his methods much more easily, this was not, however, to be its chief significance.

"Decisive above all," Courant later wrote, "is the ordering and clarifying effect of such a general function theory on the entire methodology and conceptual development in analytic investigation."

At the same time that Hilbert was developing this very advanced and abstract mathematical theory, he was also teaching calculus to first year students.

His calculus class of 1906, although typical of his teaching technique at this level, was in some ways different from those which preceded it and followed it in that it took place during the great ski winter. Inspired by Runge, who had an English mother and was thus automatically the sportsman of the faculty, Hilbert and some of the younger teachers had decided to learn to ski. Equipment was imported from Norway, since none was manufactured in Germany at that time. The sessions, coached by Runge, took place on the gentle slope below Der Rohns, a popular inn.

"Well, you know, it is very nice but it is also very strenuous," Hilbert confided to Minkowski at the weekly meeting of the Mathematics Club.

"This afternoon I found myself in a situation where I really did not know that I had fallen into a ditch. My two skis were up in the air and I was on my back. Then one of my skis came off and slid down the hill. So I had to take off the other ski and carry it through the deep snow. You know, that's all not so simple."

"Well," said Minkowski, who had not taken up the new sport, "why didn't you just let the second ski run down the same path as the first one? It would have landed next to it."

"Oh," said Hilbert. "Runge never thought of that!"

There was a slight incline between Hilbert's house and the Auditorienhaus, where the calculus class met; and when there was sufficient snow on the ground, Hilbert liked to ski to class. On such days he would dash breathlessly into the lecture hall, still wearing his enormous Norwegian ski boots with points on the toes and buckles on the backs, and leap to the rostrum, already talking.

It was still his custom to open a lecture by carefully reviewing what he had presented in the previous lecture. If he had spent 40 minutes on a subject the last time, he now spent 20 minutes on it. Only after that review did he take up the new subject.

"Last time we saw such and such. Now it seems that this is not applicable in the new case. How is that possible? Why does the old method break down? What is the matter? What can we do? How can we possibly get around this difficulty?"

He would go on in this vein for some time. He would also bring in ideas from other fields and mention very advanced results and recent work. The students would be fascinated by the glimpses of concepts and areas of mathematics which in the normal course of events they would not meet for several years, but at the same time they would find themselves growing increasingly impatient to get on to the subject of the day. Then, just as they were ready to give up hope, came the necessary new concept — "like a marble statue lighted up in a dark park."

"It was wonderful," says Paul Ewald, who was a student in the calculus class of 1906. "When it finally came, we felt as if we had actually seen Hilbert *create* the new concept that was needed."

By this time Hilbert's assistants were paid, and it was even sometimes possible to arrange a special "Ausarbeiter" for a large class. This particular year Hilbert had managed to wangle some non-mathematical funds for mathematics, and Ewald was hired for the job of "Ausarbeiter" and paid as "a forestry worker" for a nearby village. It was his duty to prepare a clean copy of the notes he took on Hilbert's lecture and then have it approved by the assistant, who was now Born's friend Hellinger.

Even in an elementary class like calculus it often happened that Hilbert got things garbled. Then Hellinger, looking sadly at Ewald's notes, would say, "Well, he's muddled it again and we'll just have to sit down and work it out." When the notes finally satisfied Hellinger, they were deposited in the Lesezimmer, where the students could consult them.

Ewald was to become a distinguished physicist, but he was always to say that he learned almost all the analysis he ever needed in Hilbert's calculus class and in the after-sessions with Hellinger.

In the spring semester of the calculus class, Hilbert bought a bicycle, a method of transportation which was just beginning to become popular in Göttingen, and at the age of 45 started to learn to ride.

Skiing was a temporary enthusiasm; but bicycling, like walking and gardening, became a regular accompaniment to his creative activity. He still preferred to work outdoors. Now the bicycle was always nearby. He would work for a while at the big blackboard which hung from his neighbor's wall. Then suddenly he would stop, jump on the bicycle, do a figure-eight around the two circular rose beds, or some other trick. After a few moments of riding, he would throw the bicycle to the ground and return to the blackboard. At other times he would stop what he was doing to pace up and down under his covered walk-way, his head down, his hands clasped behind him. Sometimes he would interrupt his work to prune a tree, dig a little, or pull some weeds. Visitors arrived constantly at the house and the housekeeper directed them to the garden, saying, "If you don't see the professor, look up in the trees." Usually the first word that Hilbert spoke revealed that in spite of appearances he was working toward the solution of some specific mathematical problem with the greatest ardor. He would continue to pursue his train of thought, but aloud now, unless the visitor had come with a problem of his own. In that case he would talk with interest and enthusiasm about that.

Richard Courant, who had recently joined the Breslau group of Born, Hellinger and Toeplitz, often observed the activity of Hilbert in the garden from the balcony of his nearby room. "A fantastic balance," it seemed to him, "between intense concentration and complete relaxation."

The year after Hilbert presented his theory of infinitely many variables to the scientific society, Erhard Schmidt, by this time a Privatdozent at

Berlin, published his own very simple and beautiful solution-method which, like his dissertation, again carried forward his teacher's work.

Thus, in Göttingen, the life of science proceeded, unforgettable to those who lived through it, but unnoted by visiting journalists.

XIV

Space, Time and Number

It was a halcyon time. Göttingen seemed a half forgotten kingdom of the past. The state of Hannover, of which it was a part, had been defeated and annexed by Prussia in 1866; but forty years later the remnants of the Hannoverian nobility still quietly resisted the rule of the victor. The mark of George II's adopted country remained on his German university as thoroughly as his own guttural accent had remained on his English. The houses that lined Prinzenstrasse were those of dukes and princes of Hannover, but their titles were given in English rather than in German. The scientific society of Göttingen was officially known in the English fashion as "die Königliche Gesellschaft der Wissenschaften" — *the Royal Society of Science*. The British minister of war regularly spent his summers in Göttingen. To the young people who had been drawn there from all over the world because of their love of mathematics, it seemed that they had "all time". But it did not always seem so to their elders.

By 1908 Hilbert and Minkowski had been friends for a quarter of a century. Hilbert was 46; Minkowski, 44.

For a while, during the anniversary summer, Hilbert's marvelously good health and natural optimism failed him. He became very nervous and depressed.

The breakdown did not seem to have been triggered by any specific experience. Some, like Blumenthal, thought it was the consequence of the reckless physical and mental exertions of the past few years. Others saw it as characteristic of the creative worker.

"Almost every great scientist I have known has been subject to such deep depressions," Courant says. "Klein, of course, but many others too. There are periods in the life of a productive person when he appears to himself and perhaps actually is losing his powers. This comes as a great shock."

In any case, Hilbert analyzed his illness with intelligence, then set about in a deliberate manner to get well. After several months of rest at a sanitorium in the nearby Harz mountains, he was lecturing again as usual in the fall.

In contrast to Hilbert, Minkowski was at a point of great creativity during the summer of 1908. In September he presented some of his new electrodynamical results at the annual meeting of the Society of German Scientists and Physicians, which was held in Cologne. The title he chose for this talk was "Space and Time."

"The views of space and time which I wish to lay before you," he began in his quiet, hesitant voice, "have sprung from the soil of experimental physics, and therein lies their strength. They are radical. Henceforth space by itself, and time by itself, are doomed to fade away into mere shadows, and only a kind of union of the two will preserve an independent reality."

He had often told his students in Göttingen, "Einstein's presentation of his deep theory is mathematically awkward — I can say that because he got his mathematical education in Zürich from me."

In his special theory of relativity, Einstein had shown that when mechanical events are described by the use of clocks and measuring rods, the description depends on the motion of the laboratory in which the instruments are used, and had stated the mathematical relations that connect the different descriptions of the same physical event.

Now came, in Minkowski's talk at Cologne, what has been called "the great moment of geometrization." In the period of a few moments Minkowski introduced into relativity theory his own beautifully simple mathematical idea of Space-Time by means of which the different descriptions of a phenomenon can be represented mathematically in a very simple manner.

"Three-dimensional geometry becomes a chapter in four-dimensional physics."

"Now you know why," he told his listeners, "I said at the outset that space and time are to fade away into shadows, and only a world in itself will subsist."

Among the members of the audience was Max Born, whose interest in relativity had been re-aroused by the recent works of Einstein. Minkowski wanted Born to return to Göttingen as his collaborator. He needed someone with the knowledge of optics which Born had. But first he wanted his former pupil to become more familiar with his own new ideas in the field. He sent Born back to Breslau with his latest electrodynamical work.

In Minkowski's work the young man found laid out "the whole arsenal of relativity mathematics . . . as it has been used every day since then by every theoretical physicist." Not until the beginning of December did he consider himself prepared to return to Göttingen.

"There followed several weeks when I saw Minkowski every day and talked with him. It was a happy time, full of scientific excitement, but also rich in experience of a personal sort, the beginning of a true friendship so far as the difference of age and experience permits this word to be used."

After they had finished discussing the relativity problems, they talked about number theory: "For Minkowski, as for Hilbert, number theory was the most wonderful creation of the human mind and spirit, equally a science and the greatest of arts."

It happened that at this time, just when Minkowski had deserted number theory for electrodynamics, Hilbert, recovering from his breakdown of the summer, had been ensnared by a notorious problem in the classical theory of numbers. In 1770, Edmund Waring, an otherwise undistinguished English mathematician, had asserted, apparently without any proof, that every ordinary whole number can be represented as the sum of four squares, nine cubes, nineteen fourth powers, and so on — in general, a finite number for every nth power. At about the same time it had been proved in connection with another theorem that every such number can in fact be represented as the sum of four squares. It did not follow, however, that Waring, having been right in the case of the squares, was also right for the other powers. It did not even follow that every number could be represented by some finite number of cubes, fourth powers, and so on. The number needed for some or all powers greater than 2 might increase indefinitely as the numbers to be represented increased in size. Since 1770, little progress had been made toward a proof of this statement of Waring's. Recently, however, mathematicians had begun to show new interest in the problem because of the possibility of successfully applying certain analytic methods to it. Hurwitz had worked in this direction; but, like all the other mathematicians who had in the past tried to prove Waring's theorem, he too had had to give up, defeated. But Hurwitz's work had aroused Hilbert's interest in the problem. For the moment he turned away from integral equations. He began where Hurwitz had left off, even using as his starting point an identity of the kind that had been set up by Hurwitz. At the end of 1908, exactly 138 years since Waring first made his conjecture, Hilbert produced a proof of Waring's theorem.

Typically, Hilbert's proof aimed at existence rather than actual construction. However, it differed from his proof of Gordan's Theorem in that while it did not actually establish the number of nth powers needed for the desired representation, it did provide a method by which at least in principle an estimate could be made in each case.

It was not by any means a simple proof. In fact, as the Russian number theorist Khinchin has pointed out, it was "not only ponderous in its formal aspect and based on complicated analytical theories... but also lacked transparency in conceptual respects." But in view of the previous intractability of the problem, it was a remarkable achievement.

"It would hardly be possible for me to exaggerate the admiration which I feel for the solution of this historic problem," G. H. Hardy wrote when, later, he and J. E. Littlewood had produced another proof of the theorem. "Within the limits which it has set for itself, it is absolutely and triumphantly successful... one of the landmarks in the modern theory of numbers."

Hilbert himself was tremendously pleased and proud. "He had fought with a master of the high degree of Hurwitz," Blumenthal noted, "and gained victory with weapons from Hurwitz's own armory at a point where Hurwitz saw no possibility of success." He thought happily of communicating his result to his old friend in his next letter. But before he told Hurwitz about it, he would present the proof of Waring's Theorem to Minkowski and the members of the joint seminar at the first meeting of the new year.

Minkowski had been gone from Göttingen during the Christmas holidays, but he returned on Wednesday, January 6. The next day being Thursday, the four mathematics professors made their weekly hike, promptly at three o'clock, to the Kehrhotel on the Hainberg. In spite of the wintry hills and leafless trees, it was a pleasant excursion. The cold air rang with loud cheerful voices and laughter. Minkowski recounted "with special liveliness" his latest results in his electrodynamical work. Hilbert astonished everyone with the announcement that he would present a proof of Waring's Theorem at the next meeting of the seminar.

On Friday Minkowski delivered a regular lecture. After that he conducted a doctoral examination.

Then, on Sunday afternoon, following dinner, he was suddenly stricken with a violent attack of appendicitis. That night the decision was made to undertake the difficult operation to remove the ruptured organ.

Through Monday Minkowski's condition worsened. He was conscious and quite clear about the hopelessness of his situation. On the hospital bed he studied the proof-sheets of some of his latest work and considered whether it would be possible to turn the still unfinished part of the work to good account.

Hilbert later recalled: "He spoke his regrets upon his fate, since he still could have accomplished much; but he decided that it would be good to correct the proof-sheets so that his latest electrodynamical works could be more easily read and better understood." He said that perhaps after his death the opposition to his new ideas could be more easily overcome.

"Even on the hospital bed, lying mortally afflicted, he was concerned with the fact that at the next meeting of the seminar, when I would talk on my solution of Waring's Problem, he would not be able to be present."

At noon on Tuesday, January 12, 1909, Minkowski asked to see his family and Hilbert again. Hilbert left home as soon as he received the message; but by the time he reached the hospital, Minkowski was dead. Not having yet attained his forty-fifth year, he had been taken "in the full possession of his vital energy, in the middle of his most joyful work, at the height of his scientific creativity."

Later that afternoon Hilbert wrote a letter to Hurwitz. "My Dear Old Friend," he began. "Now you alone are that to me . . . ."

The handwriting, larger than usual, became blurred as the short note progressed. He had planned, he said, to write about "a good idea" he had obtained for the solution of Waring's Problem "from your beautiful work" in the Göttingen *Nachrichten*, "but instead you receive this sad letter." He signed himself, "Your Old Friend." Then, as if to convince them both of what had happened, he repeated in a postscript the activities of Minkowski during the week, the return from the Berlin trip on Wednesday, the happy hike to the Kehrhotel on Thursday afternoon, the lecture on Friday and the doctoral examination, the attack on Sunday, the operation on Sunday night.

"The doctors themselves stood around his bed with tears in their eyes."

On Wednesday morning the announcement was made to the students.

"I was in class when Hilbert told us about Minkowski's death, and Hilbert wept," a former student recalls. "Because of the great position of a professor in those days and the distance between him and the students it was almost more of a shock for us to see Hilbert weep than to hear that Minkowski was dead."

On Thursday afternoon there was no mathematical walk. Instead, the mathematics professors provided Minkowski's body with a final escort. Again, Hilbert noted, it was exactly three o'clock.

"The strong mathematicians were like men confounded," another student wrote to his parents after the funeral. "To all appearances Klein himself found it difficult to speak calmly. Hilbert and Runge seemed disfigured, their eyes were so red with tears."

The solution of Waring's Problem — "over the proof-sheets of which his sure eye has never passed" — was published shortly afterwards and inscribed by its author:

"To the memory of Hermann Minkowski."

XV

Friends and Students

No amount of clever negotiation at the ministry in Berlin could obtain a replacement for the friendship and scientific stimulation which Hilbert had received from Minkowski. But life had to go on.

The strain under which Hilbert worked is revealed in an incident which occurred during a lecture a short time after Minkowski's death. Among those in the audience was a young man who, in spite of the fact that the professor was obviously distraught, persisted in interrupting him with questions. Finally Hilbert snapped:

"We are not here to give you information."

"But that is what you are paid for, Herr Geheimrat!"

In the shocked, embarrassed silence that followed, Hilbert, obviously shaken and angry, waited for the offender to leave the lecture hall. The young man remained stubbornly in his seat. Finally it was the professor, his face white, who turned and walked out.

"Such an incident would never have occurred," says a man who was present, "if Hilbert had been himself."

But for the most part Hilbert succeeded in accepting his loss with the same philosophical calm with which he had seen his friend accept death. There was no recurrence of the deep depression of the summer.

He took an active part in the selection of Minkowski's successor. He and Klein agreed that they were looking for a young man, someone whose achievements were still ahead of him. Hurwitz was thus ruled out. Various young mathematicians were considered. The choice finally narrowed down to Oskar Perron and Edmund Landau, and there was considerable discussion of the respective merits of the two among the members of the mathematical faculty.

"Oh, Perron is such a wonderful person," Klein said at last. "Everybody loves him. Landau is very disagreeable, very difficult to get along with. But we, being such a group as we are here, it is better that we have a man who is not easy."

This was the decisive statement. In the spring following Minkowski's death, the 32-year-old Landau came to Göttingen as professor of mathematics.

Landau's specialty was the application of analytic methods to the theory of numbers. Already, while a Privatdozent in Berlin, he had proved a very general theorem on the distribution of the prime ideals in an arbitrary algebraic number field, corresponding to the classical Prime Number Theorem. He had also done important work in function theory and had been able to extend Picard's famous theorem in such a wholly unexpected way that even he himself had not at first believed that he was correct and had delayed publishing his paper for more than a year.

His book on the distribution of primes, which is the central problem in the analytic theory of numbers, appeared the same year that he arrived in Göttingen. "In it," G. H. Hardy wrote many years later, "the analytic theory of numbers is presented for the first time, not as a collection of a few beautiful scattered theorems, but as a systematic science." It transformed the subject, "hitherto the hunting ground of a few adventurous heroes," into one of the most fruitful fields of mathematical research.

Although most German professors at that time came from the upper middle class and were comfortably well off, Landau was very rich. When people asked him how to get to his house in Göttingen, he said simply: "You will have no difficulty in finding it. It is the finest house in town."

Soon after his arrival at the University, the Landau stories began to compete in number with the Hilbert stories.

A student consulted Landau regarding the quality of a piece of amber, *Bernstein* in German. In replying, Landau gave at the same time his opinion of the relative merits of two mathematicians named Bernstein who were in Göttingen then. "Felix," he said. If he had said, "Serge," it would have meant that he considered the amber to be of superior quality. (This judgment was not so damning as it sounds. Felix Bernstein was a very good mathematician, famous for his work in actuarial theory and statistics; but Serge Bernstein was one of the greatest Russian mathematicians of the time.)

Landau had none of Minkowski's interest in geometry or mathematical physics, and he had an absolute contempt for applied mathematics.

One time Steinhaus was describing his doctoral examination to Landau. For this he had had to be examined by an astronomer. Landau appeared to be very impressed that a student of pure mathematics could successfully answer questions put to him by an applied mathematician. "What did he ask you?" he demanded. Pleased to have gained the professor's interest in his experience, Steinhaus explained that the astronomer had asked him for the differential equations for the movements of three celestial bodies.

"Ah, so he knows *that*!" Landau exclaimed. "So he knows that."

That was Landau.

Colleagues and students disliked his arrogance and feared his wit and ruthless honesty. They gave him their allegiance and respect, however, for his fantastic diligence and the unexpected impersonality of his devotion to mathematics. "Most of us are at bottom a little jealous of progress by others," Hardy once commented, "[but] Landau seemed singularly free from such unworthy emotions."

Landau was scarcely settled in Göttingen before he suggested to a young Dane named Harald Bohr, who had solved a problem mentioned in one of his works, that he should come down and study with him.

Bohr was an unusual mathematician. In 1908, he had been a member of Denmark's runner-up Olympic soccer team, and he was probably the only person in the history of mathematics to have his examination for the doctoral degree reported on the sports page. In the years to come he was to attach his name inextricably to what are known as "almost periodic functions." But he was never to be able to go by a ball without kicking it.

To Bohr a spirit of genuine international brotherhood seemed to reign among the young mathematicians of Göttingen when he arrived there, a few months after Minkowski's death. The transfer of foreign currency was as easy as possible. No one was ever asked for a passport. The German students, especially the slightly older ones, looked after the young foreign students "with touching care."

"The grand old man ... was Felix Klein. His imposing and powerful figure filled all, young and old, with great respect — one could almost say with awe.... But over the whole life in Göttingen shone the brilliant genius of David Hilbert, as if binding us all together.... Almost every word he said, about problems in our science and about things in general, seemed to us strangely fresh and enriching."

That spring Hilbert sent no communication on integral equations to the Göttingen Scientific Society. He and Käthe spent a great deal of time with Guste Minkowski and the little girls. He took over the general editorship of Minkowski's papers, and he began to plan a memorial talk. In preparation for this, he read over the ninety-odd letters that he had received from Minkowski since their university days. "It is like living our whole life over," he wrote to Hurwitz, "and I see how important a part you played in it."

Almost coincidentally with Minkowski's death an opportunity offered itself to bring to Göttingen distinguished visitors for the personal scientific exchange that was so necessary for Hilbert's own productivity and which in the past he had obtained from Minkowski. A Darmstadt mathematics professor named Paul Wolfskehl left in his will a prize of 100,000 Marks for the first complete proof of Fermat's Last Theorem. Until the prize was claimed, the interest accruing from this sum was to be used at the discretion of a committee of the Göttingen Scientific Society. Hilbert became the chairman of the committee, and in the April following Minkowski's death he was able to arrange that 2500 Marks be used to bring Henri Poincaré to Göttingen.

Socially and mathematically, the situation was delicate. The breakdown which had changed the entire course of Klein's career had been brought about by his competition with the young Poincaré. Now the leading mathematicians in the world were Hilbert and Poincaré, but the Bolyai Prize had gone to Poincaré. To many people in Göttingen, the Frenchman's presence was an unwelcome reminder that the mathematical world was not a sphere, with its center at Göttingen, but an ellipsoid.

Poincaré's choice of subjects for his lectures did not help the situation. He decided to speak on integral equations and relativity theory, both areas in which he had made substantial contributions, and he probably chose these topics because he knew that the Göttingen mathematicians were interested in them. But a foreign mathematician who was present was very surprised at the coolness with which the famous guest was received. "*We* were surprised," one of the Göttingen docents explained, "that Poincaré would come and talk to *us* about integral equations!"

Hilbert, however, always addressed Poincaré as "My Dear Friend" and referred to him in lectures and papers as "the most splendid mathematician of his generation." He and Käthe gave a large reception for the Frenchman and for Klein, whose sixtieth birthday fell during the visit.

Minkowski too had had the highest admiration for Poincaré; and the little Minkowski girl, seeing the great man on the steps of the house at Wilhelm Weber Strasse, dropped him a deep curtsy, as befits a little girl when she sees a king.

"What a joy it is to be a mathematician today!" Hilbert told his guests in a little speech. "Everywhere mathematics is budding, and new sprigs are blossoming. For in its applications to the natural sciences and in connection with philosophy, mathematics is becoming ever more important and is in the process of regaining its former central position!"

But in a letter to Hurwitz, thanking him for his friendly appreciation of the proof of Waring's Theorem, he wrote that it was "a bright ray in the darkness."

On the first of May he delivered his address in memory of Minkowski before a special meeting of the Göttingen Scientific Society.

With loving care he described his friend's work, recounted his successes and the appreciation which he had received from mathematicians like Hermite and Dedekind. "In spite of the fact that he was genuinely modest and willingly kept his person in the background, he nevertheless had the inner conviction that many of the works produced by him would survive those of other contemporary authors and would be admitted eventually to general appreciation. He rated the theorem which he discovered on the solvability in integers of linear inequalities, his proof of the existence of ramification numbers in number fields, and the reduction of the cubic inequality which expresses the maximum property of the sphere to a quadratic inequality equivalent to the finest achievements of the great classical mathematicians in the field of number theory combined with geometry."

He had accomplished a great deal in a short time. "Diligent he must have been!" His science had accompanied him wherever he went. "It was at all times interesting to him and in no part fatigued him, whether on an excursion or during the summer holiday, in an art gallery or a railway carriage or on the sidewalks of a big city."

Since their school days, Hilbert told his colleagues, Minkowski had been his best and most trustworthy friend:

"Our science, which we loved above everything, had brought us together. It appeared to us as a flowering garden. In this garden there were well-worn paths where one might look around at leisure and enjoy oneself without effort, especially at the side of a congenial companion. But we also liked to seek out hidden trails and discovered many an unexpected view which was pleasing to our eyes; and when the one pointed it out to the other and we admired it together, our joy was complete."

To Hilbert it seemed that his friend's nature had been like the sound of a bell, "so clear in the happiness in relation to his work and the cheerfulness of his disposition, so full in steadiness and trustworthiness, so pure in his idealistic aspirations and his life conception."

"He was for me a gift of the gods — such a one as would seldom fall to a person's lot — and I must be thankful that I so long possessed it."

In the coming months and years Hilbert tried to find congenial companionship among the advanced students and docents at Göttingen. He was quite clear about the necessity of contact with youth for his own scientific creativity.

"I seat myself with the young," he announced at one scientific meeting, "from whom there is still something to get."

One young friend of long standing was Leonard Nelson, a philosophy docent 20 years Hilbert's junior. Their acquaintance had begun several years earlier when Nelson, having received his doctor's degree at Berlin, was attempting to habilitate at Göttingen. Nelson was a young man with a strong leaning toward controversy, personal, philosophical, and political. He had incurred the dislike of Husserl, the philosophy professor; and his habilitation thesis was rejected by the majority of the Philosophical Faculty, which also included the mathematicians. Afterwards, when Nelson had received the bad news and was sitting dejectedly in his room, there was a knock on his door. "And to my astonishment," he wrote to his parents, "there was Hilbert in his own person. He invited me to supper at his house...." In the next letter he reported, "Hilbert is 'racking his brains' how we can get my habilitation paper accepted." As it turned out, this project had taken even Hilbert several years; but now Nelson was a docent and he and the professor were frequently seen "walking the wall" together, deep in discussion of that area of knowledge where philosophy, mathematics and logic meet.

Another young friend, also not a mathematician, was Theodor von Kármán, who was an assistant in Prandtl's applied mechanics institute. Von Kármán was working on a project called the Zeppelin which the government wanted to test under a variety of atmospheric conditions. Many years later, when he was an influential figure in aviation and space research in the United States, he called Hilbert "the greatest mathematician in the history of science... for he developed the theory of integral equations into a tool which enabled scientists to make break-throughs in regions once muddy with confusion."

After Minkowski's death Hilbert revived the custom of taking a group of the young people for a long walk following the weekly meetings of the Mathematics Club.

"He was not young... but he had retained all his full strength and youthfulness," it seemed to the 22-year-old Bohr, "[and] his profound originality, his total lack of prejudice, one might even say of convention, made each of these meetings with him a real experience."

Several of the gifted students of the past few years were now working their way up the academic ladder.

At Hilbert's suggestion, Max Born was entrusted by Mrs. Minkowski with the editing of her husband's physics papers. One of these Born had to reconstruct from the barest notes. He also carried on his teacher's work with a paper of his own in which he presented a new and rigorous method for calculating the electromagnetic self-energy of the electron. A talk presented on this paper so impressed Voigt that he offered Born a position as a Privatdozent in the Institute of Theoretical Physics.

Hermann Weyl also became a Privatdozent about this time. Although he had already shown his mathematical abilities, he was still too shy for the socially "in-group" of mathematics. Thus it was something of a surprise to everybody when he shortly won the hand of a much sought-after young lady whose charms were such that when her father threatened to withdraw her from the University a petition begging him to reconsider was signed even by professors.

It was also at this time that the friendship between Hilbert and Richard Courant began.

Already it was clear that here was a young man who would go far, and not only in mathematics. From the age of fourteen he had lived alone, supported himself by tutoring the students of a girls' school, and ultimately accomplished the near-impossible by passing the Abitur without having completed the gymnasium course. Unlike most university students of that time, he was completely dependent upon himself for his support.

One day, after attending a lecture by Hilbert, Courant was surprised to be invited to tea by the professor. When he arrived, he learned that the Hilberts had a request to make of him. Franz Hilbert, who was now in his teens, was not doing well in the school in Göttingen.

("My son gets his mathematics from his mother," Hilbert sometimes said. "All else is from me.")

Mrs. Hilbert thought that perhaps Franz would do better in another school. To make sure that he would be accepted, would young Courant be willing to tutor him?

"So it happened that I spent quite a bit of time with Franz Hilbert. He was not an unintelligent or untalented boy. He was accessible. He learned a little bit and he was accepted by the new school, a very well known country school. But I was always impressed with the fact that here was a boy whose mind was like a photographic plate that you put in the developer and something very nice comes out and then, after a short while, there's a veil; it becomes increasingly cloudy and finally there is nothing left on the plate."

Already "the little Courant," as he was affectionately known, had a deep feeling for the broad scientific tradition of Göttingen. He also had a flair for the dramatic. When he received his degree in February 1910, he was not content with giving the customary kiss to the little goose girl in the fountain in the Rathaus Square. Instead, two friends hired a Droschke and circled the city, trumpeting the announcement to the citizens that Richard Courant was now Doctor of Philosophy *summa cum laude*!

During the year 1910 Courant was Hilbert's assistant.

That same year, for the first time since 1906, Hilbert sent a communication on integral equations to the Göttingen Scientific Society, his sixth and last.

"One can really say that it was through Hilbert's investigations that the true significance of integral equation theory was first exposed," Courant later wrote. "Their various relations to the most different fields of mathematics, the many-sided applicability and the inner harmony and simplicity of their structure, their unifying power in relation to numerous previously isolated investigations first became truly evident in Hilbert's work."

Since Fredholm, mathematicians from all over the world, but especially from Germany and the United States, had taken up the subject of integral equations.

But the day had quite definitely gone to Hilbert.

In Göttingen, life went on.

XVI

Physics

In the fall of 1910, the Hungarian Academy of Science announced the award of its second Bolyai Prize "to David Hilbert, who by the profundity of his thought, the originality of his methods, and the rigorous logic of his demonstrations has already exercised considerable influence on the progress of the mathematical sciences."

It fell to Poincaré, as secretary of the prize committee, to prepare the general summary of Hilbert's work to be presented to the Academy and then to be published.

Qualities he chose to single out for special mention were the variety of the investigations, the importance of the problems attacked, the elegance and the simplicity of the methods, the clarity of the exposition, and the care for absolute rigor. He appreciated the readability of the work. He also noted that the influence Hilbert had had on the progress of mathematics came, not only from his personal investigations, but also from his teaching, "by the assistance which he gives to his students and which permits them in turn, utilizing the methods created by their teacher, to contribute to our understanding."

He described in detail Hilbert's achievements (devoting the most space to the work on the foundations of geometry) and tried to place them in relation to the achievements of other mathematicians.

On the proof of Gordan's Theorem — "One cannot better measure the progress accomplished by M. Hilbert than by comparing the volume which Gordan has devoted to his demonstration to the lines with which M. Hilbert has been able to content himself."

On the new proof of the transcendence of e and π — "The ability to simplify what at first seems complex presents itself as one of the characteristics of M. Hilbert's talent."

On the work in algebraic number fields — "The introduction of ideals by Kummer and Dedekind was a considerable progress: it generalized and at the same time clarified the classical results of Gauss on quadratic forms and their composition. The papers of M. Hilbert . . . constitute a new step forward which is no less important than the former."

On the investigations on the foundations of geometry — "There are in the history of the philosophy of geometry, three principal epochs: the first is that in which the thinkers, at the head of whom we must cite J. Bolyai, developed non-euclidean geometry; the second is that in which Helmholtz and Lie revealed the role of the idea of motion and the group in geometry; the third has been opened up by M. Hilbert."

On the salvaging of the Dirichlet Principle — "It is needless to lay stress on the importance of the discoveries which extend beyond the special problem of Dirichlet [and] we should not be surprised at the number of investigators who are now engaged on the path opened up by M. Hilbert."

On the proof of Waring's Theorem — "We do not doubt that these considerations . . . when they are fully understood, will be applied to problems that extend far beyond that of Waring."

On the recent work in the theory of integral equations — "This discovery of M. Fredholm is certainly one of the most remarkable of recent times, . . . M. Hilbert had made important improvements, . . . of which one has to admire the simplicity, the sureness, and the generality."

Poincaré's report for the Bolyai Prize appeared in *Acta Mathematica* in 1911. Unsuspected by anyone at the time, he had summed up in it what was to be the totality of Hilbert's contribution to constructive mathematics. The following year Hilbert — now fifty years old — became, as far his fellow mathematicians were concerned, a physicist.

The recent work on integral equations (published as a book in 1912) had brought him to the boundary of mathematics and physics. In it he had combined many theories under one comprehensive viewpoint. The result had been much greater abstraction, unification, clarity and rigor than had existed in the past. Practically speaking, however, physicists had gained little, since in the majority of cases the old methods utilizing differential equations still remained most usable. But in the introduction to his book on integral equations Hilbert had expressed his delight in finding a field of physics where choice was not possible, the physical concepts definitely led to an integral equation as the only valid expression of the data. This was kinetic gas theory; and the presentation of a paper on the foundations of this theory in the spring of 1912 announced that Hilbert, the mathematician, had now turned his attention to physics.

In retrospect, it seemed to him that the contemporary era in physics had opened during his docent days when Hertz had established the existence of the electromagnetic waves predicted by Maxwell. Then had followed in rapid succession the discovery of x-rays by Roentgen — radioactivity by the Curies — electrons by J. J. Thomson. Max Planck had put forth the quantum theory. Einstein had enunciated the special theory of relativity. In years, there had been as many great discoveries as there had been in centuries. "And not a single one," Hilbert exulted, "has to yield in magnificence to the achievements of the past!"

But as a mathematician he was disturbed by a certain lack of order in the triumphs of the physicists. In this he was not alone. Walther Lietzmann, one of his students of an earlier day, has recalled "what discomfort we mathematicians felt in the lectures on theoretical physics when sometimes this, sometimes that principle, without proof, was placed before us and all sorts of propositions and conclusions derived from it. We perceived the pressing necessity for investigation to determine whether these diverse principles were compatible with one another and in what relation they stood."

Questions similar to these had been investigated by Hilbert in connection with his work on the axioms of geometry — the questions of the sufficiency, independence and consistency of the axioms. Now it seemed to him that the time had arrived for the project which he had proposed at Paris as the sixth problem for the twentieth century — the axiomatization of physics and the other sciences closely allied to mathematics. A few fundamental physical phenomena should be set up as the axioms from which all observable data could then be derived by rigorous mathematical deduction as smoothly and as satisfyingly as the theorems of Euclid had been derived from his axioms. But this project required a mathematician.

"Physics," Hilbert announced, "is much too hard for physicists."

It seemed a rather arrogant remark, but the physicists knew what he meant.

"Although he was only joking," one Nobel Prize winner later said, "he expressed thus something completely genuine: the respect for the difficulty of the problems which are posed in this field of pure thought, recognized only by one who has actually put all his intellectual power to overcoming such problems."

At Paris, Hilbert had specifically mentioned that in his opinion the investigation of the axioms of the theory of probability should be accompanied by a rigorous and satisfactory development of the method of mean values in mathematical physics, and in particular in the kinetic theory of gases. This is where he first attacked the new project.

Kinetic gas theory had developed on the principle that because the motion of molecules in a gas appears completely disordered it can be described statistically and effects related to pressure, density, temperature and such can be predicted on the basis of average motions. But the theory had not developed in a unified manner: different aspects were treated individually and without connection. By applying the axiomatic method and his theory of integral equations, Hilbert was able to achieve a beautifully simple, unified system and thus transform the theory into a usable and acceptable mathematical tool. ("It is interesting to note," von Kármán wrote many years later, "that this work of sixty years ago, when space flight was a science fiction dream, is today the basis of most of our engineering calculations on the behavior of man-made satellites.") The value of his investigations in this area lay not so much in the derivation of the physical theorems, which were already known, as in the insight which was gained into their structure, assumptions, and range of validity.

But in spite of Hilbert's belief in the power of the axiomatic method to bring order out of disorder, he recognized that he could not solve the problems of physics by sheer mathematical power alone. He would have to inform himself on current developments. One way to do this would be to read and study the reports of the new discoveries as they appeared in print. But this was not his way. Instead, he turned for help to his old friend, Arnold Sommerfeld.

Now in Munich, Sommerfeld was the center of the most fertile group of young physicists in Germany. It was customary at German universities that each physics professor had his own "institute" which then had its own faculty, docents, assistants and students. At Munich the largest, best equipped institute was that of Roentgen, the professor of experimental physics; the smallest, that of Sommerfeld. But when Sommerfeld had come to Munich, he had insisted that in addition to the library and desks customary for an institute of theoretical physics, there must also be facilities for experiments. After his arrival he had created a rare spirit of comradeship in his institute. While Roentgen's students worked independently and "even too much communication from door to door was not encouraged," Sommerfeld's students frequently joined him in the nearby Alps for skiing in the winter and mountain climbing in the summer — "going up and going down, talking physics all the time." During the week in Munich they gathered after lunch for "cake and physics" at a cafe near the University where formulas and diagrams of important discoveries were frequently noted down on marble tabletops and later wiped away by grumbling waitresses.

Hilbert now asked his old friend to find him a young man to be his special assistant for physics. Sommerfeld offered the job to his student, Paul Ewald, who had recently completed his dissertation on the passage of light through a crystal.

When Ewald returned to Göttingen in the spring of 1912, he was welcomed as "Hilbert's physics tutor." That, it seemed, was Hilbert's conception of the new position. He promptly assigned Ewald various topics in physics that he himself wanted to learn about.

"I remember that one thing he assigned me was the following. There was a long-standing controversy about the number of constants of elasticity in a crystal — it went back to the founders in the field — and Hilbert wanted me to read up on it and tell him who was right. So I went to the Lesezimmer and got out all the old volumes and found it all very interesting. I saw that both sides had good arguments. In fact, I couldn't find a flaw on either side, as indeed these great men had never managed to find themselves, nor the many others who had studied the problem. So I went back and reported that to Hilbert. A few years later, the whole problem — which had held up crystal physics for more than fifty years — was solved by Max Born."

Hilbert's scientific program at this time, according to Ewald, could be succinctly summarized as follows: "We have reformed mathematics, the next thing is to reform physics, and then we'll go on to chemistry." The chemistry of the day was "somewhat like cooking in a girls' high school." That was the way that Hilbert described it.

Now he planned to take up one physical theory at a time and bring it into an acceptable mathematical formulation. From the kinetic gas theory, he moved on to another field in which the concepts also led directly to integral equations. This was elementary radiation theory. During the next couple of years he published a series of papers in which he derived the fundamental theorems with the help of his theory of linear integral equations, laid down the axiomatic foundation for the theorems, and demonstrated the consistency of his axioms. The treatment of this one theory was in essence a model of the approach to physics as a whole which he had proposed at Paris.

Ewald recalls that he personally could not seem to get very "warm" about the radiation problem. He felt that Erich Hecke, Hilbert's mathematics assistant, was really much better than he in recognizing the nature of Hilbert's difficulty in regard to the various physics papers being discussed, perhaps because his mind was essentially mathematical like Hilbert's.

Hecke was to become one of the great mathematicians of his time, but he was always to look on the days when he was assistant to Hilbert as the high point of his career. For his efforts he received the sum of 50 Marks a month, approximately $12.50 in the American money of the time. One day Hilbert himself decided that this sum was inadequate, and he told Hecke that when he went to Berlin next he would take up the matter with the Minister of Culture. But after finishing his business with the Minister, who had the final say in almost all university matters, Hilbert realized that he had forgotten something. Without further ado he stuck his head out of the window of the Ministry and shouted down to Mrs. Hilbert, who was waiting for him in the park below, "Käthe, Käthe! What was that other matter I wanted to talk about?" "Hecke, David, Hecke!" Hilbert pulled in his head, turned back to the startled official, and proceeded to demand that Hecke's salary be doubled, as it was.

In May of 1912, Sommerfeld came to Göttingen under the sponsorship of the Wolfskehl Prize Commission to talk on some recent discoveries in physics. At this time he reported that Max von Laue and others had recently succeeded in passing x-rays through a crystal. This achievement revealed the true nature of x-rays and opened up a new way of studying matter. When Ewald heard about it, he recalled a conversation he had had with von Laue just before coming to Göttingen. He had gone to consult the older man about something in his dissertation but after a few minutes had found him strangely distracted.

"What would happen if you assumed much shorter waves to travel in the crystal?" von Laue wanted to know.

"It is all here in this formula," Ewald said. "You are welcome to discuss the formula, which I will copy out for you. But I have to get my thesis delivered within the next couple of days and also have to do some reviewing for my oral examination, and I don't have time."

Ewald had thought no more of this incident until he heard Sommerfeld's report on von Laue's discovery. Hilbert's conviction of the value of direct, personal scientific communication could have no more dramatic support! That afternoon Ewald hurried back to his room. He had all the formulas necessary to discuss von Laue's discovery already in his dissertation and he worked straight through the night.

But for the most part the semester as Hilbert's assistant was an easy, leisurely time and Ewald took advantage of the opportunity to observe Hilbert more closely than he had been able to do when he was a student in the calculus class in 1906.

One day Otto Toeplitz, now a Privatdozent, came to Hilbert with a paper which he had received from an auditor in his seminar.

"Most doctoral dissertations contain half an idea," he told the professor. "The good ones have one idea. But this paper has two good ideas!"

There was a problem, however. The author of the paper, whose name was Jakob Grommer, was not eligible to stand for the doctoral degree. He had never received the necessary leaving certificate from a gymnasium; in fact, he had never attended a gymnasium but had studied in a Talmudic school because he had intended to become a rabbi. It was customary in the part of eastern Europe from which he came that the new rabbi marry the daughter of the old rabbi; but when the rabbi's daughter saw the grotesque hands and feet of Grommer, who suffered from acromegaly, she refused to marry him and so ended his hopes of a rabbinical life. The rejected lover then turned to mathematics.

Hilbert took on Grommer's case "with a gleam in his eye," according to Ewald.

"If I can get a doctor's degree for this young man, who is a Lithuanian and a Jew and does not have a gymnasium diploma, then I really shall have done something!"

(Needless to say, Grommer did indeed eventually receive his degree as a doctor of philosophy.)

In spite of Ewald's liking and admiring Hilbert, he found him "a bit of an arrested juvenile." If the day was warm, Hilbert came to the lecture hall in a short-sleeved open shirt, an attire inconceivably inappropriate for a professor in that day. He peddled through the streets with bouquets from his garden for his "flames." He was also quite as likely to bear as his gift a basket of compost balanced on the handlebars. At a concert or restaurant, no matter how elegant, if he felt a draft, he borrowed a fur or a feather boa from one of the ladies present. To some people, like Ewald, it seemed that he did these things because he thought they were shocking to more conventional citizens. Others thought that he did them because they were reasonable and he was not bothered by the fact that they were contrary to accepted behavior. In any case, he had such natural dignity in whatever he did that no one laughed.

He still loved to dance, much preferring the Rector's annual ball to the formal banquet which that official gave each year for the professors and their wives. He liked pretty young ladies and delighted in explaining mathematical ideas to them. "But, my child," he would say, "you *must* understand that!" On one occasion he composed a little verse to his "beloved angel" expressing the hope that some of his favorites would receive invitations to the ball:

> Lieber
> Engel,
> Mach mit Eile,
> Dass Mareille,
> Kar--, Ils--, und Wei--,
> Diese drei
> Auf jeden Fall
> Kommen zum Rektorenball.

Then, writing the verse on a sheet of paper cut to resemble an angel, he deposited it anonymously at the Rector's office.

He liked to fancy himself as a dashing man of the world. He wore a panama hat to cover his baldness and announced that his idea of a good vacation was going on a trip with a colleague's wife. But all of the time, recalls George Pólya, who was a student in Göttingen during this period, "he looked *so innocent*."

Käthe Hilbert's reaction to her husband's numerous "flames" is contained in an anecdote of Hilbert's fiftieth birthday party. To honor the professor, some of his students composed what they called a Love Alphabet. For every letter there was a verse about one of Hilbert's loves. For "I" — "Wenn sich unsere Haare lichten / Lieben wir die kleinen Nichten. / Das ist menschliche Natur / Denkt an Ilschen Hilbert nur." (*When our hair gets thinner, we love the little nieces. Such is human nature, just think of Ilse Hilbert.*) But for "K" — no one was able to think of one of Hilbert's loves that began with "K." Then Käthe Hilbert said: "Now, really, you could think of me just for once!" Delighted, the young people immediately composed the following verse:

> Gott sei Dank, nicht so genau
> Nimmt es Käthe, seine Frau.
>
> *God be thanked that Käthe, his wife,*
> *Takes not too seriously his life.*

"Without Käthe," Ewald says, "Hilbert would have been truly lost." Courant adds, "Without her, he could not have lived the life he lived."

It was during that summer — the summer of Hilbert's fiftieth year — that Henri Poincaré died. He was 59 years old; for 33 years he had been incredibly productive in almost all branches of mathematics. The year before, however, he had asked the editor of a mathematical journal to accept an unfinished paper on a problem which he considered of the highest importance:

"At my age, I may not be able to solve it, and the results obtained, susceptible of putting researchers on a new and unexpected path, seem to me too full of promise, in spite of the deceptions they have caused me, that I should resign myself to sacrificing them."

It was a poignant reminder to his contemporaries that time was short. They found themselves now filled with a certain fear of death, the special characteristic of which was expressed by Vito Volterra, the leading Italian mathematician of the time, in an address on Poincaré's work:

"Among the various ways of conceiving man's affection for life, there is one in which that desire has a majestic aspect. It is quite different from the way one usually regards the feeling of the fear of death. There come moments when the mind of a scientist engenders new ideas. He sees their fruitfulness and utility, but he knows that they are still so vague that he must go through a long process of analysis to develop them before the public will be able to understand and appreciate them at their just value. If he believes then that death may suddenly annihilate this whole world of great thoughts, and that perhaps ages may go by before another discovers them, we can understand that a sudden desire to live must seize him, and the joy of his work must be confounded with the fear of having it stop forever."

With Poincaré gone, there was no longer any question about who was the greatest living mathematician — and he was up to his ears in physics.

After Ewald left Göttingen, Sommerfeld sent Alfred Landé to serve as Hilbert's physics assistant. Hilbert had moved in his lectures from radiation theory to the molecular theory of matter — the next semester he planned to devote to the theory of the electron. His treatment of these subjects was similar to that of kinetic gas theory and radiation theory, but never published.

By this time he had worked out what was to be for him a more efficient method of utilizing his physics assistant. At their first meeting he handed Landé a sheaf of separate reprints of various recently published papers on physics and instructed him to read them.

133

"All kinds of subjects, the physics of solid bodies and of spectra and fluids and heat and electricity, everything that came to him I was to read and, when I found it interesting, to report on it to him."

Each morning Landé came to the house on Wilhelm Weber Strasse and explained to Hilbert the subject matter of the papers that he had selected as interesting.

"That was really the beginning of my whole career as a scientist. Without Hilbert, I would probably never have read all those papers, certainly not have digested them. When you have to explain something to someone else, then you first must really understand it and be able to put it into words."

What was it like, teaching physics to Hilbert?

"Well, sometimes he wasn't an easy student at all, and I had to tell him things several times before he got them. He always tried to repeat what I told him, but in a more organized way, more simple and clear. Sometimes, right after our meeting, he would have a lecture scheduled on the subject we had been discussing. I remember often walking side by side with him from his house on Wilhelm Weber Strasse to the Auditorienhaus, still explaining things to him in the last minutes. Then in the lecture he would try to present what I had said but in his own way — his way as a mathematician, which is something quite different from that of a physicist."

In his spare time Landé studied Hilbert's book on integral equations — "a wonderful book." In the evenings he went to parties and danced with professors' daughters. His social position, he found, was much improved by his being Hilbert's physics assistant. There was only one unpleasant aspect to the job. At the Hilbert parties it was his duty as assistant to select and change the phonograph records. This was a chore which he still, after fifty years, recalls with distaste. Hilbert, who continued to receive the latest model phonograph as a gift from the manufacturer, had few classical records at this time, preferring the latest music hall "Schlagers." It was hard for Landé to find a record which he himself cared to listen to. To make matters worse, Hilbert liked his music loud. At this time volume was determined by the size of the needle, and Hilbert insisted on having the large needle. Once he went with great expectations to a concert by Caruso. But he was disappointed. "Caruso," he said, "sings on the small needle."

During 1913 Paul Scherrer came to Göttingen as a student. He found under the still surface "an intellectual life of unsurpassed intensity." It was the time when the Quantum Theory of Light was at last being taken seriously "although it could by no effort be straightened out with wave theory." This was also the year that Niels Bohr, the elder brother of Harald, put forth his planetary theory of the atom and "one tried hard to become convinced of the reality of Bohr's electron orbits in the atoms in spite of all the hesitation the physicist felt in accepting the hypothesis that the electron on its stationary orbit about the atomic nucleus does not radiate."

Niels Bohr, like his younger brother Harald, was frequently in Göttingen. People there saw Harald as *L'Allegro*; Niels, as *Il Penseroso*. But their father, a professor of medicine who was tremendously proud of both his sons, summed them up differently: "Harald is silver," he said affectionately, "but Niels — Niels is pure gold."

Hilbert enjoyed the opportunity of talking informally with Niels Bohr. Communication to others of his own discoveries and the working through in his own mind of their ideas — that was vital. Particularly now that the mathematical sciences embraced such an extensive complex of human knowledge and were in such a state of rapid and intensive progress, it seemed to him that a scientist could not be expected to acquire the information he needed through the mere reading of scientific works. The papers of the day, because of the abstractness of their thought, required in his opinion the addition of "a strong display of spirit and lively power." How valuable it would be, he thought, to bring together the leading physicists for a week of lectures and conversation!

It was long before the day of foundations and grants, but — Fermat's Last Theorem being still unproved — the interest from the bequest of the Darmstadt mathematics professor was at hand. In 1910 the money had been used to bring H. A. Lorentz to Göttingen to talk on relativity and radiation theory. In 1911 visiting lecturers had been dispensed with so that a prize of 5000 Marks could be awarded to Zermelo "for his achievements in set theory and as an aid in the full recovery of his health." In 1912 Sommerfeld had given his lectures on the recent advances in physics. Now Hilbert arranged for a week-long Wolfskehl Conference on the Kinetic Theory of Matter in the spring of 1913.

"No one who participated can forget the impression of this gathering of outstanding learned men, freely discussing the problems of their science," F. W. Levi later wrote. "Hilbert presided.... The young men who filled the hall were almost all to make their own marks later.... The prosaic lecture hall with the black iron stove at one side was the arena for an assemblage of the crown princes and the kings of science."

In the course of the Gas Week, as it was of course immediately nicknamed, Hilbert met Peter Debye, a young physics professor from Holland, who had been Sommerfeld's first assistant at Munich. Hilbert was impressed with Debye and wished that there was an appropriate place at Göttingen which could be offered to him. He proposed to the Wolfskehl Commission that the following year's interest on the prize be used to bring guest professors in the mathematical sciences to Göttingen during the summer semesters. The summer of 1914 saw the first Darmstadt Professors at Göttingen: one of these was Hilbert's former student, Alfred Haar, now a professor at Klausenburg; the other was Peter Debye.

(When people asked Hilbert why he didn't prove Fermat's Last Theorem and win the Wolfskehl Prize, he said, "Why should I kill the goose that lays the golden egg?")

That same summer it seemed also that at last Klein's plan for a separate building — *an institute* — for mathematics was about to be carried out. The land had been obtained, the funds set aside, construction scheduled to begin.

It was the summer that, in Sarajevo, the Archduke Ferdinand of Austria was assassinated by a Serbian student.

XVII

War

The long vacation began at Göttingen on the first of August. Already, Austria-Hungary had declared war on Serbia. The French army mobilized. The German army began to march through Belgium. By the end of August a dozen countries were at war.

Hilbert thought the war was stupid, and said so.

Letters came from his former students in the United States assuring him of their continued love and respect.

The enemy, recoiling before the "atrocities of the Hun" and finding itself hard put to reconcile German "barbarism" with respected German achievements in the arts and sciences, rationalized that there must be two Germanys — the military Germany of the Kaiser and the cultural Germany of Goethe, Beethoven and Kant. Germany responded with a declaration by a group of its most famous artists and scientists that they, like all the German people, were solidly behind the Kaiser. Addressed "To the Cultural World," the declaration listed the "lies and slanders of the enemy" and, beginning with the statement, "It is not true that Germany caused this war," categorically denied each one.

Those who had drawn up the declaration recognized that mathematicians, no matter how great, are not as a rule well known except to other mathematicians. The international reputations of Klein and Hilbert were such, however, that they were both asked to sign.

Klein had always been an extremely patriotic man — in 1870 he had rushed home from Paris to volunteer for the army — and now he gave permission for his name to be used without questioning the statements made in the declaration. Hilbert, on the other hand, examined the list of sentences, each one beginning, "It is not true that . . .," and, since he could not ascertain whether they were in fact true, refused to sign.

On October 15, 1914, the Declaration to the Cultural World was publicized by the German government. Those signing included such famous scientists as Ehrlich, Fischer, Nernst, Planck, Roentgen, Wasserman, Wien. One name which was conspicuously missing was that of Einstein, who was now at the Kaiser Wilhelm Institute in Berlin. According to Einstein's friend and biographer Philipp Frank, only the fact that Einstein had become a Swiss citizen saved him from being considered a traitor. Hilbert had no such protection. His refusal to sign was the more unforgivable because he was, not merely German, but Prussian. When classes resumed at the beginning of November, many people turned away from him as if he were indeed a traitor.

Most of Hilbert's mathematical colleagues were sympathetic with his action, however; even Klein shortly regretted the excess of patriotism that had led him to sign a document without first ascertaining the validity of the statements it contained. As it happened, the Declaration did not have the effect which had been hoped for. The Cultural World was shocked that respected men would put their names to such statements as "It is not true that Germany violated the neutrality of Belgium." Klein was expelled from the Paris Academy. Hilbert was allowed to remain a member.

In spite of the war, the Thursday afternoon mathematical walks continued in Göttingen. There were now more participants than there had been in Minkowski's day. Landau and Ludwig Prandtl, the professor of applied mechanics, had been added. Another participant was Carathéodory, who had returned to relieve Klein after the older man had suffered another breakdown in 1911. The easy, cultured Greek, whose family motto was "No Effort Too Much," had become a mathematician in the traditional Göttingen style. The physicist Peter Debye, who had so impressed Hilbert during the Wolfskehl Conference, had also become a regular member of the faculty by this time.

Almost all the younger men, students and docents, were gone, or soon to go. The Lesezimmer, which before the war had been crowded at all times, was almost empty. There was no such thing as an educational deferment. Brains, good grades, letters of recommendation from one's professors, great expectations carried no weight. Only a few young men had not been called. One of these was Hilbert's assistant Landé, whose bad eyesight disqualified him for the army in the beginning.

Franz Hilbert was 21 years old the year that the war began, but the army did not take him. For a long time Hilbert had held hopes for his son. There was a period when he was apprenticed to a gardener in Göttingen. "But you never can tell," Hilbert said to Ewald, who was his assistant at the time. "I was also in my youth a bit *dammelig*." Later a small job in a bookstore in Frankfurt was obtained for Franz. He did not do well at all. It became increasingly clear that he was a disturbed boy. Mrs. Hilbert worried a great deal about him and got regular reports from friends in Frankfurt.

One night, before the war, when Courant was at the house, she received a message that her son had failed to appear for work that day and no one knew where he was. Courant, who was just leaving for Berlin, volunteered to go instead with Mrs. Hilbert to Frankfurt and help her search for Franz. While they were sitting and talking to Hilbert, waiting for train time, there was a great commotion outside and Franz suddenly appeared, covered with mud and very excited. He had left the train at a village along the way and walked home. He had come to save them, he announced, from evil spirits that were after them.

"I can still see the scene before me," Courant says today. "Hilbert said to Franz, 'Oh, you stupid boy, there is nothing — there are no ghosts or devils.' Then Franz became even more upset. There was much shouting back and forth. Franz kept haranguing us about these invisible creatures that wanted to harm us. Hilbert kept hitting his hand on the table and saying, 'There are no ghosts.' It was a very weird scene. Something obviously had to be done right away. So I called the professor of psychiatry, who came and gave Franz a little injection to quiet him. Then we took him in a taxi to a clinic for mental diseases which was near the University, and he was admitted right away."

By the time they left the clinic it was morning. Courant and Hilbert went for a short walk.

"From now on," Hilbert said quietly, "I must consider myself as not having a son."

"It was very sad the way he said it, but very determined."

The tragedy of Franz Hilbert was personally disturbing to mathematicians and mathematics students at Göttingen. To explain how two such great and gifted people could have had this unhappy offspring, they began to say that the Hilberts were first cousins. This was not true, although they were cousins by marriage.

Her husband's attitude toward Franz caused Käthe Hilbert a great deal of sorrow. She, unlike Hilbert, could not consider that she no longer had a son; and the young mathematicians quickly learned that they could always get on the good side of Mrs. Hilbert with a kind word about Franz. During the war years, however, she did not let either the personal nor the general tragedy hinder her husband from functioning as a scientist. Under her skillful management, the combination of fellowship, comfort and order necessary for Hilbert to work continued to be maintained in the house on Wilhelm Weber Strasse.

Klein, in spite of poor health, also managed to maintain what Hilbert called "the mathematical arrangements." The war had ended many of the older man's activities, such as the International Commission on Schools, and had curtailed others. Several years earlier, he had brushed away the suggestion that he write the history of the mathematics of the nineteenth century: "I am too old. It needs a young man who could devote years to its preparation. No, all that I could do would be to give a few lectures on the great events; but now I am too occupied even to prepare those."

The war gave him time.

The lectures on the mathematics of the nineteenth century, delivered in the dining room of his home, were to seem to Courant, who later helped to edit them, "the perfect sweet fruit of the wisdom of Klein's old age." Courant himself never heard the lectures. He was at the front.

Hilbert, still absorbed in physics, had few students, most of them foreign. But with Debye's arrival in the summer of 1914 he had been all set to learn about the structure of matter, and he saw no reason to let the war change his plans. He asked Debye to organize a seminar on the subject. He himself opened each session with the only half-humorous request, "Now, meine Herren, you tell me, just what is an atom?"

Scherrer, working very closely with Debye at this time and himself a member of the seminar, later recalled Hilbert as "by far the most intelligent person I have ever known."

Hilbert was now primarily interested in the fundamental problems of physics and their mathematical formulation. Sometimes in the seminar he would throw out a question with the comment, "That is a purely mathematical problem." Other times he would say, "For that problem, the physicist has the great calculating machine, Nature." It was Hilbert's opinion, according to Debye, that Maxwell's equations did not attack the essence of the problem of the structure of matter — at that time the electron was the only known fundamental particle — and that equations were needed from which *it should follow* that such a particle exists.

In their daily sessions, Landé was presenting to Hilbert "in a distilled form for mathematicians" the quantum mechanics of randomly scattered events, which was at that time still in a quite primitive stage. Then in December 1914, although still not drafted, Landé decided to volunteer for the Red Cross. When Hilbert heard that his assistant was going to leave him, he was most annoyed. To Landé, his response was another example of his extreme egocentricity:

"He thought only about mathematics; and since he was considered, now that Poincaré was gone, the greatest mathematician of the time, he thought every ease was due to him, from his wife and everybody else. He squeezed me out for my physics. That was all I meant to him."

(To Landé's teacher Sommerfeld, however, Hilbert's "naive and imperative egoism" was always "egoism in the interest of his mission, never of his own person.")

Just before Christmas, Landé left Göttingen. He was in the Red Cross for two years. Then he was drafted, "because by that time they would take anybody."

In Göttingen, in the weekly Hilbert-Debye seminars, it seemed to those few students who were left that the "living pulse" of physical research was at their finger tips. The work of Einstein as he pressed forward toward a general theory of relativity was followed with great interest. Also followed was the work of others who were trying to reach the same goal. Hilbert was especially fascinated by the ideas of Gustav Mie, then in Greifswald, who was attempting to develop a theory of matter on the fundamentals of the relativity principle; and in his own investigations he was able to bring together Mie's program of pure field theory and Einstein's theory of gravitation. At the same time, while Einstein was attempting in a rather roundabout way to develop the binding laws for the 10 coefficients of the differential form which determines gravitation, Hilbert independently solved the problem in a different, more direct way.

Both men arrived at almost the same time at the goal. As the western front settled down for the winter, Einstein presented his two papers "On general relativity theory" to the Berlin Academy on November 11 and 25; Hilbert presented his first note on "The foundations of physics" to the Royal Society of Science in Göttingen on November 20, 1915.

It was a remarkable coincidence — reminiscent of Minkowski's work on special relativity and electrodynamics in the joint seminar of 1905 — but even more remarkable, according to Born (who was now in Berlin with Einstein), was the fact that it led, not to a controversy over priority, but to a series of friendly encounters and letters.

Hilbert freely admitted, and frequently stated in lectures, that the great idea was Einstein's.

"Every boy in the streets of Göttingen understands more about four-dimensional geometry than Einstein," he once remarked. "Yet, in spite of that, Einstein did the work and not the mathematicians."

On another occasion, in a public lecture, he demanded: "Do you know why Einstein said the most original and profound things about space and time in our generation? Because he had learned nothing at all about the philosophy and mathematics of time and space!"

Each man, however, was essentially a man of his own science. Originally Einstein had believed that the most primitive mathematical principles would be adequate to formulate the fundamental laws of physics. Not until much later did he see that the opposite was the case. Then it turned out that it was Minkowski, whose lectures he had found so uninteresting, who had created the mathematical conception of Space-Time which made possible his own formulation of general relativity.

"The people in Göttingen," Einstein once wryly observed, "sometimes strike me, not as if they want to help one formulate something clearly, but as if they want only to show us physicists how much brighter they are than we."

To Hilbert, the beauty of Einstein's theory lay in its great geometrical abstraction; and when the time arrived for the awarding of the third Bolyai Prize in 1915, he recommended that it go to Einstein "for the high mathematical spirit behind all his achievements."

Klein also contributed to the development of relativity theory. He was greatly impressed by Hilbert's papers on the foundations of physics. Now, almost 70, he thought he saw a way to clarify the fundamental laws of relativity theory with the old ideas of his Erlangen Program. With his knowledge of infinitesimal transformations, he was able to achieve an important abbreviation of Hilbert's calculations.

The war went on.

While the fate of Verdun hung in the balance, a young woman arrived in Göttingen. She was the daughter of the mathematician Max Noether and had been a student of his friend Gordan, the one-time "king of the invariants," now dead. She had published half a dozen papers and had lectured to her father's classes from time to time when he was ill. Now her father had retired, her mother had recently died, her brother Fritz — who had earlier been a mathematics student in Göttingen — had gone into the army. It was a time of change, and she had decided to take advantage of it.

Emmy Noether had little in common with the legendary "female mathematician" Sonya Kowalewski, who had bewitched even Weierstrass with her young charms as well as her mind. She was not even feminine in her appearance or manner. This is the first thing, even today, that the men who knew her recall. "She had a loud and disagreeable voice." "She looked like an energetic and very nearsighted washerwoman." "Her clothes were always baggy." And they still quote with delight the gentle remark of Hermann Weyl that "the graces did not preside at her cradle." But she was to be much more important to mathematics than the bewitching Sonya. Even at this time, she had an impressive knowledge of certain subjects which Hilbert and Klein needed for their work on relativity theory, and they were both determined that she must stay in Göttingen. But in spite of the fact that Göttingen had been the first university in Germany to grant a doctoral degree to a woman, it was still not an easy matter to obtain habilitation for one. The entire Philosophical Faculty, which included philosophers, philologists and historians as well as natural scientists and mathematicians, had to vote on the acceptance of the habilitation thesis. Particular opposition came from the non-mathematical members of the Faculty.

They argued formally: "How can it be allowed that a woman become a Privatdozent? Having become a Privatdozent, she can then become a professor and a member of the University Senate. Is it permitted that a woman enter the Senate?" They argued informally, "What will our soldiers think when they return to the University and find that they are expected to learn at the feet of a woman?"

Hilbert had heard what to him were similarly irrelevant arguments when he had been attempting to have Grommer's dissertation approved by the same faculty members. "If students without the gymnasium diploma will always write such dissertations as Grommer's," he had told them, "it will be necessary to make a law forbidding the taking of the examination for the diploma." Now he answered their formal argument against habilitating Emmy Noether with equal directness:

"Meine Herren, I do not see that the sex of the candidate is an argument against her admission as a Privatdozent. After all, the Senate is not a bath-house."

When, in spite of this rejoinder, he still could not obtain her habilitation, he solved the problem of keeping her at Göttingen in his own way. Lectures would be announced under the name of Professor Hilbert, but delivered by Fräulein Noether.

The war went on.

German submarines were sinking one out of every four ships that left an English port, but the English blockade was beginning to be felt in

Germany. Food was extremely scarce. Nineteen-sixteen saw the worst famine of the war — "the Turnip Winter," as it was called. Hilbert went as often as possible to Switzerland. In the past it had seemed to him that his old friend Hurwitz had been passed over by the German universities in favor of men often "not fit to hold a candle to him." Now, in the tranquillity of Zürich, he thought that perhaps this had been for the best: the ailing Hurwitz would never have been able to stand the privations of wartime Germany.

With his native shrewdness where his own needs were concerned and with the indispensable help of his wife, Hilbert himself managed all during the war — sometimes to the amusement of his friends and colleagues — to maintain the domestic comfort which was necessary for his work.

Food was a problem. He considered meat and eggs an absolute necessity if his brains were to function at their best for mathematics. He always had great scorn for the arguments of the vegetarians: "If they had their way, we should then have to pension the oxen." His garden furnished him with fruits and vegetables. Getting meat was more difficult. But one day the Rector of the University called all the professors to the Great Hall.

"Ah, I wonder what it will be this time!" Hilbert said with anticipation to his neighbor. The last time such a meeting had been called, the University had obtained a number of geese from a peasant and had distributed these among the professors. "Perhaps now we get swine!"

The Rector began to speak. He had great news. "Our highest commander, his Majesty the Kaiser, has just declared unrestricted submarine warfare upon our enemy!"

While most of the professors clapped and cheered at this announcement, Hilbert turned back disgustedly to his neighbor: "And I thought we would get swine!" he said. "But, you see, the German people are like that. They don't want swine. They want unrestricted submarine warfare."

The lack of contact with foreign mathematicians was extremely frustrating to Hilbert. Just before the war Bertrand Russell, with A. N. Whitehead, had published his *Principia Mathematica*. Hilbert was convinced that the combination of mathematics, philosophy and logic represented by Russell should play a greater role in science. Since he could not now bring Russell himself to Göttingen, he set about improving the position of his philosopher friend Leonard Nelson.

Nelson was also a champion of the axiomatic method. His philosophical work treated two main problems: the laying of a scientific foundation for philosophy and the systematic development of philosophical ethics and a "philosophy of right." He was still firmly opposed by Husserl, the philosophy professor; and Hilbert's files contain an extremely bulky item labeled "The Nelson Affair," recording his efforts to obtain an associate professorship for Nelson during this time.

Nelson (who finally did become an associate professor, but not until after the war) later dedicated to Hilbert the three volumes of his *Lectures on the foundations of ethics* — "an attempt to open up for the sovereign domain of exact science a new province."

In the spring of 1917 the United States at last entered the war against Germany.

That same year news arrived in Göttingen that Gaston Darboux had died. Hilbert admired Darboux, not only for his mathematical work, but also for the influence he had had on mathematics in France as a man and as a teacher. He immediately prepared a memorial for publication in the *Nachrichten*. When it appeared in print, an outraged mob of students gathered in front of Hilbert's house and demanded that the memorial to the "enemy mathematician" be immediately repudiated by its author and all copies destroyed. Hilbert refused. Instead he went to the Rector of the University and threatened to resign unless he received an official apology for the behavior of the students. The apology was immediately forthcoming; the memorial to Darboux — one of the four that Hilbert wrote during his career — remained in print. (The others were to Weierstrass, Minkowski, and Hurwitz.)

At the beginning of 1918, as new men fought in the Kremlin, Germany concluded a separate peace with the Ukraine; and Alexander Ostrowski, a young Ukrainian who had been a civil prisoner in Marburg during the war, was now able to come to Göttingen. During his enforced stay in Marburg, Ostrowski had thoroughly studied Hilbert's work and also the work of Klein. Upon his arrival in Göttingen, he paid the traditional calls upon the famous mathematicians — "a right as well as a duty."

He found Klein friendly. "He spoke with me about different things and was very much astonished that I knew so much about his work." Hilbert was polite but distant. "I believe he was rather distrustful of people whom he saw for the first time."

At the beginning of the spring semester, the Central Powers, led by Germany, launched a great offensive. For a moment, to the Germans, victory seemed at hand.

Klein's friends had recently urged him to edit his collected works. At first he had refused, saying that he could not do so without the help of some younger mathematician who would know the modern point of view. After meeting Ostrowski, he felt that he had found such a person; and he started work on the project.

Klein had always had a great intuitive gift. "In his youth," Carathéodory once wrote, "he used to look at the most difficult problems and *guess* their solutions." But he had never possessed the patience to provide the logically perfect demonstration for the theorems which he was convinced were true. "He did not want to admit that the exercise of such a demonstration could be elevated into an art and that the right execution of that art is the real essence of mathematics."

This quality of Klein's made Ostrowski's work extremely difficult.

"It happened now and then that we had to discuss results in his papers which he gave without, in my opinion, sufficient proof. Then I tried to get from him a proof. I asked him, 'Well, now, how is this and this? It is a point I don't understand.' He explained it. I still did not understand. Finally I said, 'Herr Geheimrat, may I ask questions?' Now the problem was to put the questions as sharply as possible. Well, he felt himself awfully maltreated, as if somebody had pinned him up to the wall. It happened sometimes that he just stood up and went to the window for a minute to cool off. He was never disagreeable, but he really had to apply quite a lot of control."

During these months, Ostrowski also had more contact with Hilbert. He found it fascinating to study at firsthand the personality of the man whose mathematical work he had studied so thoroughly. He was particularly impressed with the way in which Hilbert had solved the problem of "how a man of exceptional quality has to arrange himself for living among people of lesser quality."

"The problem obviously offered itself to him very early ... and he probably saw very early what the difficulties were. He was a great friend of Minkowski's, and Minkowski was a blazing star — a student in the university who wins the great prize of the Paris Academy! People must have admired Minkowski, but in a small university like Königsberg, a lot of people must have disliked him too. Minkowski was obviously a Jew and not even a Jew of German origin. At that time, I suppose, Hilbert made his first observations on the problem of a superior individual having to live with lesser beings. It is a problem that comes up quite often, and I would say that most people do not get it solved in time. They fail to recognize the existence of the problem, or else they need their superiority to overcome some complex they have. Hilbert, in my opinion, avoided the difficulties very well."

By summer the situation at the front had changed drastically. In July the German army began to retreat. News of the true state of the military affairs now reached even beyond the Rhine. The poet Richard Dehmel issued a plea to old men and young boys to stage a last resistance against the enemy. Hilbert's childhood friend, Käthe Kollwitz, now a great and also exceptionally popular artist (who had lost one of her two sons in Belgium), replied with a stirring letter to the press:

"There has been enough of dying! Let not another man fall! Against Richard Dehmel, I ask that the words of an even greater poet be remembered: *The seed for the planting must not be ground.*"

Almost exactly four years after the Declaration to the Cultural World, a new chancellor asked for an Armistice. In the early morning of November 9, 1918, the Kaiser crossed the frontier into the Netherlands.

XVIII

The Foundations of Mathematics

With a marked stiffness, a face scarred less dashingly than by a sabre, an empty sleeve or trouser leg, the young men who had been in the trenches began to return to class.

Mathematics lay before them, "fresh as May."

While they had been gone, Einstein had changed the conception of space, time and matter and had created a need for a whole new kind of geometry. In three papers totaling less than 17 pages, a young Dutchman named Brouwer had challenged the belief that the laws of classical logic have an absolute validity independent of the subject matter to which they are applied, and had proposed a drastic program to end the "foundations crisis" precipitated by the discovery of the antinomies in set theory at the beginning of the century.

The new ideas had swept up Hilbert's gifted pupil Hermann Weyl when he had returned to Zürich after service in the German army. Before the war he had become acquainted with Einstein. Now he gave a brilliant series of lectures on Einstein's ideas and published them as the book *Space, Time and Matter*, which became a scientific best seller. To his friends it seemed that Weyl "could take an intoxicated pleasure in allowing himself to be carried away or merely tossed about by the opposing currents which disturbed the period." The "foundations crisis" was irresistible to him. In 1918 he made his own contribution with his paper on the logical foundations of the continuum. He also carefully studied Intuitionism, as Brouwer's program was called.

Hilbert was disturbed by his former student's fascination with the ideas of Brouwer, who aroused in Hilbert the memory of Kronecker. At the end of the war, Brouwer was a few years older than Weyl and 20 years younger than Hilbert. He had made impressive contributions to mathematics. In 1911 he had opened a new era in topology with his proof that the dimensionality of a euclidean space is a topological invariant. His papers on point sets were considered by many to be the deepest since those of Cantor. But, like Kronecker before him, he was willing to jettison a great part of his mathematical achievement because of his philosophical ideas.

For Brouwer, neither language nor logic was a presupposition for mathematics, which, in his view, had its source in an intuition which makes its concepts and inferences immediately clear. To Weyl it was to seem that Brouwer had "opened our eyes and made us see how far generally accepted mathematics goes beyond such statements as can claim real meaning and truth founded on evidence."

Brouwer, for instance, refused to accept the Logical Principle of the Excluded Middle although, since the time of Aristotle, mathematicians had accepted without hesitation the idea that for any sentence A there are two possibilities only: either A or not-A. Now Brouwer insisted that there was a third possibility — in other words, a middle which could not be excluded.

His argument was the following:

Suppose that A is the statement "There exists a member of the set S having the property P." If the set S is finite, it is possible — in principle — to examine each member of S and determine either that there is a member of S with the property P or that every member of S lacks the property P. For finite sets, therefore, Brouwer accepted the Principle of the Excluded Middle as valid. He refused to accept it for infinite sets because if the set S is infinite, we cannot — even in principle — examine each member of the set. If, during the course of our examination, we find a member of the set with the property P, the first alternative is substantiated; but if we never find such a member, the second alternative is still not substantiated — perhaps we have just not persisted long enough!

Since mathematical theorems are often proved by establishing that the negation would involve us in a contradiction, this third possibility which Brouwer suggested would throw into question many of the mathematical statements currently accepted.

"Taking the Principle of the Excluded Middle from the mathematician," Hilbert said, "is the same as . . . prohibiting the boxer the use of his fists."

The possible loss did not seem to bother Weyl.

Brouwer's program was the coming thing, he insisted to his friends in Zürich.

"Hermann, that is mathematics in shirt-sleeves," George Pólya told him; in other words, not completely attired.

Weyl promptly offered to wager Pólya on the future of two specific propositions which would be eliminated from mathematics if Brouwer's ideas were to be accepted, as Weyl was convinced they would be — and within 20 years. The winner of the wager was to be decided by whether in 1938 Pólya was willing to admit that the two following propositions —

1. That each [non-empty] bounded set of real numbers has a precise upper bound,

2. That each infinite set of real numbers has a countable sub-set

— were in fact completely vague "and that one could ask of their truth or falsity as little as he could ask of the truth or falsity of the main ideas of Hegel's philosophy." If, by 1938, Pólya and Weyl could not agree between themselves as to the state of affairs then existing in mathematics, the determining opinion would be that of the majority of the full professors of mathematics at the Swiss Federal Institute and at the universities of Zürich, Berlin and Göttingen. The loser would then publish the conditions of the wager and the fact that he had lost it in the German Mathematical Society's *Jahresbericht*.

Hilbert himself never read a line of Brouwer's work. Increasingly, he avoided papers, preferring to get his information from lectures and conversation. Weyl was invited to Göttingen to talk to the Mathematics Club about Intuitionism.

It will be remembered that at the Heidelberg Congress, shortly after the discovery by Russell and Zermelo of a fundamental antinomy in set theory, Hilbert had sketched a mathematical-logical program which he believed would remove "once and forever" any doubts as to the soundness of the foundations of mathematics and the methods of mathematical reasoning. During the intervening years, absorbed first in integral equations and later in physics, he had apparently dropped this project. In fact, just before the war, Blumenthal, walking with the Hilberts and recalling the Heidelberg Congress, had remarked that it now seemed nothing would ever come of the idea for a "theory of proof." Hilbert had made no comment, but Mrs. Hilbert (Blumenthal was later to recall) had smiled.

Since the Heidelberg Congress there had been several important developments in the study of foundations. Zermelo had proved the well-ordering theorem and had developed his axiom system for set theory. Russell and Whitehead had published their *Principia Mathematica*. But Hilbert himself did not return, publicly at least, to the foundations of mathematics until 1917.

In the spring of that year, on a visit to Zürich, he arranged for two of the young mathematicians in the circle around Hurwitz to accompany him on a walk. One of these was Weyl's friend Pólya. The other was a reserved, shy and somewhat nervous man named Paul Bernays. To the surprise of Pólya and Bernays the subject of conversation on the walk to the top of the Zürichberg was not mathematics but philosophy. Neither of them had specialized in that field. Bernays, however, had studied some philosophy and, during his student days at Göttingen, had been close to Leonard Nelson. In fact, his first publication had been in Nelson's philosophical journal. Now, in spite of his quietness, Bernays had much more to say than the usually voluble Pólya. At the end of the walk Hilbert asked Bernays to come to Göttingen as his assistant. Bernays accepted.

That September Hilbert returned to Zürich to deliver a lecture before the Swiss Mathematical Society. It was a week or so after the third anniversary of the beginning of the war, and his first words were timely:

"As in the life of nations, the single nation can prosper only when all goes well with its neighbors and the interest of the states requires that order prevails, not only in each one of the separate states, but in the relations among the states as well — so is it also in the life of science."

The talk was devoted to a favorite subject — the importance of the role of mathematics in the sciences — and might have been entitled "In praise of the axiomatic method."

"I believe," Hilbert said firmly, "that all which is subject to scientific thought, as soon as it is ready for the development of a theory, comes into the power of the axiomatic method and thus of mathematics."

But it was also in the course of this talk that he brought up certain questions which revealed for the first time since 1904 in public utterance his continued interest in the subject of the foundations of his science:

The problem of the solvability in principle of every mathematical question.

The problem of finding a standard of simplicity for mathematical proof.

The problem of the relation of content and formalism in mathematics.

The problem of the decidability of a mathematical question by a finite procedure.

To investigate these questions — he pointed out — it would be necessary first to examine the concepts of mathematical proof.

But he himself was still not ready to enter the foundations crisis personally. There were problems at home, both personal and professional.

Franz had been released from the hospital. Little jobs were obtained for him through the connections of the University, but he was not able to keep them for long; then Mrs. Hilbert would have to bring her son home, and the peace in the house on Wilhelm Weber Strasse would be disrupted.

"Hilbert suffered very much because he couldn't work in an atmosphere of this kind of thing," Courant says. "So it was rather poisonous for him. He needed an easy, protected life. His wife of course didn't want to give up her only son, couldn't, so that was the basis of some tension between husband and wife. But Hilbert was so intelligent that there was no real danger."

The Göttingen Association had been disbanded. The Lesezimmer had huge gaps in its collection. Almost all the German publishers of scientific journals and books were pulling back. The construction of the Mathematical Institute had had to be abandonned. Klein was 70 in 1919, Hilbert approaching 60. Carathéodory had left Göttingen for Berlin, where he was again in the company of his old friend Erhard Schmidt. Peter Debye had accepted a position in Switzerland. The Mark had been steadily declining. Food was scarce, living conditions crowded. Hilbert complained to Bernays that his salary was now worth less than it had been when he was a Privatdozent in Königsberg. The future looked very bleak.

In the summer of 1919, Hilbert, vacationing in Switzerland, let it be known that he "would perhaps consider favorably," "would not be totally adverse to," and "might even be inclined to accept" a position in Bern. Under normal conditions Bern could not hope to entice Hilbert from Göttingen, but conditions were not normal. Bern saw an opportunity to add the most famous mathematician in the world to its faculty; and, by-passing the canton rule that all openings were to be advertised in the press, the University eagerly extended an offer to the great German mathematician.

It seems clear now that Hilbert never had any real intention of accepting. At the end of his career he did not even list the offer from Bern among those which he had received. He apparently wanted it only as a negotiating lever with which to improve the situation at home "for mathematics."

Basically, his personal desires were very modest. Courant recalls his saying on his fiftieth birthday, when he was at the height of his fame and influence, "From now on, I think I will indulge myself in the luxury of traveling first class on the train."

The offer from Bern seems to have had the desired effect. By August 1919 Hilbert was corresponding with the new Minister for Science, Art and Popular Education about bringing foreign guest professors to Göttingen. His earlier request for 5000 Marks for this purpose was now raised to 10,000 and, in view of the progressive inflation, "perhaps we should have at least 15,000 Marks."

It was that fall, on November 18, 1919, that Hurwitz died. Since his student days, Hilbert had admired Hurwitz and his mathematical abilities without reservation. Once, in conversation with Ostrowski, he had mentioned that in his opinion there were just two kinds of mathematicians, those who tackled and solved worthwhile problems and those who did not.

"I was surprised how definite it was for him, how there were a few people who were really good and then there were the other ones that he just did not care about. I was also surprised that he would say such a thing to me. It was almost the only time when he did not behave like a 'sage.' If he had said something like that in public, other people would begin to wonder in what group he considered them. But there was no question in which group he considered Hurwitz. At that time he mentioned to me a paper of Hurwitz's which he said had completely supplanted a paper of his own. No one else would have said that about Hilbert's work. But he said it."

For the second time Hilbert went before the Göttingen Scientific Society to deliver a memorial to a lost friend of his youth. For eight and a half years he and Hurwitz had explored "every corner" of mathematics on daily walks in Königsberg. Hurwitz, he now told his colleagues, had been "a harmoniously developed, philosophically enlightened spirit, willingly prepared to acknowledge and appreciate the achievements of others and filled with sincere joy at every scientific advance." He took comfort in the fact that, passing away in a coma, Hurwitz had not had to take leave of his family. To be spared that had been his last wish.

After Hurwitz's death there was a rumor that Hilbert had been offered Hurwitz's chair in Zürich. A group of students went to see him with a poetic petition that he remain in Göttingen: "Hilbert, gehen Sie nicht nach Zürich. / Leben da ist auch recht 'schwürich.'" (*Hilbert, don't go to Zürich, life is hard there too.*) However, no offer was forthcoming from the Swiss.

The relative weight of Hilbert's scientific interests during this period was being gauged by his assistants, Bernays for mathematics and Adolf Kratzer for physics. On the day before a lecture, both men came to Hilbert's house. As his interest moved from physics back to mathematics, so did the role played by the assistants.

"In the summer of 1920, he was concerned primarily with problems of atom mechanics," Kratzer says. "His goal here was still axiomatization. Questions were directed to me. I seemed to do most of the talking while Bernays listened. But by the winter of 1920–21 his interest had begun to change. Now his chief goal was the formalization of the foundations of mathematics on a logistic basis, and Bernays talked while I listened."

Although in his own work he was moving toward the most abstract and formal conception of mathematics, Hilbert delivered at this time a series of lectures on geometry based on an approach of strictly visual intuition. They were quite frankly designed by him to popularize mathematics with the young men returning to the University after the war.

"For it is true," he conceded, "that mathematics is not, generally speaking, a popular subject."

He saw the reason for the lack of popularity in "the common superstition that mathematics is . . . a further development of the fine art of arithmetic, of juggling with numbers . . . ." He thought he could bring about a greater enjoyment of the subject he himself enjoyed so thoroughly if he could make it possible for his listeners "to penetrate to the essence of mathematics without having to weigh themselves down under a laborious course of studies." He planned instead "a leisurely walk in the big garden that is geometry so that each may pick for himself a bouquet to his liking."

The next summer Hilbert lectured on relativity theory as part of a University series for all the Faculties. Here he demonstrated, according to Born, "that only one to whom the logical structure of a difficult, complicated territory is completely clear can discourse on it successfully to a lay audience.

He enjoyed these excursions into popularization, and during the twenties frequently delivered such lectures on various subjects.

But now Hilbert was becoming increasingly alarmed by the gains that Brouwer's conception of mathematics was making among the younger mathematicians. To him, the program of the Intuitionists represented quite simply a clear and present danger to mathematics. Many of the theorems of classical mathematics could be established by their methods, but in a more complicated and lengthy way than was customary. Much — including all pure existence proofs and a great part of analysis and Cantor's theory of infinite sets — would have to be given up.

"Existential" ideas permeated Hilbert's thinking, not only in mathematics, but also in everyday life. This is illustrated by an incident which Helmut Hasse observed at this time. The Society of German Scientists and Physicians was holding its first meeting after the war in Leipzig. In the evenings at the Burgkeller there was much questioning of the type, "What about Professor K. from A., is he still alive?" The 24-year-old Hasse was seated with other young mathematicians at a table quite near to the table shared by Hilbert and his party.

"I heard him put exactly this type of question to a Hungarian mathematician about another Hungarian mathematician. The former began to answer, 'Yes, he teaches at — and concerns himself with the theory of —, he was married a few years ago, there are three children, the oldest . . . .' But after the first few words Hilbert began to interrupt, 'Yes, but . . . .' When he finally succeeded in stopping the flow of information, he continued, 'Yes, but all of that I don't want to know. I have asked only *Does he still exist?*'"

According to Brouwer, a statement that an object exists having a given property means that, and is only proved when, a method is known which in principle at least will enable such an object to be found or constructed. Thus Brouwer would not accept Hilbert's youthful proof of the existence of a finite basis of the invariant system, or many others.

Hilbert naturally disagreed.

". . . pure existence proofs have been the most important landmarks in the historical development of our science," he maintained.

The fact that Weyl was moving closer to Brouwer's position disturbed Hilbert considerably.

In 1919 Weyl had published some "long held" thoughts of his own on foundations. Then in 1920 he had delivered several lectures on Brouwer's program. In the course of one of these he had declared: "I now give up my own attempt and join Brouwer." He was never called "Brouwer's Bulldog," but he might have been. In 1921 he proceeded to use his literary gifts to put Brouwer's ideas into even wider circulation.

This was too much for Hilbert.

At a meeting in Hamburg in 1922 he came roaring back to the defense of mathematics.

The state of affairs brought about by the discovery of the antinomies in set theory was intolerable — he conceded — but "high-ranking and meritorious mathematicians, Weyl and Brouwer, seek the solution of the problem through false ways."

Weyl heard "anger and determination" in the voice of his old teacher.

"What Weyl and Brouwer do comes to the same thing as to follow in the footsteps of Kronecker! They seek to save mathematics by throwing overboard all that which is troublesome . . . . They would chop up and mangle the science. If we would follow such a reform as the one they suggest, we would run the risk of losing a great part of our most valuable treasures!"

Down the list he went of some of the treasures which would be lost if the Intuitionists' program were to be adopted:

The general concept of the irrational number.
The function. "Even the number theory function."
Cantor's transfinite numbers.
The theorem that among infinitely many whole numbers there is a smallest.
The logical principle of the excluded middle.

Hilbert refused to accept such a "mutilation" of mathematics. He thought he saw a way in which he could regain the elementary mathematical objectivity which Brouwer and Weyl demanded without giving up any of the treasures that would have to be sacrificed in their program. This was essentially the "theory of proof" which he had sketched at Heidelberg in 1904.

Characteristically, his approach was a frontal assault on the problem. As Weyl himself later had to concede, Hilbert now gave "a completely new turn to the questions of the foundations and the truth content of mathematics."

The Intuitionists had objected that "much generally accepted mathematics goes beyond such statements as can claim real meaning." Hilbert countered their argument, so Weyl claimed, by relinquishing meaning altogether.

He proposed to formalize mathematics into a system in which the objects of the system — the mathematical theorems and their proofs — were expressed in the language of symbolic logic as sentences which have a logical structure but no content. These objects of the formal system would be chosen in such a way as to represent faithfully mathematical theory as far as the totality of theorems was concerned. The consistency of the formal system — that is, mathematics — would then be established by what Hilbert called finitary methods. "Finitary" was defined as meaning that "the discussion, assertion or definition in question is kept within the boundaries of thorough-going producibility of objects and thorough-going practicality of methods and may accordingly be carried out within the domain of concrete inspection."

In this way, through methods as severely limited as any Brouwer and Weyl could demand, Hilbert believed that he could surmount the new foundations crisis and dispose of the foundations questions *once and for all*.

In the year of his sixtieth birthday he set out to save the entirety of classical mathematics by what his former student was to call "a radical reinterpretation of its meaning without reducing its inventory."

The ghost of Kronecker seemed to rise before him in the program of the Intuitionists, and the violence with which he slashed out at it was (as Weyl was quick to point out) oddly in contrast to the confidence with which he predicted its ultimate failure:

"I believe that as little as Kronecker was able to abolish the irrational numbers... just as little will Weyl and Brouwer today be able to succeed. Brouwer is not, as Weyl believes him to be, the Revolution — only the repetition of an attempted *Putsch*, in its day more sharply undertaken yet failing utterly, and now, with the State armed and strengthened..., doomed from the start!"

XIX

The New Order

Hilbert was 60 years old on January 23, 1922.

Naturwissenschaften, the German equivalent of the British scientific weekly *Nature*, dedicated its last issue in January to the birthday. The frontispiece was a photograph of Hilbert sitting in a wide-armed wicker chair. He had not changed much over the years, but time had honed the intelligence and concentration in his face until, in age, he was a more impressive looking man than he had been in his youth.

Otto Blumenthal led off the issue with a sketch of Hilbert's career and of his character. As the "oldest student," Blumenthal had observed his Doctor-Father carefully for almost a quarter of a century. Now the pattern of Hilbert's life seemed to stand out in relief. His career had developed at the hand of problems. Then, with the work on the foundations of geometry, the axiomatic method had become so much a part of him that it, like the problems, had always accompanied him and led him. To Blumenthal, now, the most striking aspect of Hilbert's life seemed to be the remarkable continuity of progress. Immediately upon solving one problem, he had applied himself to the next. Indeed, it might seem to some who did not know him that he had been all mathematician, a logical machine and problem solver, a creature of pure thought.

"But I believe that Hilbert himself would wish to be judged differently," Blumenthal wrote. "The longer I know him and the more I learn about him, I see him as a wise human being who, after he first became conscious of his power, has unfailingly kept before himself a supreme goal toward which he strives on a well-plotted course: the goal of a unified view of life, at least in the specific field of the exact sciences."

There were other articles by former pupils on the five main areas in which Hilbert had worked — algebra, geometry, analysis, mathematical physics, and the philosophy of mathematics. (An article entitled "Hilbert and Women" was prepared by Courant and his friend Ferdinand Springer but, Courant recalls, "we didn't get it finished in time.")

There was also a birthday banquet, and the 73-year-old Klein, now confined to a wheel chair, presented the honored professor with the copy of the Vortrag which young Dr. Hilbert had given in 1885 in Klein's seminar at Leipzig.

The celebration marked in a sense the passing of the old order at Göttingen. After the war Richard Courant had returned to academic life, first in Münster and then, when Erich Hecke left, as a professor in Göttingen. In the coming years he was to fill the place of Klein.

Courant was an entirely different personality from the old Jupiter. Little, gnomish-faced, soft-voiced, he was never described as "olympian." Rather, his students remember "how he could present a picture of utter helplessness and indecision, how he could grumble almost inaudibly, how he could interfere and guide by non-interference and finally obtain the unfailing attachment and devotion of all his associates."

Courant was unusually democratic for a German professor. Even his books were in part the result of group effort. What were known as "Proofreading Festivals" regularly took place at a long table with all the assistants participating. Among these at various times were Willy Feller, Kurt Friedrichs, Hans Lewy, Otto Neugebauer, Franz Rellich.

"Red ink, glue and personal temperament were available in abundance," recalls Otto Neugebauer, who held the important and influential post of "head assistant." "Courant had certainly no easy time in defending his position and reaching a generally accepted solution under the impact of simultaneously uttered and often widely diverging individual opinions about proofs, style, formulations, figures, and many other details. At the end of such a meeting he had to stuff into his briefcase galleys, or even page proofs, which can only be described as Riemann surfaces of high genus; and it needed completely unshakable faith in the correctness of the uniformization theorems to believe that these proofs would ever be mapped on *schlicht* pages."

Like Klein, however, Courant was in the broad mathematical-physical tradition of Göttingen. The real core of his work was to be (as Neugebauer later saw it) "the conscious continuation and ever-widening development of the ideas of Riemann, Klein and Hilbert and the insistence on demonstrating the fundamental unity of all mathematical disciplines."

When Courant took over from Klein, the students of mathematics and theoretical physics still attended classes in the only classroom building of the University, the three-story Auditorienhaus where Weender Strasse crossed the old city wall. The third floor of this building remained the heart of the mathematical life: the common room where the Mathematics Club held its weekly meetings — the Lesezimmer with the mathematical books and journals on open shelves as Klein had decreed — the Room of the Mathematical Models where the students gathered outside the main lecture hall. A large wooden cabinet contained the entire administrative apparatus of mathematics at Göttingen — the stamps and the stationery. This was where Courant took his first revolutionary step. He applied to the Minister of Culture for permission to change the heading on the stationery from "Universität Göttingen" to "Mathematisches Institut der Universität Göttingen." After an appropriate delay, he received permission for this change.

"They don't know how much that will cost them," the head of the new Institute said softly.

Thus, at Göttingen, the new order began.

The problem of publication, so important for the progress of science, was already being solved by Courant. During the war a personal bond had developed between the Göttingen mathematicians and the publisher Ferdinand Springer. After the war (as Hilbert later described it) "under the impact of Klein's personality and my active influence, Dr. Springer placed his energy and his firm at the disposal of mathematics." Courant and Springer became close friends; and as a result of their combined efforts, scientific publishing in Germany began to return to normal.

In addition to Courant, there was another former pupil whom Klein and Hilbert wanted to see back at Göttingen. This was Hermann Weyl. In 1922 — the same year in which Hilbert had delivered his polemic against the Intuitionists at Hamburg — an offer went to Weyl.

Like Courant, Weyl was still in his thirties. As a result of the popularity of his book on relativity theory, which had gone into five printings in five years, and of his active participation in the controversy over foundations, he was perhaps the most generally known of his generation of mathematicians. But he also had already behind him impressive solid achievements in mathematics and mathematical physics. He was now at the height of his creative powers. A great stream of papers gushed forth, not only on his main mathematical themes, but on any mathematical topic that interested him. And it was not just mathematics that interested Weyl. There was philosophy. Art. Literature. He was convinced that the problems of science could not be separated from the problems of philosophy; also convinced that mathematics — like art, music and literature — was a creative activity of mankind. He loved to write, and wrote well. It has been said that no mathematical papers of the century express as vividly their author's personality. "Expression and shape are almost more to me than knowledge itself," he said once. And another time: "My work has always tried to unite the true with the beautiful; and when I had to choose one or the other, I usually chose the beautiful."

Weyl respected and loved Klein and Hilbert. He was dedicated to the Göttingen tradition. But he did not immediately agree to return to his old university. Still debating his decision at the last possible moment, he marched his wife around and around the block of their home in Zürich. When it was nearly midnight, he decided that he would accept the offer from Göttingen. Hurrying off to send the telegram, he returned a few hours later, having wired instead his refusal.

"I could not bring myself," he explained, "to exchange the tranquillity of life in Zürich for the uncertainties of post-war Germany."

Life in Germany was indeed uncertain. A period of violent unrest had followed the surrender. Then the people had elected a national assembly, which had met at Weimar and drawn up a republican constitution; but the new government was constantly under attack. Monarchists wanted to restore the empire. Communists wanted an experiment in the Russian style. Nationalistic groups demanded a dictatorship, the rearming of Germany, and the tearing-up of the Treaty of Versailles. "Germans will have to get used to politics just as the cavemen had to get used to soap and water," Hilbert observed.

It was during this uncertain time that Courant began to bring to physical reality Klein's old dream of a great Mathematical Institute at Göttingen.

The Mark had been steadily declining in value. By 1922 the new government was issuing paper money to meet its needs, and inflation was well under way. The price of a volume of the *Annalen*, which had been 64 Marks in 1920, had doubled by the beginning of 1922. By the end of the year it was 400 Marks. By 1923 it had gone to 800 Marks; by the end of 1923, 28,000 Marks. The fees which the students paid at the beginning of the semester for lectures were virtually valueless by the end of the semester when the University turned them over to the Privatdozents. The Wolfskehl Prize of 100,000 Marks was shortly to be worth no more than a few scraps of paper (but in 1921 the interest from the Prize could still be used to bring Niels Bohr for a series of lectures — "the Bohr Festival Week").

Courant, to emphasize his Kleinian interest in applied as well as pure mathematics, had equipped his new institute with one of the early electrically driven desk computers. Its range of 19 digits now turned out to be just about right to handle the inflationary currency. Salaries and prices were expressed by basic numbers which were then multiplied by a rapidly increasing coefficient $c(t)$ such that the result represented the value expressed in Marks at a given moment t. Salaries were then computed each week on the basis of the current value of $c(t)$, which was obtained confidentially from the government. Now Courant offered to lend his computer to the University for the privilege of being given the value $c(t)$ when it was received, hours before its publication in the newspapers. By this simple method he greatly increased the purchasing power of the funds budgeted for mathematics. For the most part he used the extra money so obtained to fill the great gaps which had developed in the Lesezimmer collection during the war. For with Courant, as with Klein, the Lesezimmer was the center around which mathematics at Göttingen revolved.

What the Lesezimmer meant to the students has been described by B. L. van der Waerden, who after completing the university course at Amsterdam had come to Göttingen on the recommendation of Brouwer. Van der Waerden was a gifted young man. His father, a high school teacher, had once taken away his mathematical books because he thought the boy should be out playing with the other boys in the fresh air. He returned the books, however, when he discovered that his son had invented trigonometry on his own and was using the names and notations he had made up instead of the traditional ones.

In Göttingen, van der Waerden spent much of his time in the Lesezimmer. There had been nothing at all like it in Holland. Today he recalls how his regular lunching and walking companion, Helmuth Kneser, "used to start on a certain subject and make a few remarks which I couldn't understand at all. Then I would say to him that I would like to learn about that subject. Where could I find out about it? So he would give me the names of some books which I could find in the Lesezimmer. A day or so later I would be able to answer his questions and also make some significant remarks of my own, and then I learned much more." Sometimes, while van der Waerden was looking for a book "by Author A," he would find next to it a book "by Author B" which was even more interesting and useful. "In this way, I learned more in weeks or months in the Lesezimmer than many students learn in years and years."

In 1923 the inflation ended abruptly through the creation of a new unit of currency called the Rentenmark. Although Hilbert remarked sceptically, "One cannot solve a problem by changing the name of the independent variable," the stability of conditions was gradually restored.

Students again began to come to Göttingen from all over the world.

The University was, thanks to Landau, the center of the great number theory activity of the 1920's — "the beginning," it has been said, "of an era of arithmetic comparable to that inaugurated by Gauss in 1801." Two problems seemed to attract the most interest. One of these was the hypothesis of Riemann concerning the zeros of the zeta function, which Hilbert had listed as the eighth of his Paris Problems. The other was the determination of the exact values for the number of nth powers in Waring's Theorem, work on which had been opened up by Hilbert's proof of the theorem in 1909. Waring's conjecture had turned out to be, according to mathematical historians, "one of those problems that have started epochs in mathematics."

Harald Bohr and G. H. Hardy were frequent visitors in Göttingen, usually on their way to Denmark or to England to visit one another. When Hardy would leave Bohr, to return home over the choppy North Sea channel, he always mailed him a card announcing, "I have a proof for the Riemann hypothesis!" — confident, Hardy said, that God — with whom he waged a very personal war — would not let Hardy die with such glory.

There is a Hilbert story in connection with the Riemann hypothesis which, although perhaps apocryphal, must be included. According to this story, Hilbert had a student who one day presented him with a paper purporting to prove the Riemann hypothesis. Hilbert studied the paper carefully and was really impressed by the depth of the argument; but unfortunately he found an error in it which even he could not eliminate. The following year the student died. Hilbert asked the grieving parents if he might be permitted to make a funeral oration. While the student's relatives and friends were weeping beside the grave in the rain, Hilbert came forward. He began by saying what a tragedy it was that such a gifted young man had died before he had had an opportunity to show what he could accomplish. But, he continued, in spite of the fact that this young man's proof of the Riemann hypothesis contained an error, it was still possible that some day a proof of the famous problem would be obtained along the lines which the deceased had indicated. "In fact," he continued with enthusiasm, standing there in the rain by the dead student's grave, "let us consider a function of a complex variable...."

During this period there came to Göttingen a big, shy boy who was to be an outstanding number theorist — in this area of mathematics, the Minkowski of the new generation. He had refused to serve in the army and had been confined in a mental institution which was located next to the clinic owned by Landau's father. Thus young Carl Ludwig Siegel, extremely gifted but without money, had become acquainted with the Göttingen professor. To Siegel, Landau presented a quite different picture from the spoiled cherub whom Norbert Wiener saw at about this same time.

"If it had not been for Landau," Siegel says simply, "I would have died."

When, however, Siegel came to Göttingen as a student in 1919, he worked almost entirely alone. "I was very eager to show what I could do by myself." He had no direct personal contact with Hilbert, but he was always to remember a lecture on number theory which he heard from Hilbert at this time. Hilbert wanted to give his listeners examples of the characteristic problems of the theory of numbers which seem at first glance so very simple but turn out to be incredibly difficult to solve. He mentioned Riemann's hypothesis, Fermat's theorem, and the transcendence of $2^{\sqrt{2}}$ (which he had listed as his seventh problem at Paris) as examples of this type of problem. Then he went on to say that there had recently been much progress on Riemann's hypothesis and he was very hopeful that he himself would live to see it proved. Fermat's problem had been around for a long time and apparently demanded entirely new methods for its solution — perhaps the youngest members of his audience would live to see it solved. But as for establishing the transcendence of $2^{\sqrt{2}}$, no one present in the lecture hall would live to see that!

The first two problems which Hilbert mentioned are still unsolved. But less than ten years later a young Russian named Gelfond established the transcendence of $2^{\sqrt{-2}}$. Utilizing this work, Siegel himself was shortly able to establish the desired transcendence of $2^{\sqrt{2}}$.

Siegel wrote to Hilbert about the proof. He reminded him of what he had said in the lecture of 1920 and emphasized that the important work was that of Gelfond. Hilbert was frequently criticized for "acting as if everything had been done in Göttingen." Now he responded with enthusiastic delight to Siegel's letter, but he made no mention of the young Russian's contribution. He wanted only to publish Siegel's solution. Siegel refused, certain that Gelfond himself would eventually solve this problem too. Hilbert immediately lost all interest in the matter.

After a semester in Hamburg with Hecke, who was now a professor there, Siegel returned to Göttingen as assistant to Courant and later became a Privatdozent. The money he earned was so little that Courant, who wanted a cycling companion, had to arrange an extra stipend so that Siegel could afford to buy a bicycle.

Courant liked to keep Klein and Hilbert in touch with the gifted young people. It was through him that Siegel had his first personal contacts with the famous mathematicians of Göttingen. Because of the post-war housing shortage he lived for a while at Klein's house. But even living under the same roof he felt the distance which people had always felt between themselves and Klein. He worried constantly that he "would say the wrong thing." Later he was taken by Courant to swim in that part of the Leine river which was roped off for the faculty. He met Hilbert in the little shed where the professors changed into their bathing suits. Young Siegel, Courant explained to Hilbert, had recently found another proof of a theorem of Hecke's connected with the Riemann hypothesis. Hilbert was very enthusiastic. "He always liked to make young people feel hopeful." In the bathhouse there with Hilbert, Siegel felt none of the constraint he had felt in Klein's house.

Soon after this meeting with Hilbert, Siegel was asked by Courant to referee a paper for the *Annalen*, of which Hilbert was still one of the principal editors. The young man found the paper inaccurate in many places and, even where accurate, unnecessarily roundabout in its methods. He reported to Hilbert that in his opinion the paper was not publishable.

"No, no, I must publish it!" Hilbert insisted. "In 1910 this man was a member of the committee that gave me the Bolyai Prize, and now I simply cannot refuse to publish his paper! Take it and change whatever should be changed. But I must publish it!"

The paper appeared in an improved form in the *Annalen*. Several months later, when Siegel was sure that Hilbert had forgotten all about the matter, a package was delivered to his rooms. It contained the two volumes of Minkowski's collected works, inscribed "With friendly thoughts from the editor."

One of the most fertile circles of research in post-war Göttingen revolved around Emmy Noether. The desired position of Privatdozent had at last been obtained for her in 1919. This was still the lowest possible rank on the university scale, not a job but a privilege. But Emmy Noether was delighted with the appointment. In the thirteen years which had passed since she had had to defend her doctoral dissertation before Gordan, she had come a long way. Already she had achieved important results in differential invariants, which the Soviet mathematician Paul Alexandroff was to consider sufficient to secure her a reputation as a first-rate mathematician, "hardly less a contribution to mathematical science than the notable researches of Kowalewski." She herself was always to dismiss these works as standing to the side of her main scientific path, on which at last she was now, at the age of 39, taking her first step — the building up on an axiomatic basis of a completely general theory of ideals. This work would have its source in the early algebraic work of Hilbert, but in her hands the axiomatic method would become no longer "merely a method for logical clarification and deepening of the foundations [as it was with Hilbert] but a powerful weapon of concrete mathematical research." Gordan's picture still hung over her desk in Göttingen; but although she had been so thoroughly under his influence in her youth that her dissertation had concluded with a table of the complete system of covariant forms for a given ternary quartic and had contained more than three hundred forms in symbolic representation — a maiden work which she in later years dismissed as "Formelgestrüpp!" — *a jungle of formulas* — she was destined in the next decade to make Hilbert's "theology" look like mathematics.

In 1922 she became a "nicht beamteter ausserordentlicher Professor" — an unofficial extraordinary, or associate, professor. There were no obligations connected with this new title — and no salary, such an extraordinary professor being considered more than usually inferior to an ordinary professor. The title could be explained only by a Göttingen saying to the effect that "an extraordinary professor knows nothing ordinary and an ordinary professor knows nothing extraordinary." By this time, however, inflation had so reduced the students' ability to pay fees that if the Privatdozents were not to starve away they had to be given some small sums by the University for delivering lectures in their specialties. Such a "Lehrauftrag" for algebra was now awarded to Emmy Noether, the first and only salary she was ever to be paid in Göttingen.

She and her work were not on the whole much admired in her native land. She was never even elected to the Göttingen Scientific Society. "It is time that we begin to elect some people of real stature to this society," Hilbert once remarked at a meeting. "Ja, now, how many people of stature have we indeed elected in the past few years?" He looked thoughtfully around at the members. "Only — zero," he said at last. "Only zero!"

A Dutchman, attending one of Emmy Noether's lectures for the first time, remembers that she greeted him, "Ah, another foreigner! I get only foreigners!" But among the foreigners who came to her were van der Waerden from Holland, Artin from Austria, Alexandroff from Russia.

It was Alexandroff who christened her "der Noether", *der* being the definite article which precedes all masculine nouns in German. But he later said: "Her femininity appeared in that gentle and subtle lyricism which lay at the heart of the far-flung but never superficial concerns which she maintained for people, for her profession, and for the interests of all mankind."

She was not a good lecturer and her classes usually numbered no more than five or ten. Once though, she arrived at the appointed hour to find more than a hundred students waiting for her. "You must have the wrong class," she told them. But they began the traditional noisy shuffling of the feet which, in lieu of clapping, preceded and ended each university class. So she went ahead and delivered her lecture to this unusually large number of students. When she finished, a note was passed up to her by one of her regular students who was in the group. "The visitors," it read, "have understood the lecture just as well as any of the regular students."

It was true, she had no pedagogical talents. Her mind was open only to those who were in sympathy with it. Her teaching approach, like her thinking, was wholly conceptual. The German letters which she chalked up on the blackboard were representatives of concepts. It seemed to van der Waerden that "her touching efforts to clarify these, even before she had quite verbalized them . . . had the opposite effect." But of all the new generation in Göttingen, Emmy Noether was to have the greatest influence on the course of mathematics.

While these widening circles of varying mathematical activity were forming themselves around Courant, Landau and Emmy Noether, a group of exceptionally gifted young physicists were gathering around Max Born, who (like Courant, still in his thirties) had become professor of theoretical physics after the war. From the beginning, it was Born's goal to establish at Göttingen a physics institute comparable to Sommerfeld's institute at Munich. When the opportunity arose, he arranged that his best friend, James Franck, join him in Göttingen as professor of experimental physics. It was a stroke reminiscent of Hilbert's bringing Minkowski to the University in 1902. But even before Franck's arrival in 1922, the spectacular series of students who would make their way to Göttingen during the 1920's had begun; and Born's first assistants were Wolfgang Pauli and Werner Heisenberg.

Since the war, the Germans had been barred from most international scientific gatherings; but now it seemed once again that in Göttingen an international congress was permanently in session.

167

XX

The Infinite!

The highpoint of the mathematical week at Göttingen during the 1920's was the regular session of the Mathematics Club.

The club was a very informal kind of organization without officers, members or dues. Anyone with a doctor's degree could come to meetings, and because of the quality of mathematics at Göttingen it was always "a very high class affair." Sometimes the speaker was a distinguished visitor who reported on his own recent work or that of his students. More often he was a member of the Göttingen circle — professor, docent or student.

The bright young newcomers who saw the famous Hilbert in action for the first time at these events were struck by his slowness in comprehending ideas which they themselves "got" immediately. Often he did not understand the speaker's meaning. The speaker would try to explain. Others would join in. Finally it would seem that everyone present was involved in trying to help Hilbert to understand.

"That I have been able to accomplish anything in mathematics," Hilbert once said to Harald Bohr, "is really due to the fact that I have always found it so difficult. When I read, or when I am told about something, it nearly always seems so difficult, and practically impossible to understand, and then I cannot help wondering if it might not be simpler. And," he added, with his still childlike smile, "on several occasions it has turned out that it really was more simple!"

Some of the young people were irritated by the precious time consumed by Hilbert's questions; others found it fascinating to watch Hilbert's mind in action.

"Scientifically, he did not grasp complicated things at a flash and absorb them. This kind of talent he did not have," Courant explains. "He had to go to the bottom of things."

Hilbert still set high standards of simplicity and clarity for the talks to the Mathematics Club. His guiding rule for the speaker was "only the raisins out of the cake." If computations were complicated, he would interrupt with, "We are not here to check that the sign is right." If an explanation seemed too obvious to him, he would reprove the speaker, "We are not in *tertia*" — *tertia* being the level of the gymnasium in which the student is 12 to 14 years old.

The brutality with which he could dispose of someone who did not meet his standards was well known. There were important mathematicians in Europe and America who dreaded a speech before the Mathematics Club in Göttingen. It seemed now sometimes to Ostrowski that Hilbert was unnecessarily rough on speakers—as if he no longer attended so carefully as in the past to the problem of the superior individual living among lesser individuals.

One young Scandinavian, today highly esteemed, came to Göttingen and spoke about his work — "really important and beautiful and very difficult" in Ostrowski's opinion. Hilbert listened and, when the visitor was through, demanded only, "What is it good for?"

On another occasion he interrupted the speaker with, "My dear colleague, I am very much afraid that you do not know what a differential equation is." Stunned and humiliated, the man turned instantly and left the meeting, going into the next room, which was the Lesezimmer. "You really shouldn't have done that," everyone scolded Hilbert. "But he *doesn't* know what a differential equation is," Hilbert insisted. "Now, you see, he has gone to the Lesezimmer to look it up!"

Still another time the speaker was the young Norbert Wiener. The importance of his talk in Göttingen can be gauged by the fact that many years later in his autobiography he devoted more than a dozen pages to it. After Wiener's talk to the Mathematics Club, everyone hiked up to Der Rohns, as was the custom, and had supper together. Hilbert began in a rambling way during supper to talk about speeches which he had heard during the years he had been at Göttingen.

"The speeches that are given nowadays are so much worse than they used to be. In my time there was an art to giving speeches. People thought a lot about what they wanted to say and their talks were good. But now the young people cannot give good talks any more. It is indeed exceptionally bad here in Göttingen. I guess the worst talks in the whole world are given in Göttingen. This year especially they have been very bad. There have been — no, I have heard no good talks at all. Recently it has been especially bad. But now, this afternoon, there was an exception —"

The young "ex-prodigy" from America prepared himself to accept the compliment.

"This afternoon's talk," Hilbert concluded, "was the worst there ever has been!"

In spite of this remark (which was not reported in the autobiography), Wiener continued to see Hilbert as "the sort of mathematician I [would like] to become, combining tremendous abstract power with a down-to-earth sense of physical reality."

The presence of Klein was still felt in Göttingen during the early twenties, but now like the sunset rather than the high noon-day sun. The editing of the collected works was completed, each paper accompanied by detailed notes on the historical context in which it had originated — a history of the mathematics of his time as well as of his own career. It often seemed to Courant that Klein felt his own life was also completed. He continued to take on projects, such as the editing of his war-time lectures on the history of nineteenth century mathematics, "but with the knowledge that these would have to be finished by others."

When a young mathematician did not immediately follow up a suggestion, Klein dismissed him with, "I am an old man, I can't wait."

Young Norbert Wiener went to pay a call on Klein in the spring of 1925.

"The great man sat in an armchair behind a table, with a rug about his knees. He . . . carried about him an aura of the wisdom of the ages . . . and as he spoke the great names of the past ceased to be the mere shadowy authors of papers and became real human beings. There was a timelessness about him which became a man to whom time no longer had a meaning."

The 1920's were "the beautiful years" when modern physics was developing at an almost magical rate within a triangle which had as its vertices Cambridge, Copenhagen and Göttingen. The 20-year-old Werner Heisenberg, still wearing the khaki shorts of the Youth Movement, came from Munich to Göttingen in 1921. He recalls himself as "much impressed" by the number of young physicists who were interested in the particular problem that was currently interesting Hilbert — "a problem which at that time exceeded by far my own mathematical and physical knowledge." Hilbert had recently returned to his war-time ideas on relativity; and for a while, according to Weyl, hopes ran high in the Hilbert circle of a unified field theory. But, on the whole, it was Hilbert's spirit rather than his person which was felt in physics at this time.

From 1922 on, Hilbert was no longer a physicist. The seminar on the Structure of Matter, which he had instituted with Debye during the war, was now carried on by Born and Franck. The members included at various times during the twenties Heisenberg, Wolfgang Pauli, Robert Oppenheimer, K. T. Compton, Pascual Jordan, Paul Dirac, Linus Pauling, Fritz Houtermans, P. M. S. Blackett among others. Hilbert rarely attended.

His own personal achievement in physics had been a disappointment, "in no way comparable," Weyl later said in summary, "to the mathematical achievement of any single period of his career." The axiomatization of physics, which had been his goal when he first began the joint study with Minkowski, always eluded him.

To Weyl, who himself made important contributions to mathematical physics, it seemed that "the maze of experimental facts which the physicist has to take into account is too manifold, their expansion too fast, and their aspect and relative weight too changeable for the axiomatic method to find a firm enough foothold, except in the thoroughly consolidated parts of our physical knowledge. Men like Einstein or Niels Bohr grope their way in the dark toward their conceptions of general relativity or atomic structure by another type of experience and imagination than those of the mathematician, altough no doubt mathematics is an essential ingredient."

Hilbert's real contribution to physics was to lie in the mathematical methods which he had created in his work on integral equations and in the unification which this work had brought about. When, at the end of 1924, Courant published the first volume of his *Methods of Mathematical Physics*, he placed the name of Hilbert on the title page with his own. This act seemed justified, Courant wrote in the preface, by the fact that much material from Hilbert's papers and lectures had been used as well as by the hope that the book expressed some of Hilbert's spirit, "which had such decisive influence on mathematical research and education."

"Actually it is more than a mere act of dedication that Hilbert's name stands next to that of Courant on the title page," Ewald pointed out in a review of the book for *Naturwissenschaften*. "Hilbert's spirit radiates from the entire book — that elemental spirit, passionately seeking to grasp completely the clear and simple truths, pushing trivialities aside and with masterful clarity establishing connections between the high points of recognition — a spirit that filled generations of searchers with enthusiasm for science."

"Courant-Hilbert," as the book immediately became known, represented a tremendous advance over previous classics of applied mathematics. There had, in fact, really been nothing like it. In the past theoretical physicists had for the most part had to obtain their mathematics from the work of Rayleigh and other physicists. Now they welcomed "Courant-Hilbert."

Hilbert continued to have an assistant to keep him informed on the latest developments in physics. Beginning in 1922, this position was held by Lothar Nordheim, who like all of the other assistants was chosen for Hilbert by Sommerfeld.

Hilbert, in Nordheim's opinion, still had hopes at this time for the achievement of his goal of the axiomatization of physics. To his assistant, however, he was no longer the legendary "great thinker." He was not well. He seemed to live much in the past, had difficulty accepting changes, was prejudiced in many things, his egoism having become more marked. "He could not imagine any greater privilege for a young man than to be his assistant." Nordheim would have preferred a position in Born's institute. Working with Hilbert now in his home, he felt very much out of the mainstream of physics.

But in spite of these signs of apparently early aging, Hilbert continued to maintain his close contacts with youth.

At the same time that Nordheim was coming regularly to Hilbert's house, another young man was also a frequent visitor. John von Neumann had studied in Berlin with Erhard Schmidt, Hilbert's former student who, at the beginning of the century, had so significantly forwarded the Hilbert work on integral equations. He was a young man who was in one respect at least the exact opposite of Hilbert. Whereas Hilbert was "slow to understand," von Neumann was equipped with "the fastest mind I ever met," according to Nordheim. He frequently expressed his opinion that the mathematical powers decline after the age of 26, but that a certain prosaic shrewdness developing from experience manages to compensate for this gradual loss. (During his own life, he slowly raised the limiting age.)

Von Neumann was 21 in 1924, deeply interested in Hilbert's approach to physics and also in his ideas on proof theory. The two mathematicians, more than forty years apart in age, spent long hours together in Hilbert's garden or in his study.

Hilbert's real collaborator during these days, however, was Bernays. To some people it seemed that he was even exploiting his logic assistant. Bernays was no young student but a man in his middle thirties, a mature mathematician. As Hilbert's assistant, he received a salary and, having habilitated shortly after his arrival in Göttingen, also received fees from the students who attended his lectures. He could live on what he received, but certainly he could not marry.

Hilbert was very opposed to marriage for young scientists anyway. He felt that it kept them from fulfilling their obligations to science. Later, when Wilhelm Ackermann, with whom he had worked and collaborated on a book, married, Hilbert was very angry. He refused to do anything more to further Ackermann's career; and as a result, not obtaining a university position, the gifted young logician had to take a job teaching in a high school. When, sometime later, Hilbert heard that the Ackermanns were expecting a child, he was delighted.

"Oh, that is wonderful!" he said. "That is wonderful news for me. Because if this man is so crazy that he gets married and then even has a child, it completely relieves me from having to do anything for such a crazy man!"

In addition to preparing his own lectures, Bernays helped Hilbert prepare his lectures, accompanied him to class and often took over the teaching for part of the hour, supervised Hilbert's students who were working for the doctoral degree, studied and digested the literature necessary for their work, and did a great deal of writing on their joint book, which was to be entitled *Grundlagen der Mathematik*. In Bernays, Hilbert had found someone as interested in the foundations of mathematics as he was. He had no compunction about working his assistant as hard as he worked himself. "Genius is industry," he liked to tell his students and his assistants, quoting Lichtenberg. He himself was, as Weyl later recalled, "enormously industrious."

The two men sometimes, however, got into rather violent arguments over the subject of foundations. Bernays attributes the emotional quality of these arguments to a fundamental "opposition" in Hilbert's feelings about mathematics.

"For Hilbert's program," he explains, "experiences out of the early part of his scientific career (in fact, even out of his student days) had considerable significance; namely, his resistance to Kronecker's tendency to restrict mathematical methods and, particularly, set theory. Under the influence of the discovery of the antinomies in set theory, Hilbert temporarily thought that Kronecker had probably been right there. But soon he changed his mind. Now it became his goal, one might say, to do battle with Kronecker with his own weapons of finiteness by means of a modified conception of mathematics . . . .

"In addition, two other motives were in opposition to each other — both strong tendencies in Hilbert's way of thinking. On one side, he was convinced of the soundness of existing mathematics; on the other side, he had — philosophically — a strong scepticism."

An example was Hilbert's attitude toward the question of the solvability of every definite mathematical problem. At Paris he had spoken in ringing tones of the axiom of the solvability of every problem, "the conviction which every mathematician shares, although it has not yet been supported by proof." He was convinced that in mathematics at least "there is no *ignorabimus*." Yet, at Zürich, he listed among the epistemological questions which he felt should be investigated the question of the solvability in principle of every mathematical question.

"The problem for Hilbert," Bernays explains, "was to bring together these opposing tendencies, and he thought that he could do this through the method of formalizing mathematics."

Bernays did not always agree with Hilbert about their program, but he appreciated the fact that, passionate though Hilbert was in his disputation, he never held it against his assistant personally when he took the opposite side.

After their work was finished, Hilbert and Bernays often argued about politics. Hilbert enjoyed expressing his views on the subject in extreme and paradoxical ways.

Although he was generally considered conservative, he surprised everybody by proposing Käthe Kollwitz, who was known to be very strongly oriented to the left, for the Star of the Order of Merit, Peace Class. She had become one of the great women artists of all time. ("I have never seen such a drawing by the hand of a woman," the sculptor Constantin Meunier said.) Her subject matter reflected her feeling for the sufferings of humanity.

"Of course what she draws is horrible to look at," Hilbert told his fellow wearers of the Star. "But when we were young in Königsberg and used to dance, she was one of the first girls to dance without her corset!"

In spite of his conservative background, Hilbert was always a liberal in the sense that he never considered himself bound to any certain political view. In his arguments with his assistant, he often criticized "liberals" for seeing things as they wished them to be and not as they were.

"Sometimes," he said, "it happens that a man's circle of horizon becomes smaller and smaller, and as the radius approaches zero it concentrates on one point. And then that becomes his point of view."

He liked to remind his younger assistant: "Mankind is always the constant."

Music often brought peace after the arguments, logical or political. Bernays loved music and had played "four hands" with Hurwitz when he was in Zürich. He was impressed by how much Hilbert's musical knowledge and appreciation had developed over the years as a result of his love for his phonograph, a new model of which was still being supplied regularly to him by the manufacturer. Now he was a member of a group of professors and their wives who attended concerts together in Göttingen and traveled to Leipzig or Hannover for special musical events.

Hilbert sometimes seemed to have very little appreciation of the arts other than music. Yet he was drawn to literature and, as Courant says, "wanted to be aware." He appreciated Goethe and Homer, but he insisted on action in novels. One of his "flames" once set out to educate him in literature. She began by giving him a historical novel about the Swiss civil wars, quite a bloody tale. Hilbert shortly returned it. "If I am given a book to read," he said, "it should be one in which something really happens. Describing the soul and the variations of mood — that I can do for myself!"

There is one story about his attitude toward literature which also reveals a great deal about his feeling for mathematics. It seems that there was a mathematician who had become a novelist. "Why did he do that?" people in Göttingen marvelled. "How can a man who was a mathematician write novels?" "But that is completely simple," Hilbert said. "He did not have enough imagination for mathematics, but he had enough for novels."

The current creation of Hilbert's own mathematical imagination was his proof theory. At Zürich, in 1917, he had announced the general idea and aims of the theory — but not the means of investigation. "For indeed," Bernays was later to comment, "the theory was not to rely on the current mathematical methods." In the first communication on the new theory (the attack on Brouwer and Weyl in Hamburg in 1922), Hilbert had argued that mathematicians could regain elementary objectivity by formalizing the statements and proofs of mathematics in the language of symbolic logic and then taking the represented formulas and proofs directly as objects for study. That same year, at Leipzig, he had added further refinements which reduced the problem of proving the consistency of a formalized domain of arithmetic — the task which he had set for the new century in Paris in 1900.

"Thus it seemed," Bernays later wrote, "that carrying out proof theory was only a matter of mathematical technique."

On the occasion of a celebration at Münster in honor of Weierstrass, Hilbert chose to talk "On the Infinite." He felt that the occasion was appropriate for the fullest exposition of his program of Formalism up to that time. The analysis of Weierstrass and the concept of the infinite as it appeared in the work of Cantor had been prime targets for Kronecker. In the current program of Brouwer, many of the achievements of Weierstrass and Cantor would be among the sacrifices required.

Hilbert was not at all well at the time of his talk in Münster. Recently it had become clear that the deterioration noted by Nordheim was not that of age alone, but the nature of the illness was still undetermined. In spite of his poor health, however, Hilbert spoke as enthusiastically and optimistically as ever.

He began his talk by pointing out that the present "happy state of affairs" in analysis was entirely due to Weierstrass and his penetrating critique of its methods. And yet — disputes about the foundations of analysis did continue up to the present day. This was, in his opinion, because the meaning of *the infinite*, as that concept was used in mathematics, had not yet been completely clarified.

The infinite was nowhere to be found in reality; yet it existed in a very real sense, in his opinion, as an "over-all negation." From time immemorial the idea of the infinite had stirred men's emotions as no other subject. Therefore, he felt, the definitive clarification of its nature went far beyond the sphere of specialized scientific interest: it was needed for the dignity of the human intellect itself!

The deepest insight into the nature of the infinite to date had been obtained by a theory which came closer to a general philosophical way of thinking than to mathematics. This theory, created by Georg Cantor, was set theory.

"It is, I think, the finest product of mathematical genius," Hilbert said, "and one of the supreme achievements of purely intellectual human activity."

But it was in Cantor's set theory, simply as a result of employing definitions and deductive methods which had become customary in mathematics, that the catastrophic antinomies had begun to appear.

"... the present state of affairs ... is intolerable. Just think, the definitions and deductive methods which everyone learns, teaches, and uses in mathematics, the paragon of truth and certitude, lead to absurdities! If mathematical thinking is defective, where are we to find truth and certitude?"

There was, however, "a completely satisfactory way of avoiding the paradoxes of set theory without betraying our science." Mathematicians must establish throughout mathematics the same certitude for their deductions as exists in the ordinary arithmetic of whole numbers, "which no one doubts and where contradictions and paradoxes arise only through our own carelessness."

But if men were to remain within the domain of such purely intuitive and finitary statements — as they must — they would have to have, as a rule, more complicated logical laws. The logical laws which Aristotle had taught and which men had used since they began to think, would not hold.

"We could, of course, develop logical laws which do hold for the domain of finitary statements. But ... we do not want to give up the use of the simple laws of aristotelian logic.... What then are we to do?

"Let us remember that *we are mathematicians* and that as mathematicians we have often been in precarious situations from which we have been rescued by the ingenious method of ideal elements.... Similarly, to preserve the simple formal rules of aristotelian logic, we must *supplement the finitary statements with ideal statements.*"

Mathematics, under this view, would become a stock of two kinds of formulas: first, those to which the meaningful communications correspond and, secondly, other formulas which signify nothing but which are the ideal structures of the theory.

"But in our general joy over this achievement, and in our particular joy over finding that indispensable tool, the logical calculus, already developed without any effort on our part, we must not forget the essential condition of the method of ideal elements — *a proof of consistency.*"

For the extension of a domain by the addition of ideal elements is legitimate only if the extension does not cause contradictions to appear.

This problem of consistency could be "easily handled." It was possible, in his opinion, to obtain in a purely intuitive and finitary way — the way in which truths are obtained in elementary number theory — the insights which would guarantee the validity of the mathematical apparatus. Then the test of the theory would be its ability to solve old problems, for the solution of which it had not been expressly designed. He cited as an example of such a problem the continuum hypothesis of Cantor, which he had listed as the first of the Paris Problems. He now devoted the last part of his talk to sketching an attack on this famous problem.

"No one," he promised his fellow mathematicians, "will drive us out of this paradise that Cantor has created for us!"

XXI

Borrowed Time

On a soft warm evening in June 1925, Felix Klein died.

Everyone at Göttingen had long been prepared for Klein's death.

"But the event after it happened touched us all deeply and affected us painfully," Hilbert said in a little speech to his colleagues the next morning. "Up until yesterday Felix Klein was still with us, we could pay him a visit, we could get his advice, we could see how highly interested he was in us. But that is now all over."

Everything they saw around them in Göttingen was the work of Klein, the collection of mathematical models in the adjoining corridor, the Lesezimmer with all the books on open shelves, the numerous technical institutes that had grown up around the University, the easy relation they had with the education ministry, the many important people from business and industry who were interested in them.... They had lost "a great spirit, a strong will, and a noble character."

An era had come to an end.

A few months later at a memorial meeting of the Göttingen Scientific Society, Courant recalled the dramatic story of the great Felix: the meager beginnings, the spectacular successes ("If today we are able to build on the work of Riemann, it is thanks to Klein."), the tragic breakdown, and then — "the wonderful turning point" — the seemingly broken man who had lived another 43 years and displayed the most varied sides as researcher, teacher, organizer and administrator.

And yet Klein's life had not been without its inner tragedy. The power of synthesis had been granted to him to an extraordinary degree. The other great mathematical power of analysis had been to a certain extent withheld. His ability to bring together the most distant, abstract parts of mathematics had been remarkable, but the sense for the formulation of an individual problem and the absorption in it had been lacking. "He was like a flier who, soaring high over the world, discovers and looks over new fields ... but cannot land his plane in order to take actual possession, to plow and to harvest." Perhaps Klein had himself been unaware of this deep schism but, in Courant's opinion, it had been one of the causes of the decisive breakdown during his competition with Poincaré. Certainly he had perceived "that his most splendid scientific creations were fundamentally gigantic sketches, the completion of which he had to leave to other hands."

Sometimes he had failed to preserve purely human relationships. "Many who knew him only as an organizer ... found him too harsh and violent, so he produced much opposition to his ideas ... which a gentler hand would easily have overcome." Yet his nearest relatives and colleagues and the great majority of his students had known always that behind the relentlessly naive drive, a good human being stood.

They placed on his grave a simple inscription: "Felix Klein, A Friend, Sincere and Constant."

The same year that Klein died Runge retired; his place was taken by Gustav Herglotz.

The ailing Hilbert's condition steadily worsened. In the fall of 1925 it was at last recognized that he was suffering from pernicious anemia. The disease, which was generally considered fatal, had gone so long undetected because the first symptoms, occurring in someone of his age, seemed merely an early failing of powers. Now the doctors gave him at best a few months, maybe even weeks.

In spite of what the doctors said, Hilbert remained optimistic about his condition. He insisted that he actually did not have pernicious anemia — it was some other, less serious disease which merely had all the same symptoms.

It was Hilbert's sheer luck that earlier in the year 1925 G. H. Whipple and F. S. Robscheit-Robbins had discovered the beneficial effects of raw liver on blood regeneration, and by 1926 their work was being applied to the treatment of pernicious anemia by G. R. Minot in America. A pharmacologist friend in Göttingen now chanced to read about Minot's work in the *Journal of the American Medical Association* and showed the article to Hilbert. In addition to describing the new treatment — still, it stressed, in a highly experimental state — the article also described vividly the mortal seriousness of "P. A." But Hilbert reading it completely ignored all the distressing details. He concentrated only upon the hopes raised by Minot's work.

Mrs. Landau was the daughter of Paul Ehrlich, who had discovered salvarsan, the "magic bullet" treatment for syphilis; and she had many contacts in the medical world. With the help of Courant, she drafted a long telegram to Minot, who was at Harvard. "It was the longest telegram I ever sent," Courant says. At the same time another telegram went to Oliver Kellogg who, in 1902, had been the first Hilbert student to write his doctoral dissertation on integral equations. Now a mathematics professor at Harvard, Kellogg rallied support among the mathematicians for the request from Göttingen.

At first Minot and his associates were not too receptive. They had very little of the treatment substance, which would have to be administered to the patient for the rest of his life. People were dying of pernicious anemia within a few miles of Harvard University . . . .

George Birkhoff, the leading American mathematician and also a Harvard mathematics professor, had recently seen a play in which a doctor was able to save but 10 men. How should he choose the 10? "On the basis of their value to mankind" was the playwright's answer. In conversation with Minot, Birkhoff quoted *The Doctor's Dilemma* of George Bernard Shaw.

Mathematics makes mathematicians persistent men. Minot gave in. Instructions were wired to the pharmacologist in Göttingen for concocting large amounts of raw liver which would serve for treatment until the more concentrated experimental substance arrived from the United States. E. U. Condon, visiting Göttingen in the summer of 1926, heard Hilbert complaining that he would rather die than eat that much raw liver.

Eventually, though, Minot's preparation arrived.

At this late stage it was probably not possible to reverse completely the progress of the disease; however, everyone in Göttingen noticed that Hilbert's condition began to improve almost immediately. All during his illness he had continued to work to the best of his ability — even turning his dining room into a lecture hall when he was not well enough to go to the University. Now when a former student inquired after his health, he replied firmly, "That illness — well, it no longer *exists*."

The two years, beginning in 1925, were the "Wunderjahre" of what was known in Göttingen as "boy physics" because so many of the great discoveries were being made by physicists still in their twenties. Early in 1925 Heisenberg came to Born with the seemingly weird mathematics that had developed in a new theory of quantum mechanics which he had created. Heisenberg thought that this was the one thing that still had to be corrected in his theory. In actuality, it was his great discovery. Born promptly identified the weird mathematics as matrix algebra, the germ of which had existed in the quaternions developed by William Rowan Hamilton more than three-quarters of a century before.

In matrix algebra multiplication is not commutative: $a \times b$ may not equal $b \times a$ but something entirely different. Prior to Heisenberg's work, matrices had rarely been used by physicists, although one exception had been Born's earlier work on the lattice theory of crystals. But now even Born had to consult his old friend Otto Toeplitz about certain properties of matrices and considered himself fortunate to obtain as his assistant Pascual Jordan, whom he just happened to meet when Jordan overheard him talking about matrices to a companion in a train compartment and proceeded to introduce himself. Jordan had been one of Courant's assistants in the preparation of Courant-Hilbert and was therefore very familiar with matrix algebra.

The Heisenberg paper was followed just 60 days later by the great Born-Jordan paper, which provided the necessarily rigorous mathematical foundation for the new matrix mechanics. The next year saw the publication of Born's famous statistical interpretation, for which he later received the Nobel Prize.

Hilbert never went as deeply into quantum mechanics as he had gone into relativity, but he still demanded that his physics assistant teach him the new theory.

"Generally he tried to give a course on what he was learning," says Nordheim. "He was a person for whom it was difficult to understand others. He always had to work things through for himself. That seemed to be his only way of really understanding. So when there was a new development, he tried to give a course on it. Usually this also contained some old material, for nothing grows up entirely of itself. For the new parts we had to make drafts. After that he would try to put the ideas into his own words."

In the spring of 1926 Hilbert announced his first lectures on quantum mechanics. Nordheim recalls how he had to extract "rather laboriously" the essence from the papers of Born and his collaborators for Hilbert, who was still at this time not well.

"Of course, he knew a lot about matrix algebra and differential equations and so on, and all of these things are the mathematical tools of quantum mechanics. In this respect my job was made easier. I went to his home two or three times a week, as required, and we discussed the general situation. Then he would ask for writeups on specific points or the development of formulas on particular applications. The next time we would talk these over — whether everything was correct and understood."

The matrix mechanics of Heisenberg was followed in short order by the wave mechanics of Erwin Schrödinger. The two papers, although they were on the same subject and led to the same results, astonished physicists; for, as one of them marvelled, "they started from entirely different physical assumptions, used entirely different mathematical methods, and seemed to have nothing to do with each other."

The equivalence of Heisenberg's and Schrödinger's theories, however, was soon established.

The whole development gave Hilbert "a great laugh," according to Condon:

"... when [Born and Heisenberg and the Göttingen theoretical physicists] first discovered matrix mechanics they were having, of course, the same kind of trouble that everybody else had in trying to solve problems and to manipulate and to really do things with matrices. So they had gone to Hilbert for help and Hilbert said the only times he had ever had anything to do with matrices was when they came up as a sort of by-product of the eigenvalues of the boundary-value problem of a differential equation. So if you look for the differential equation which has these matrices you can probably do more with that. They had thought it was a goofy idea and that Hilbert didn't know what he was talking about. So he was having a lot of fun pointing out to them that they could have discovered Schrödinger's wave mechanics six months earlier if they had paid a little more attention to him."

As a result of his almost miraculous recovery Hilbert lived to see what has been called "one of the most dramatic anticipations in the history of mathematical physics."

The Courant-Hilbert book on mathematical methods of physics, which had appeared at the end of 1924, before both Heisenberg's and Schrödinger's work, instead of being outdated by the new discoveries, seemed to have been written expressly for the physicists who now had to deal with them. Hilbert's own work at the beginning of the century on integral equations, the theory of eigenfunctions and eigenvalues of 1903–04 and the theory of infinitely many variables of 1905–06, turned out to be the appropriate mathematics for quantum mechanics (as was first established by Born in a joint paper with Heisenberg and Jordan).

"Indirectly Hilbert exerted the strongest influence on the development of quantum mechanics in Göttingen," Heisenberg was later to write. "This influence can be fully recognized only by one who studied in Göttingen during the twenties. Hilbert and his colleagues had created there an atmosphere of mathematics, and all the younger mathematicians were so trained in the thought processes of the Hilbert theory of integral equations and linear algebra that each project which belonged in this field could develop better in Göttingen than in any other place. It was an especially fortunate coincidence that the mathematical methods of quantum mechanics turned out to be a direct application of Hilbert's theory of integral equations...."

To Hilbert himself this was yet another example of that pre-established harmony which seemed to him almost the embodiment and realization of mathematical thought.

"I developed my theory of infinitely many variables from purely mathematical interests," he marvelled, "and even called it 'spectral analysis' without any presentiment that it would later find an application to the actual spectrum of physics!"

What happened next was also impressive to Hilbert, for it underlined the continuity of mathematical effort. Hilbert's theory of infinitely many variables — which had become known as "Hilbert Space" theory — now turned out to be in several respects not quite equal to the task of handling quantum mechanics. At this point young John von Neumann, inspired by Erhard Schmidt, formulated Hilbert's concept of a quadratic form more abstractly so that the extended Hilbert theory was able to meet completely the needs of the physicists.

Hilbert's last publication in physics was a collaboration with Nordheim and von Neumann on the axiomatic foundations of quantum mechanics, in which, although he did almost none of the work, the spirit was quite definitely that of Hilbert. While the effort turned out to be not mathematically rigorous, it served to introduce von Neumann to quantum mechanics and inspired him to create his later famous analysis of the foundations of the subject.

In 1927 Nordheim left Göttingen and Eugene Wigner became Hilbert's special assistant for physics. He recalls that he saw Hilbert "only about five times." When Wigner left in 1928, his place was taken by a mathematics student named Arnold Schmidt and the position became a second assistantship in logic. Hearing a lecture by Schrödinger on the new physics in 1928 or 1929, Hilbert grumbled to his former student Paul Funk: "I don't see how anybody understands what is happening in physics today. Even I don't understand much which I would like to learn from physics books. But with me, if I don't understand something, then I go to the telephone and call up Debye or Born, and they come and explain it to me. And then I understand it — but what do other people do?"

He himself was still, after his illness, deep in the work on the foundations of mathematics.

The enthusiasm for Brouwer's Intuitionism had definitely begun to wane. Brouwer came to Göttingen to deliver a talk on his ideas to the Mathematics Club.

"You say that we can't *know* whether in the decimal representation of π ten 9's occur in succession," someone objected after Brouwer finished. "Maybe we can't know — but God knows!"

To this Brouwer replied dryly, "I do not have a pipeline to God."

After a lively discussion Hilbert finally stood up.

"With your methods," he said to Brouwer, "most of the results of modern mathematics would have to be abandoned, and to me the important thing is not to get fewer results but to get more results."

He sat down to enthusiastic applause.

The feeling of most mathematicians has been informally expressed by Hans Lewy, who as a Privatdozent was present at Brouwer's talk in Göttingen:

"It seems that there are some mathematicians who lack a sense of humor or have an over-swollen conscience. What Hilbert expressed there seems exactly right to me. If we have to go through so much trouble as Brouwer says, then nobody will want to be a mathematician any more. After all, it is a human activity. Until Brouwer can produce a contradiction in classical mathematics, nobody is going to listen to him.

"That is the way, in my opinion, that logic has developed. One has accepted principles until such time as one notices that they may lead to contradiction and then he has modified them. I think this is the way it will always be. There may be lots of contradictions hidden somewhere; and as soon as they appear, all mathematicians will wish to have them eliminated. But until then we will continue to accept those principles that advance us most speedily."

Hilbert's program, however, also received its share of criticism. Some mathematicians objected that in his formalism he had reduced their science to "a meaningless game played with meaningless marks on paper." But to those familiar with Hilbert's work this criticism did not seem valid.

"... is it really credible that this is a fair account of Hilbert's view," Hardy demanded, "the view of the man who has probably added to the structure of significant mathematics a richer and more beautiful aggregate of theorems than any other mathematician of his time? I can believe that Hilbert's philosophy is as inadequate as you please, but not that an ambitious mathematical theory which he has elaborated is trivial or ridiculous. It is impossible to suppose that Hilbert denies the significance and reality of mathematical concepts, and we have the best of reasons for refusing to believe it: 'The axioms and demonstrable theorems,' he says himself, 'which arise in our formalistic game, are the images of the ideas which form the subject-matter of ordinary mathematics.'"

By 1927 Hilbert was well enough to go again to Hamburg "to round out and develop my thoughts on the foundations of mathematics, which I expounded here one day five years ago and which since then have kept me most actively occupied." His goal was still to remove "once and for all" any question as to the soundness of the foundations of mathematics. "I believe," he said, "I can attain this goal completely with my proof theory, even though a great deal of work must still be done before it is fully developed."

In the course of his talk, he took up various criticisms of his program, "all of which I consider just as unfair as can be." He went back even as far as Poincaré's remarks on the Heidelberg talk. "Regrettably, Poincaré, the mathematician who in his generation was the richest in ideas and the most fertile, had a decided prejudice against Cantor's theory that kept him from forming a just opinion of Cantor's magnificent conceptions." As for the most recent investigations, of which the program advanced by Brouwer formed the greater part, "the fact that research on foundations has again come to attract such lively appreciation and interest certainly gives me the greatest pleasure. When I reflect on the content and the results of these investigations, however, I cannot for the most part agree with their tendency; I feel, rather, that they are to a large extent behind the times, as if they came from a period when Cantor's majestic world of ideas had not yet been discovered."

The whole talk had a strongly polemical quality: "Not even the sketch of my proof of Cantor's continuum hypothesis has remained uncriticized!" Hilbert complained, and took up this proof again at length.

The formula game "which Brouwer so deprecates," he pointed out, enabled mathematicians to express the entire thought-content of the science of mathematics in a uniform manner and develop it in such a way that, at the same time, the interconnections between the individual propositions and the facts become clear. It had, besides its mathematical value, an important general philosophical significance.

"For this formula game is carried out according to certain definite rules, in which the *technique of our thinking* is expressed. These rules form a closed system that can be discovered and definitively stated. The fundamental idea of my proof theory is none other than to describe the activity of our understanding, to make a protocol of the rules according to which our thinking actually proceeds .... If any totality of observations and phenomena deserves to be made the object of a serious and thorough investigation, it is this one — since, after all, it is a part of the task of science to liberate us from arbitrariness, sentiment and habit and to protect us from the subjectivism that already made itself felt in Kronecker's views and, it seems to me, finds its culmination in Intuitionism...."

It was true, Hilbert conceded, that the consistency proof of the formalized arithmetic which would "determine the effective scope of proof theory and in general constitute its core" was not yet at hand. But, as he concluded his address, he was thoroughly optimistic: such a proof would soon be produced.

"Already at this time I should like to assert what the final outcome will be: mathematics is a presuppositionless science. To found it I do not need God, as does Kronecker, or the assumption of a special faculty of our understanding attuned to the principle of mathematical induction, as does Poincaré, or the primal intuition of Brouwer, or, finally, as do Russell and Whitehead, axioms of infinity, reducibility, or completeness...."

When Hilbert finished, Hermann Weyl rose to make a few remarks. Weyl's love for his old teacher had not been affected by the five years of controversy. Also his enthusiasm for Brouwer's ideas had abated. He nevertheless felt that at this point he should defend Brouwer:

"Brouwer was first to see exactly and in full measure how [mathematics] had in fact everywhere far exceeded the limits of contentual thought. I believe that we are all indebted to him for this recognition of the limits of contentual thought. In the contentual considerations that are intended to establish the consistency of formalized mathematics, Hilbert fully respects these limits, and he does so as a matter of course; we are really not dealing with artificial prohibitions by any means. Accordingly, it does not seem strange to me that Brouwer's ideas have found a following; his position resulted of necessity from a thesis shared by all mathematicians before Hilbert proposed his formal approach and forms a new, indubitable fundamental logical insight that even Hilbert acknowledges.

"That from this point of view only a part, perhaps only a wretched part of classical mathematics is tenable is a bitter and inevitable fact. Hilbert could not bear this mutilation. And it is again a different matter that he succeeded in saving classical mathematics by a radical reinterpretation of its meaning without reducing its inventory, namely by formalizing it, thus transforming it in principle from a system of intuitive results into a game with formulas that proceeds according to fixed rules.

"Let me now by all means acknowledge the immense significance and scope of this step of Hilbert's, which evidently was made necessary by the pressure of circumstances. All of us who witnessed this development are full of admiration for the genius and steadfastness with which Hilbert, through his proof theory of formalized mathematics, crowned his axiomatic life work. And, I am very glad to confirm, that there is nothing that separates me from Hilbert in the epistemological appraisal of the new situation thus created."

In contrast to Weyl, Brouwer had become, like Kronecker, a fanatic in the service of his cause. He looked upon Hilbert as "my enemy," and once left a house in Amsterdam where he was a guest when van der Waerden, who was also a guest, referred to Hilbert and Courant as his friends.

The ill-feeling was undoubtedly intensified by the fact that circumstances placed Hilbert constantly in opposition to Brouwer.

Both men were on the editorial staff of the *Annalen*. Hilbert was one of the three principal editors, a position which he had held since 1902; Brouwer, a member of the seven-man editorial board. At about this time Brouwer began to insist that all papers by Dutch mathematicians and all papers on topology be submitted directly to him. Everyone objected, especially Dutch topologists, since it was well known that when a paper got into Brouwer's hands, it did not get out for several years. Although personally unaffected, Hilbert was repelled by Brouwer's dictatorial demands. When he had been in good health, he had been confident of his ability to protect the integrity of the *Annalen*. Since his illness, however, he had begun to fear that if anything happened to him, Brouwer would take over the journal to the detriment of mathematics. So he now called together his friends to devise a way of removing Brouwer from the editorial board.

Carathéodory, who was himself a member of the board, came up with a solution. Since Brouwer alone could not be asked to resign, the entire seven-man board should be dismissed. Hilbert promptly acted. The change is reflected in the covers of Vol. 100 of the *Annalen* and Vol. 101, on which only the names of Hilbert, Hecke and Blumenthal remain.

(It should be mentioned that Einstein, disturbed by the controversy, resigned from his position as one of the three principal editors. "What is this frog and mouse battle among the mathematicians?" he asked a friend.)

Hilbert and Brouwer were then placed in opposition by another circumstance.

Since the war, the German mathematicians had not been invited to any international meetings. In 1928, however, the Italians, planning the first official International Congress since 1912, determined to make it truly international. Once again, invitations were sent to all German schools and mathematical organizations. Many Germans did not want to accept. The leader of this group was a professor at the University of Berlin named Ludwig Bieberbach. In his opposition to accepting the invitation of the Italians, he was seconded by Brouwer, who although Dutch was an ardent German nationalist. In the spring of 1928 Bieberbach sent a letter to all German secondary schools and universities urging them to boycott the congress at Bologna. Hilbert responded by sending out a letter of his own:

"We are convinced that pursuing Herr Bieberbach's way will bring misfortune to German science and will expose us all to justifiable criticism from well disposed sides . . . . The Italian colleagues have troubled themselves with the greatest idealism and expense in time and effort . . . . It appears under the present circumstances a command of rectitude and the most elementary courtesy to take a friendly attitude toward the Congress."

In August, although suffering from a recurrence of his illness, Hilbert personally led a delegation of 67 mathematicians to the Congress. At the opening session, as the Germans came into an international meeting for the first time since the war, the delegates saw a familiar figure, more frail than they remembered, marching at their head. For a few minutes there was not a sound in the hall. Then, spontaneously, every person present rose and applauded.

"It makes me very happy," Hilbert told them in the familiar accents, "that after a long, hard time all the mathematicians of the world are represented here. This is as it should be and as it must be for the prosperity of our beloved science.

"Let us consider that we as mathematicians stand on the highest pinnacle of the cultivation of the exact sciences. We have no other choice than to assume this highest place, because all limits, especially national ones, are contrary to the nature of mathematics. It is a complete misunderstanding of our science to construct differences according to peoples and races, and the reasons for which this has been done are very shabby ones.

"Mathematics knows no races . . . . For mathematics, the whole cultural world is a single country."

Hilbert's scientific paper for presentation at the Congress concerned again the fundamental problems of mathematics. Recently there had been signs that his expectation that carrying out proof theory was only a matter of mathematical technique was perhaps overly optimistic. The first attempt at a substantial consistency proof (in Ackermann's dissertation) had required an essential restriction of the formal system not in the original plan. Similarly, in a paper of von Neumann's, the consistency proof, following Hilbert's line of reasoning, had not applied to the full system. Now, though, Ackermann's proof had been revised and simplified; and it seemed at the moment that the consistency of formalized number theory, at least, had been proved.

Hilbert now added to the problem of consistency another problem, that of the *completeness* of the formal system.

When Hilbert went to pay his hotel bill, he was informed that it had already been paid for him by the committee in charge of planning the Congress.

"Ah, if I had only known that," he said, "I would have eaten a great deal more."

The Hilbert career was almost over.

The year after the Bologna Congress, he was permitted to see what Felix Klein had not lived to see — the dedication of a handsome building to house the Mathematical Institute of Göttingen.

The new institute had been made possible by Courant's friendship with the Bohr brothers and the entrée which they had provided to the Rockefeller Foundation. The funds given by the Foundation were then matched by the German government. Thus the Institute was a joint project of German and American money and effort.

"There will never be another institute like this!" Hilbert exulted. "For to have another such institute, there would have to be another Courant — and there can never be another Courant!"

XXII

Logic and the Understanding of Nature

The mandatory age of retirement for a professor was 68, an age Hilbert would attain on January 23, 1930; in Göttingen a bittersweet feeling of anticipation and regret was in the air.

During the winter semester 1929–30, Hilbert delivered his "Farewell to Teaching." For his subject he went back to the foundations of his fame and lectured for almost the first time in 40 years on the invariants. Professors crowded into the lecture hall with the students. A street was named Hilbert Strasse. "A street named after you!" Mrs. Hilbert exclaimed. "Isn't that a *nice* idea, David?" Hilbert shrugged. "The idea, no, but the execution, ah — that is nice. Klein had to wait until he was dead to have a street named after him!"

He saw another student through the doctoral. Appropriately this was Haskell Curry, an American. Curry had little contact with Hilbert though. He remembers he came to class on a warm spring day wearing a fur-lined coat. He was always accompanied by Bernays, who sometimes had to step in and lecture for a bit. Curry had most of his conferences with Bernays; but since Bernays was not a full professor, he had to take the final examination with Hilbert.

"I rather enjoyed my final examination with him . . . . He did not ask me any questions having to do with logic, but only to do with general mathematics. One question was about the uniformization of algebraic functions. It happened that I had just had a course on that subject with Professor Osgood at Harvard. Although it was a way off from my special field, I gave as precise an answer as anyone would expect in a field so far removed from the candidate's specialty; he was quite impressed with me and turned to me and said, 'Where did you learn that?' Although he seemed rather frail, he was razor-sharp and alert."

As the time of the retirement approached, the choice of a successor was discussed. There was, it was generally agreed, but one possible choice. If Courant had shown himself to be the Klein of the new generation, Weyl was the Hilbert.

A decade before, Weyl had refused an offer from Göttingen because of the uncertainties of life in Germany after the war. Hilbert said, "It is easy to call Weyl but hard to get him." This time Weyl again had difficulty in making up his mind. He had recently returned from England; and the pessimism which he found expressed in the newspapers and letters which had piled up on his desk during his absence filled him with apprehension about returning to Germany. He was also concerned that he might not be the right choice for Göttingen at this time. By now he was 45 years old. He knew he was close to the end of his creative period. Perhaps the Institute should get someone like young Emil Artin, "from whom great results can still be expected." Yet he was tempted. He loved and revered Hilbert — the Pied Piper who had led all the young rats down into the deep river of mathematics. He knew that he was more completely in the mathematical-physical tradition of Göttingen than Artin. He would enjoy the opportunity of working with Courant, Born and Franck. The situation in Germany seemed to be improving. The Dawes Plan had helped to relieve the economic problems. The lunatic fringe which muttered about the "Jewish physics" of his friend Einstein still seemed on the fringe. In the end, this time, Weyl wired his acceptance.

"I don't have to tell you with how much joy and how much pride I was filled to be called as your successor," he wrote to Hilbert. ". . . I am looking forward most eagerly to working with the colleagues you have gathered around yourself, you to whom the mathematical-scientific faculty owes its strength and harmony." The dark clouds that hung over Germany might not disappear quickly. "But I hope it will be granted to me to live many happy years near you. . . . Please do not be angry about my tardiness in accepting."

The Göttingen which greeted Weyl in the spring of 1930 was at the height of its new glory. More than ever before, it could be said that an international congress of mathematicians was perpetually in session in the quiet little town with its linden-lined streets and the solid respectable, now old "Jugendstil" houses. On the outskirts a series of scientific industries and laboratories surrounded the city like another wall. The Mathematical Institute was housed in its handsome new building, the Lesezimmer a long well-lighted library. *Extra Gottingen non est vita*. The Latin motto was still blazoned on the wall of the Ratskeller. In the sunshine outside, students

Hilbert

and professors sat at small tables and argued about politics, love and science. The little goose girl gazed tranquilly down into her fountain. Weyl, returning to the beloved town of his college years, must have agreed. Away from Göttingen there was no life.

Of all the honors being showered on Hilbert during the retirement year, the one that seemed to please him most came from his native city. The Königsberg town council voted to present its famous son with "honorary citizenship." The presentation was scheduled to be made in the fall at the meeting of the Society of German Scientists and Physicians, which was being held that year in Königsberg.

Hilbert gave considerable thought to the selection of a topic for his acceptance address. It must be something of wide and general interest. In Königsberg, the birthplace of Kant, it must be philosophical in tone. It must also be a fitting conclusion to the career that had begun long ago at the university in Königsberg. When he thought of the university, he remembered the statue of Kant on the grounds and the laconic inscription "Kant" — so expressive in its brevity. He also remembered Jacobi, from whom the mathematical tradition of Königsberg derived as in Göttingen it derived from Gauss. He wanted a topic which would weave together these great names and all the separate strands of his career — Königsberg and Göttingen, Jacobi, Gauss, Kant, mathematics and science, science and experience, the great developments in knowledge and in thought through which he had lived.

Naturerkennen — the understanding of nature — *und Logik*. This would be his subject.

During the past decade he had become increasingly interested in reaching a greater audience with mathematical ideas. He had frequently accepted the opportunity of giving popular lectures in the Saturday morning series "for all the Faculties of the University." He took subjects like "Relativity Theory" or "The Infinite" or "The Principles of Mathematics" and tried, by finding examples from familiar fields outside mathematics, to make the fundamental concepts comprehensible to laymen.

"An enormous amount of labor was devoted to this task," Nordheim recalls. "We had to prepare preliminary outlines either of new material or from old lectures. These were then worked and re-worked practically every morning and in this process flavored with Hilbert's own inimitable brand of logic and humor."

During this period Hilbert and some of the other mathematics professors regularly attended the lectures of a zoologist. Hilbert had developed a great interest in genetics. He delighted in the laws determining the heredity of Drosophila, which could be obtained by the application of certain of the geometric axioms. He was fascinated by the Pferdespulwurm — "that creature with the most modest number of chromosomes, which corresponds therefore to the hydrogen atom with only one electron." But he was also impressed by the ability of the biologists to make their subject interesting and understandable to laymen.

"The biologists understand popular presentation especially well," he told Paul Funk one time. "In order to prevent the fatigue which straining thoughts bring forth in laymen, one must occasionally insert a little *dessin* (a French word meaning pattern, or design), and at that the biologists are pre-eminent." Pronouncing the French word sharply in his Königsberg dialect, he went on to say: "For us mathematicians, popular presentation is much harder, but still it must happen — if we go about it right — that we find a beautiful *dessin*."

Now, in the summer of 1930, among his mature fruit trees, he sought such a beautiful *dessin* as he began to prepare his speech for the Königsberg meeting, stripping away all vague generalities from his subject, putting his ideas into simple non-technical language for a general audience (that "man in the street" he had mentioned in Paris, to whom one should always be able to explain any fully realized mathematical theory).

He had returned many times over the years to his native city, but there was a special quality to this return. Kurt Reidemeister and Gabor Szegö, the mathematics professors now at the University, noted how pleased he was at the social gathering arranged in connection with his speech, so exuberant "that his wife must always again call him back." But Königsberg seemed colder than in the old days, and Hilbert had to borrow a fur coat from Szegö to keep warm.

The honorary citizenship was presented at the opening session. Then Hilbert took his place at the rostrum. His head was now almost entirely bald, the broad scholar's forehead contrasting more sharply than ever with the delicate chin; the white moustache and small beard were neatly trimmed. (Ostrowski was reminded of the head of Lenin.) He looked out at his audience through the familiar rimless glasses, the blue eyes still sharp and searching, and still so innocent. He placed his hands firmly on the manuscript in front of him and began to speak very slowly and carefully:

"The understanding of nature and life is our noblest task."

In recent times richer and deeper knowledge had been obtained in decades than had previously been obtained in the same number of centuries.

The science of logic had also progressed until there was now, in the axiomatic method, a general technique for the theoretical treatment of all scientific questions. Because of these developments — he told his audience — the men of today were better equipped than the philosophers of old to answer an ancient philosophical question: "the part which is played in our understanding by Thinking on the one side and Experience on the other."

It was a worthy question with which to conclude a career; for, fundamentally, to answer it would be to ascertain by what means general understanding is achieved and in what sense "all the knowledge which we collect in our scientific activities is truth."

Certain parallels between nature and thought had always been recognized. The most striking of these was a pre-established harmony which seemed to be almost the embodiment and realization of mathematical thought, the most magnificent and wonderful example of which was Einstein's theory of relativity.

But it seemed to him that the long recognized accord between nature and thought, experiment and theory, could only be understood when one took into account the formal element and the mechanism linked with it which exists on both sides, in nature and in thought. The extension of the methods of modern science should lead to a system of natural laws which corresponded with reality in every respect. Then we should need only pure thought — abstract deduction — in order to gain all physical knowledge. But this was not, in his opinion, the complete answer: "For what is the origin of these laws? How do we obtain them? How do we know that they correspond with reality? The answer is that we can obtain these laws only through our own experience.... Whoever wants nevertheless to deny that universal laws are derived from experience must contend that there is still a third source of understanding...."

Königsberg's great son, Immanuel Kant, had been the classical exponent of this point of view — the point of view which Hilbert 45 years ago had defended at his public promotion for the degree of doctor of philosophy. Now, before his talk, he had smilingly commented to a young relative that a lot of what Kant had said was "pure nonsense" — but that, of course, he could not say to the citizens of Königsberg.

Kant had stated that man possesses beyond logic and experience certain *a priori* knowledge of reality.

"I admit," Hilbert told his audience, "that even for the construction of special theoretical subjects certain *a priori* insights are necessary.... I even believe that mathematical knowledge depends ultimately on some kind of such intuitive insight . . . . Thus the most general basic thought of Kant's theory of knowledge retains its importance . . . . The *a priori* is nothing more or less than . . . the expression for certain indispensable preliminary conditions of thinking and experiencing. But the line between that which we possess *a priori* and that for which experience is necessary must be drawn differently by us than by Kant — Kant has greatly overestimated the role and the extent of the *a priori*."

Men now knew that many facts previously considered as holding good *a priori* were not true, the most striking being the notion of an absolute present. But it had also been shown, through the work of Helmholtz and Gauss, that geometry was "nothing more than a branch of the total conceptual framework of physics." We had forgotten that the geometrical theorems were once experiences!

"We see now: Kant's *a priori* theory contains anthropomorphic dross from which it must be freed. After we remove that, only that *a priori* will remain which also is the foundation of pure mathematical knowledge."

In essence, this was the attitude which he had characterized in his recent work on the foundations of mathematics.

"The instrument which brings about the adjustment of differences between theory and practice, between thought and experiment, is mathematics. It builds the connecting bridge and continually strengthens it. Thus it happens that our entire present culture, insofar as it is concerned with the intellectual understanding and conquest of nature, rests upon mathematics!"

The effect of Hilbert's speech on the audience has been recalled by Oystein Ore, who, as a young man on his honeymoon, was there in Königsberg:

"I remember that there was a feeling of excitement and interest both in Hilbert's lecture and in the lecture of von Neumann on the foundations of set theory — a feeling that one now finally was coming to grips with both the axiomatic foundation of mathematics and with the reasons for the applications of mathematics in the natural sciences."

In the final part of his speech, Hilbert carefully made the point that in spite of the importance of the applications of mathematics, these must never be made the measure of its value. He concluded with that defense of *pure* mathematics which he had wanted so long ago to make in answer to the speech given by Poincaré at the first International Congress of Mathematicians.

"Pure number theory is that part of mathematics for which up to now no application has ever been found. But it is number theory which was considered by Gauss [who himself made untold contributions to applied mathematics] as the queen of mathematics...."

Kronecker had compared the mathematicians who concerned themselves with number theory to Lotus-eaters who "once having consumed this food can never give it up."

"Even our great Königsberg mathematician Jacobi felt this way.... When the famous Fourier maintained that the purpose of mathematics lies in the explanation of natural phenomena, Jacobi objected, 'A philosopher like Fourier should know that the glory of the human spirit is the sole aim of all science!'.... Whoever perceives the truth of the generous thinking and philosophy which shines forth from Jacobi's words will not fall into regressive and barren scepticism."

Reidemeister and Szegö had made arrangements for Hilbert to repeat the conclusion of his speech over the local radio station; and when the session adjourned, they accompanied him to the broadcasting studio.

There, as Hilbert spoke into the unfamiliar instrument, it seemed that his voice rang out again with the enthusiasm and optimism of the vigorous man who in the prime of his life had sent his listeners out to seek the solution to 23 problems which, he was certain, would lead to progress in mathematics.

"In an effort to give an example of an unsolvable problem, the philosopher Comte once said that science would never succeed in ascertaining the secret of the chemical composition of the bodies of the universe. A few years later this problem was solved....

"The true reason, according to my thinking, why Comte could not find an unsolvable problem lies in the fact that there is no such thing as an unsolvable problem."

He denied again, at the end of his career, the "foolish *ignorabimus*" of du Bois-Reymond and his followers. His last words into the microphone were firm and strong:

"Wir müssen wissen. Wir werden wissen."

We must know. We shall know.

As he raised his eyes from his paper and the technician snapped off the recording machine, he laughed.

The record which he made of this last part of his speech at Königsberg is still in existence. At the end, if one listens very carefully, he can hear Hilbert laugh.

"Wir müssen wissen. Wir werden wissen."

We must know. We shall know.

It was in every respect a great last line.

However, lives do not always end on great last lines.

At almost the same time that Hilbert was making his speech at Königsberg, a piece of work was being brought to a conclusion which was to deal a death blow to the specific epistemological objective of the final program of Hilbert's career. On November 17, 1930, the *Monatshefte für Mathematik und Physik* received for publication a paper by a 25-year-old mathematical logician named Kurt Gödel.

XXIII

Exodus

When Hilbert first learned about Gödel's work from Bernays, he was "somewhat angry."

The young man had taken up both of the problems of completeness which Hilbert had proposed at Bologna. He had established completeness for the case of the predicate calculus. But then he had proceeded to prove — with all the finality of which mathematics is uniquely capable — the *incompleteness* of the formalized number theory. He had also proved a theorem from which it follows that a finitist proof of consistency for a formal system strong enough to formalize all finitist reasonings is *impossible*.

In the highly ingenious work of Gödel, Hilbert saw, intellectually, that the goal toward which he had directed much effort since the beginning of the century — the final unanswerable answer to Kronecker and Brouwer and the others who would restrict the methods of mathematics — could not be achieved. Classical mathematics might be consistent and, in fact, probably was; but its consistency could never be established by mathematical proof, as he had hoped and believed it could be.

The boundless confidence in the power of human thought which had led him inexorably to this last great work of his career now made it almost impossible for him to accept Gödel's result emotionally. There was also perhaps the quite human rejection of the fact that Gödel's discovery was a verification of certain indications, the significance of which he himself had up to now refused to recognize, that the framework of formalism was not strong enough for the burden he wanted it to carry.

At first he was only angry and frustrated, but then he began to try to deal constructively with the problem. Bernays found himself impressed that even now, at the very end of his career, Hilbert was able to make great changes in his program. It was not yet clear just what influence Gödel's work would ultimately have. Gödel himself felt — and expressed the thought in his paper — that his work did not contradict Hilbert's formalistic point of view; and it soon became apparent that proof theory could still be fruitfully developed without keeping to the original program. Broadened methods would permit the loosening of the requirements of formalizing. Hilbert himself now took a step in this direction. This was the replacing of the schema of complete induction by a looser rule called "unendliche Induktion." In 1931 two papers in the new direction appeared.

Although he had retired, he continued to lecture regularly at the University. He still prepared in only the most general way, still frequently got stuck. When he found himself unable to work through a proof on the blackboard, he would dismiss it with a wave of his hand as "completely elementary." He sometimes stumbled over details, rambled impossibly, repeated himself. "But still, one out of three lectures was *superb!*"

Hilbert's career having come to its official end with his retirement, plans were made to begin the collecting and editing of his mathematical works. Blumenthal, who had observed and studied the personality and achievements of his teacher since 1895, was asked to compose a biography for the final volume. Although Blumenthal had been a professor at Aachen now for many years, he had never lost his strong feeling for Göttingen, returning time after time to be (as he said) "refreshed." Wherever he was, even at the front during the first world war, he always organized the former inhabitants of Göttingen into a social club. He took on the assignment of a life history with pleasure and painstaking care.

Volume I of the collected works was to be devoted to the *Zahlbericht* and the other number theory papers. For Hilbert, as for Gauss, the first years at Göttingen had been "the fortunate years." The papers on algebraic number fields were now recognized as the deepest and most beautiful of all his mathematical works. Helmut Hasse, who with Emmy Noether, van der Waerden, Artin, Takagi and others had taken part in carrying out the program for class-fields which Hilbert had outlined in the last number theory paper, was asked to write an evaluation of Hilbert's contribution in this area.

It seemed to Hasse, in retrospect, that Hilbert's work on algebraic number theory, like so much of his work, had stood in time and content at the turn of two centuries. On one side, treating problems in great generality with new methods which far surpassed the earlier methods in elegance and simplicity, he had thrown into relief the works of the number theorists of the old century. On the other side, "with wonderful farsightedness," he had sketched out paths to the positive final treatment of the whole complex of problems and had indicated the direction for the new century.

Three young mathematicians were brought to Göttingen to assist with the editing. One of these was a young woman named Olga Taussky. She had been trained in number theory by Philipp Furtwängler, who, although he had never studied with Hilbert, had proved a number of theorems which Hilbert had conjectured. Hilbert still enjoyed talking to young women. Mostly he talked to Fräulein Taussky about his health and about his wish to return some day and live out his life in Rauschen, the little Baltic fishing village where he had spent the vacations of his youth. But one day, looking back over his career and the many fields of mathematics in which he had worked, he remarked to her that, much as he admired all branches of mathematics, he considered number theory the most beautiful.

(That same year at the International Congress of Mathematicians at Zürich, in connection with a talk by his former student, Rudolf Fueter, Hilbert stated that the theory of complex multiplication of elliptic modular functions, which brings together number theory and analysis, was not only the most beautiful part of mathematics but also of all science.)

In the course of her work on Hilbert's papers, Fräulein Taussky was astonished to discover many technical errors of varying degree. Although she recognized that, because of Hilbert's powerful mathematical intuition, the errors had not affected the ultimate results, she felt that they should be corrected in the collected works.

Fräulein Taussky never consulted Hilbert about the correction of errors, nor was she expected to. He had long ago put out of his mind his own work in the field and had not followed the work of later mathematicians who had carried out his program. At one point, with some trepidation, she pointed out in a footnote (p. 506) that two conjectures had been incorrect.

The number theory volume was to be presented to Hilbert on his seventieth birthday. A celebration was scheduled, a whole day of festivities. It was all very bothersome, Hilbert complained to Bernays, but it would be "good for mathematics."

Hermann Weyl wrote the birthday greeting that appeared in *Naturwissenschaften*. Throughout his scientific career, as he wrote to his old friend Robert König, he had kept before him a simple motto: "True to the spirit of Hilbert." The birthday of Hilbert, Weyl now noted in his greeting, was the high feast day for German mathematicians, celebrated year after year in warm personal veneration for the master but also in personal affirmation of their own beliefs and unity.

"Without doubt on the whole globe today Hilbert's name represents most concretely what mathematics means in the framework of objective spirits and how making mathematics, as a fundamental creative activity of mankind, is alive among us."

And yet, Weyl had to concede, Hilbert's own brand of optimism, his supreme confidence in the power of reason to come up with simple and clear answers to simple and clear questions, was "not popular nowadays" among the younger generation.

"Admittedly, one sentence or another of Hilbert's lecture [on logic and the understanding of nature, which he had given in Konigsberg in 1930] comes dangerously close to the opening words of Gottfried Keller's novel *Das Sinngedicht*, in which he mocks his scientist Reinhardt — 'About twenty-five years ago, when once again the natural sciences were at their highest peak ....'

"However, we do Hilbert an injustice if we confuse his rationalism with that of Haeckel .... He would rightly be called presumptious if, Faust-like, he had striven after the kind of magical knowledge which unlocks the very core of being to the intellect .... Such knowledge is different from the knowledge of reality that must prove itself by accurate prediction [and] can be advanced only by the mathematical method ....

"Hilbert seems to me to be an outstanding example of a man through whom the immensely creative power of naked scientific genius manifests itself .... I remember how enthralled I was by the first mathematics class I ever attended [at the University] .... It was Hilbert's famous course on the transcendence of e and π ....

"Woe to the youth that fails to be touched to the core by such a man as Hilbert!"

On the birthday itself, Ferdinand Springer, who was the publisher of the collected works, came to Göttingen to present personally to Hilbert the special white and gold leather-bound copy of the first volume. The beautiful cover contained, however, not the printed pages, but only the proofs of the pages; for Fräulein Taussky was still not satisfied. Hilbert made no comment on the unfinished nature of the volume. But later, in his presence, Fräulein Taussky declined a certain brand of cigarette as being too strong for her. Somebody said that one really couldn't tell one brand from another. "Aber nein!" Hilbert said. "Fräulein Taussky can tell the difference. She is capable of making the finest, the very finest distinctions." She knew he was needling her for taking so seriously errors which he considered unimportant, and suspected that he had seen the footnote on p. 506.

On the birthday evening there was a party in the magnificent new building of the Mathematical Institute. Former colleagues and students came from all over Germany and many from abroad. Although it was during the Depression, everyone managed to look very elegant in shabby formal dress. Olga Taussky remembers that she purchased a beautiful evening gown for about two dollars, and it was much admired. There was a banquet with many loving speeches and toasts. Arnold Sommerfeld read to Hilbert a little verse which he had written: "Seiner Freunde treuester Freund / Hohler Phrase ärgster Feind." (*To his friends, truest friend / To the hollow phrase, bitterest enemy.*)

Then Hilbert made a short speech. He recalled the great good luck with which he had been blessed: the friendships with Minkowski and Hurwitz, the study time in Leipzig with Felix Klein, the Easter trip of 1888 when he had visited Gordan and Kronecker and many other mathematicians, his appointment by Althoff as Lindemann's successor at an unusually early age. And in his native city, he reminded the guests, he had had the good fortune to find his wife "who since then in faithful comradeship has taken a decisive part in my whole activity and especially in my concerns for the younger generation." Minkowski's name was mentioned frequently. His sudden death, Hilbert recalled, had left a "deep emptiness, both human and scientific," but life had had to go on. Edmund Landau had come to take Minkowski's place. Now Felix Klein's great goal had at last been achieved, and he himself was celebrating his seventieth birthday "in this beautiful Institute."

There was dancing after the banquet, and the guest of honor danced almost every dance. A procession of students carrying torches marched through the snow to the entrance of the brightly lighted building on Bunsen Strasse, and shouted for Hilbert. He came out and stood on the steps, bundled in his big coat with the fur collar, and somebody took a picture. From every window of the Institute famous faces looked out.

Here at the end was the highest honor which the students could give to a professor.

"For mathematics," Hilbert exhorted the shouting students, "hoch — hoch — hoch!" In English, it would have been "Hip hip hooray!"

A few days after the birthday celebration, Hasse expressed to Mrs. Hilbert "my ardent desire to talk once in my life personally to the great man." Mrs. Hilbert invited him to come to tea and afterwards left him alone in the garden with Hilbert.

"I began talking to him about what interested me most in those days — the theory of algebraic numbers and in particular class-field theory. On this theory I had written a report, in continuation of Hilbert's celebrated *Zahlbericht*; and I began telling him what I had done in this theory, based on his own famous results of the late nineties. But he interrupted me repeatedly and insisted that I explain to him the basic conceptions and results of that theory before he could listen to what I wanted to tell him. So I explained to him the very foundations of class-field theory. About this he got very enthusiastic and said, 'But that is extremely beautiful, who has created it?' And I had to tell him that it was he himself who had laid that foundation and envisaged that beautiful theory. After that he listened to what I had to tell him about my own results. He listened attentively, but more politely than intelligently."

In the Reichstag elections in the year of Hilbert's seventieth birthday, the National Socialist Party made great gains. The following January, President von Hindenburg appointed Adolf Hitler the chancellor of Germany. Almost immediately came the first measure designed to break that "satanical power" which had "grasped in its hands all key positions of scientific and intellectual as well as political and economic life." The universities were ordered to remove from their employment almost every full-blooded Jew who held any sort of teaching position.

The Hilbert school was perhaps the hardest hit. Hilbert's devotion to his science had always been complete. No prejudice — national, sexual or racial — had ever been allowed to enter into it. In 1917 an appropriate memorial had had to be written for Darboux even though his country was at war with Germany. A position had had to be obtained for Emmy Noether even though a woman had never been a Privatdozent at Göttingen. Since the earliest friendships with Minkowski and Hurwitz, scientists had never been classified by Hilbert as Aryan and non-Aryan. There were only two kinds of scientists: those who solved problems of recognized worth, and those who did not.

Now, in the Mathematical Institute itself, to whom did the ultimatum apply? To Courant, who had replaced Klein and brought to reality Klein's great dream. To Landau, who had come to Göttingen after the death of Minkowski and made the University the center of research in the theory of numbers. To Emmy Noether, who — in spite of the fact that she still received no more than a trifling stipend — was the center of the most fertile circle of research at that time in Göttingen. To Bernays, who had been Hilbert's assistant and collaborator for almost sixteen years. In the Physics Institute, both Born and Franck were Jews. A distinction, however, was made between them by the new government. Franck, who had already received his Nobel Prize, was exempted from the order. Born, who would not receive his Nobel for some years, had to go. The ultimatum applied to many others; sometimes it seemed to everyone.

Hilbert was extremely upset when he heard that many of his friends were being put on "forced leave," as the current euphemism had it.

"Why don't you sue the government?" he demanded of Courant. "Go to the state court? It is *illegal* for such a thing to happen!"

It seemed to Courant that Hilbert was completely unable to understand that lawlessness had taken over. Since his birthday, it had been hard to get him to listen and to accept innovations at the Institute. But chiefly his difficulty seemed to be that he still believed the old system of justice prevailed. He continued to retain the deep Prussian faith in the law which had been inculcated in him by Judge Hilbert. It is exemplified by the story of how when Frederick the Great was disturbed by the sound of a peasant's mill and threatened to confiscate it, the peasant replied to the king with complete confidence, "No — *in Prussia there are still judges!*" Frederick, ashamed, had had the peasant's words inscribed across the portico of his summer palace, where they still stood in 1933.

There was at first no general agreement among those affected about what was to be done. How far would it all go? "If you knew the German people, you knew it would go all the way." Young Hans Lewy decided to leave Germany when Hitler was appointed chancellor. By the first of April he was already in Paris. Some people who did not have to go left in protest. Franck aligned himself with his fellow Jews. Others thought that something of the greatness of Göttingen could still be salvaged. Landau was allowed to stay on because he had been a professor under the Empire. Further exceptions would be made. Courant had been gassed and wounded in the stomach fighting for Germany; surely that made him a German. Letters were sent to the Minister about the case of Fräulein Noether. She held such a minor job, received so little for her services. "I don't think there was ever such a distinguished list of recommendations," Weyl later said. Hilbert's name was at the top. But all the distinguished names had no effect.

"The so-called Jews are so attached to Germany," Hilbert said plaintively, "but the rest of us would like to leave."

Otto Neugebauer, now an associate professor, was placed at the head of the Mathematical Institute. He held the famous chair for exactly one day, refusing in a stormy session in the Rector's office to sign the required loyalty declaration. The position of the head of the Mathematical Institute passed to Weyl. Although his wife was part Jewish, he was one of those who thought that something might yet be salvaged. All during the bitter uncertain spring and summer of 1933 he worked, wrote letters, interviewed officials of the government. But nothing could be changed.

By late summer nearly everyone was gone. Weyl, vacationing with his family in Switzerland, still considered returning to Göttingen in the hope that somehow he could keep alive the great scientific tradition. In America, his many friends worried about him and wrote long letters, advising, urging, begging that he leave Germany before it was too late. Abraham Flexner offered him a position at the Institute for Advanced Study. Finally Einstein, who had already been at the newly created Institute for several years, prevailed upon the younger man to come and join him there.

In Göttingen, Hilbert was left almost alone. He kept Bernays on as his assistant at his own expense. *The Foundations of Mathematics*, which he and Bernays had written in collaboration, was almost ready for publication. He put away his general mathematical books and became progressively more distant. With Bernays's help, he saw Arnold Schmidt and Kurt Schütte through the doctorate. Schütte was the last of 69 mathematicians (40 of them during the years from 1900 to 1914) to receive their degrees from Hilbert. In actuality, however, all of Schütte's contacts were through Bernays. He saw Hilbert only once.

"When I was young," Hilbert said to young Franz Rellich, one of the few remaining members of the old circle, "I resolved never to repeat what I heard the old people say — how beautiful the old days were, how ugly the present. I would never say that when I was old. But, now, I must."

Sitting next to the Nazis' newly appointed minister of education at a banquet, he was asked, "And how is mathematics in Göttingen now that it has been freed of the Jewish influence?"

"Mathematics in Göttingen?" Hilbert replied. "There is really none any more."

XXIV

Age

In the center of the town the swastika flew above the Rathaus and cast its shadow on the little goose girl. The university bulletin and publications appeared again in the traditional German script, the first page of each one bearing the statement that it appeared under Herr Goebbels's sponsorship.

A Nazi functionary became the head of the Mathematical Institute. During the winter semester 1933–34 Hilbert lectured for one hour a week on the foundations of geometry. After the end of the semester, he never again came to the Institute.

Landau continued to lecture; but when he announced a course in calculus, an unruly mob prevented his entering the lecture hall. "It is all right for you to teach advanced courses," he was told, "but these are beginners and we don't want them taught by a Jew." Siegel, now a professor at Frankfurt, attempted to get support for his old teacher from a group of professors who were safe in their positions. He was not successful.

After a while Landau too was gone from Göttingen. Unlike the others, he did not leave the country, being tied to his native land by the fact of his wealth and possessions. Hardy arranged for him to deliver a series of lectures in England: "It was quite pathetic to see his delight when he found himself again in front of a blackboard and his sorrow when his opportunity came to an end."

By the spring of 1934 the situation had become so bad for the Jews that Bernays felt he must leave Germany, and he returned to Zürich. The Mathematical Institute continued to pay the salary of Hilbert's remaining assistant, Arnold Schmidt, who worked with him in his home on problems of logic and foundations.

"There were brief failings of memory which might make strangers think he was not so sharp," Schmidt says, "but those who worked with him in this area knew differently."

Now Helmut Hasse was made the head of the Institute. This was a great improvement; for although Hasse had long been a convinced nationalist, he was a first-rate mathematician.

That summer Emmy Noether, for whom a place had been found in America at Bryn Mawr, returned to Göttingen. "Her heart knew no malice," Weyl later explained. "She did not believe in evil — indeed it never entered her mind that it could play a role among men." Things were not so clear then as they were later to seem, and she wished Hasse only success in his efforts to rebuild the great tradition of Göttingen after the exodus of the previous year. At the end of the summer she returned to Bryn Mawr. She was at the height of her powers, her imagination and her technique having reached the maximum point of perfect balance. In her hands "the axiomatic method, no longer merely a method for logical clarification and deepening of the foundations, [had become] a powerful weapon of concrete mathematical research." Already, with it, she and van der Waerden and others had laid the foundations of modern algebra.

At first both the Hilberts had spoken out in such a forthright way against the new regime that their friends remaining in Göttingen were frightened for their safety. But they did not trust many of the people who were left, nor the new people who came, and after a while they too fell silent.

"Well, Herr Geheimrat, how do you fare?" one of the now infrequent visitors inquired of Hilbert.

"I — well, I don't fare too well. It fares well only with the Jews," he replied in the old unexpected way. "The Jews know where to stand."

Von Hindenburg died in the summer of 1934, purportedly leaving a will which bequeathed the presidency of the Reich to Hitler, who would then be both president and chancellor. An election was scheduled for August with the alternatives, *yes* or *no*. The day before the election, the newspapers carried a proclamation announcing that Hitler had the support of German science. The list of signatures included the name of Hilbert. Whether Hilbert actually signed the proclamation is not known. Arnold Schmidt, who was at that time seeing him almost every day, was not aware of the existence of such a proclamation until he was shown a copy of the newspaper report of it more than thirty years later. Signing would have been contrary to everything Schmidt knew from personal experience that Hilbert believed. He had to concede, however, that "at that time it is possible that Hilbert would have signed anything to get rid of someone who was bothering him."

In 1935 the final volume of the collected works, which contained the life history written by Blumenthal, was published. Hilbert wrote a little note to his oldest student, commenting on this last great piece of luck — that he should have such a splendid interpreter of his life and work. Blumenthal placed the note in his own copy of Hilbert's collected works.

For his biographical article, Blumenthal had called up his memories of his teacher since that day when the "medium-sized, quick, unpretentiously dressed man, who did not look at all like a professor" had come to Göttingen in the spring of 1895 as the successor of Heinrich Weber. But in spite of the warmth and affection, the life history remained objective.

"For the analysis of a great mathematical talent," Blumenthal concluded, "one has to differentiate between the ability to create new concepts and the gift for sensing the depth of connections and simplifying fundamentals. Hilbert's greatness consists in his overpowering, deep-penetrating insight. All of his works contain examples from far-flung fields, the inner relatedness of which and the connection with the problem at hand only he had been able to discern; from all these the synthesis — and his work of art — was ultimately created. As far as the creation of new things is concerned, I would place Minkowski higher, and from the classical great ones, for instance, Gauss, Galois, Riemann. But in his sense for discovering the synthesis only a very few of the great have equaled Hilbert."

In the spring of 1935 Emmy Noether died in the United States following an operation.

In his office at the Institute for Advanced Study, Einstein wrote a letter to the editor of The New York Times, which had reported her death only briefly: "In the judgment of most competent living mathematicians, Fräulein Noether was the most significant creative mathematical genius [of the female sex] thus far produced . . . .

"Beneath the effort directed toward the accumulation of worldly goods lies all too frequently the illusion that this is the most substantial and desirable end to be achieved; but there is, fortunately, a minority composed of those who recognize early in their lives that the most beautiful and satisfying experiences open to human kind are not derived from the outside but are bound up with the individual's own feeling, thinking and acting . . . . However inconspicuously the lives of these individuals run their course, nonetheless, the fruits of their endeavors are the most valuable contributions which one generation can make to its successors."

The editors of the *Annalen* decided to risk publishing a memorial article by van der Waerden. After the journal appeared in print, they waited for the blow to fall; but nothing happened. Taking courage, they published a paper by Blumenthal, who was still listed on the cover of the *Annalen* as one of the editors, although as a consequence of the Nuremberg Laws he had recently been removed from his professorship in Aachen. Again, nothing happened.

But the general scientific situation in Germany was progressively deteriorating. A strong supporter of the Third Reich was Ludwig Bieberbach, who had so passionately opposed the German mathematicians' attending the International Congress at Bologna. He and others analyzed the differences in the creative styles of German mathematicians and Jewish mathematicians. Death was no protection. When Klein was listed in a Jewish encyclopedia, his antecedents were carefully examined and it was determined finally that he was "a great German mathematician." Hilbert's antecedents were also examined. There was a joke that there was only one Aryan mathematician in Göttingen and in his veins Jewish blood flowed. The joke depended upon the fact that during Hilbert's illness, he had received a blood transfusion from Courant. Now the question was seriously raised if it was not suspicious for an Aryan mathematician to have the name *David*. It finally became necessary for Hilbert to produce the autobiography of Christian David Hilbert to show that David was a family name and that the other family names indicated that the Hilberts had at one time been Pietists.

In the late summer of 1936 the mathematicians of the world met again for another International Congress, this time at Oslo. Although Hilbert did not attend, he was remembered with a telegram of greeting from the delegates. Courant, who came from the United States, where he was now teaching at New York University, telephoned from Oslo; but Hilbert did not know what to say to his former pupil and colleague, and the conversation consisted of his fumbling over and over, "Well, what should I say? What should I ask now? Let me think for a moment."

In 1937 Hilbert was 75. A newspaper reporter came to interview him and ask him about places in Göttingen connected with the history of mathematics. "I actually know none," he said without (to the reporter's surprise) a trace of embarrassment at his ignorance. "Memory only confuses thought — I have completely abolished it for a long time. I really don't need to know anything, for there are others, my wife and our maid — they will know." As the reporter then began to express "a courteous doubt" whether one could so eliminate memory and history, Hilbert put back his head and gave a little laugh.

"Ja, probably I have even been known to be especially gifted for forgetting. For that reason indeed did I study mathematics."

Then he closed his eyes.

The reporter refrained from disturbing any further the old man, "the honorary doctor of five universities, who with easy serenity could completely forget everything — house, streets, city, names, occurrences and facts — because he had the power in each remaining moment to derive and develop again a whole world."

That night there was a birthday party at the Hilberts', a comparatively large affair for the new days. While the congratulatory speeches were being made, Hilbert sat in another room with his arms around the two young nurses who came regularly to the house to give him some physical treatments. When Hecke, who had come from Hamburg for the party, reminded him that he really should listen to the glowing speeches being made about him and his work, he laughed, "This is much better!"

Elizabeth Reidemeister took a birthday picture. She reminded him of what seemed to her some important event in which they had both participated, and was surprised that he remembered nothing of it. "I am interested only in the stars," he explained.

During this period Franz was home again. With age he had come to look disconcertingly like his father. He patterned himself after him, loudly spoke out his opinion on all subjects — a tragic parody — "the sound without the substance," as the people in Göttingen observed. He never held any real job.

But he also studied various subjects very thoroughly — Goethe, theology. He was a real "Kenner," according to Arnold Schmidt — an expert on the fields of his interests. He spoke often of learning mathematics so that he could appreciate his father's work.

The next year — 1938 — saw the last birthday party in the house on Wilhelm Weber Strasse. There were only a few old friends for lunch. Hecke came from Hamburg, Carathéodory from Munich. Siegel, who was now at the Institute in Göttingen, was there. Also present was Blumenthal.

"What subjects are you lecturing on this semester?" Hilbert asked.

"I do not lecture any more," Blumenthal gently reminded him.

"What do you mean, you do not lecture?"

"I am not allowed to lecture any more."

"But that is completely impossible! That cannot be done. Nobody has a right to dismiss a professor unless he has committed a crime. Why do you not apply for justice?"

The others tried to explain Blumenthal's situation, but Hilbert became increasingly angry with them.

"I felt," Siegel says, "that he had the impression we were trying to play a bad joke on him."

Shortly afterwards, Blumenthal's name had to be removed from the cover of the *Annalen*. With the help of his friends he left Germany for Holland.

Across the ocean in America, George Pólya, who was now at Stanford University, reminded Weyl that it was 1938 and that by its terms their wager on the future of Intuitionism was now up. Weyl conceded that he had indeed lost, but he asked Pólya please not to make him concede publicly.

That same year Edmund Landau died.

Life went on in Göttingen. Mrs. Hilbert, who was gradually losing her sight, mourned that the people who used to come to visit no longer came. Sometimes, though, there were still small social gatherings around the Hilberts.

On one of these occasions there was a discussion about which German town was the most beautiful. Some of the guests said Dresden, others said Munich. But Hilbert insisted, "No, no — the most beautiful town in Germany is still Königsberg!" When his wife protested, "But, David, you can't really say that — Königsberg is not all that beautiful," he replied: "But, Käthe, after all I must know, for I have spent my whole life there." Even when she reminded him that in fact they had come to Göttingen more than forty years before, he shook his head: "Ah, a few little years — I have spent my whole life in Königsberg!"

"So," Hasse, who was present, thought sadly, "his mind has condensed all the forty fruitful years of his wonderful achievement in so many branches of mathematics into *a few little years.*"

In 1939 an agreement signed in Munich among Germany, France, England and Italy seemed for the moment to guarantee "peace in our time." The Swedish Academy of Science announced the first Mittag-Leffler Prize, which was to go to David Hilbert and Émile Picard. The old Frenchman received his prize from the emissary of the Academy, Torsten Carleman, at a large banquet in Paris. After a glowing tribute, Carleman presented Picard with a complete, beautifully bound set of Mittag-Leffler's *Acta*

Mathematica, the journal which had been one of the first to publish Cantor's works. From Paris, Carleman went to Göttingen. He expected that there he would enjoy a repetition of banquet, speech and presentation. Disappointed when such was not forthcoming, he still insisted that he wanted to award the prize to Hilbert in person. Hasse and Siegel finally located Hilbert, who was in the nearby Harz mountains with his wife, and drove Carleman to the inn where they were staying. Hilbert listened silently to Carleman's tribute. Shortly afterwards, the 72 red leather volumes of *Acta Mathematica* appeared on the shelves of the library of another mathematician, to whom Hilbert had almost immediately sold them.

It was August again. On the first day of September Germany invaded Poland. Within a week France and England had declared war on Germany.

Hilbert's assistant now was the gifted young logician, Gerhard Gentzen, who, following the new and less restrictive methods of "transfinite induction," had been able to achieve the long-sought proof of the consistency of arithmetic. The proof had been managed, however, only by substantially lowering the standards Hilbert had originally set up. Gentzen came regularly to Hilbert's house and read aloud — at Hilbert's request — the poems of Schiller. After a while Gentzen too was gone. He died in 1945 following arrest and imprisonment in Prague.

Holland was invaded. In England efforts were made by Ewald and others to get Blumenthal to safety. But it was too late.

Siegel had vowed at the end of the first war that he would not remain in Germany during another war. In March 1940, he received an invitation to deliver a lecture in Oslo. He knew that he would not see the Hilberts again, and so he went to say goodbye. They were not at the house on Wilhelm Weber Strasse, the furnace having broken down; but he found them in a shabby hotel where, Mrs. Hilbert told him, Hermann Amandus Schwarz always used to stay when he was in Göttingen. Schwarz had been responsible for bringing Klein to the University and had been dead now some twenty years. The Hilberts were having breakfast in their room, Hilbert sitting on the bed and eating from a jar of caviar that Niels Bohr had sent him from Copenhagen. Siegel said goodbye. In Oslo he found that the Bohr brothers and Oswald Veblen had already arranged passage to the United States, where a place at the Institute for Advanced Study would be waiting for him. Two days after he left Oslo, the Germans invaded Norway.

In December 1941, a month before Hilbert's eightieth birthday, the United States entered the war. Although there was no party on the eightieth birthday, a tribute to Hilbert appeared as usual. It was prepared by Walther Lietzmann, who in 1902 had headed the delegation from the Mathematics Club who had pleaded with Hilbert to refuse the tempting offer from Berlin and remain in Göttingen. The story of Hilbert's life and his achievements, everything was there in Lietzmann's tribute except the names of the many Jews (other than Minkowski and Hurwitz) who had played such an important part in his career. Blumenthal was circumspectly quoted merely as "the author of the life history in the collected works." The picture that accompanied the tribute was a recent one, and the eyes which had looked so firmly and innocently out at the world seemed now distrustful.

In Holland, Blumenthal dedicated a paper to Hilbert in honor of the eightieth birthday.

The Berlin Academy voted to commemorate the birthday with a special citation for that work which of all the influential Hilbert works had had the most pervasive influence on the progress of mathematics — the little 92-page book on the foundations of geometry.

The day that this award was voted by the Academy, Hilbert fell on the street in Göttingen and broke his arm. He died, a little more than a year later, on February 14, 1943, of complications arising from the physical inactivity that resulted from the accident.

Not more than a dozen people attended the morning funeral service in the living room of the house on Wilhelm Weber Strasse. From Munich came one of his oldest friends. Standing beside the coffin, Arnold Sommerfeld spoke of Hilbert's work.

What had been his greatest mathematical achievement?

"The invariants? The number theory, which was so loved by him? The axiomatics of geometry, which was the first great achievement in this field since Euclid and the non-euclidean geometries? What Riemann and Dirichlet surmised, Hilbert established by proof at the foundation of function theory and the calculus of variations. Or were integral equations the high point.... Soon in the new physics ... they bore most beautiful fruits. His gas theory had a fundamental effect on the new experimental knowledge, which has not yet been played out. Also his contributions to general relativity theory are of permanent value. Of his final endeavors in connection with mathematical knowledge, the last word has still not been spoken. But when in this field a further development is possible, it will not by-pass Hilbert but go through him."

Carathéodory had also planned to come from Munich for the funeral, but he had fallen ill. The tribute which he had written was read by Gustav Herglotz with tears streaming down his face.

Above all, it had seemed to Carathéodory, there had been such an absolute integrity and oneness in the conduct of Hilbert's life that "even the peculiarities of his old age, which might cause a stranger surprise, were for us, his friends, genuine manifestations of the Hilbert character."

The active influence which Hilbert had exercised on the mathematicians of his time was embodied for Carathéodory in a statement he had heard one of the most important mathematicians of the day make directly to Hilbert himself: "You have made us all think only that which you would have us think!"

Recalling what the dead man had meant to them during his lifetime, he addressed the widow directly:

"Dear Frau Hilbert, now he lies in this same room in which we spent so many joyous hours with him. In a short time we will carry him to his last resting place. But so long as our hearts beat, we shall be bound together by the memory of this great man."

They buried Hilbert in the cemetery out beyond the river, where Klein also lay. Only the name and dates were inscribed on a plaque in the grass.

The news of Hilbert's death came to the outside warring world by way of Switzerland. Attending a mathematical meeting in New York, Hermann Weyl saw a small notice from Bern in The Times. He recalled again the summer months which he had spent long ago working his way through the *Zahlbericht*, the happiest months of his life, "whose shine, across years burdened with our common share of doubt and failure, still comforts my soul." Back home in Princeton, he wrote to Auguste Minkowski, who was living in Boston with her older daughter:

"The report of Hilbert's death brought up again the whole Göttingen past for me. I had the great luck to grow up in the most beautiful years ... when Hilbert and your husband both stood in the prime of their power.... I believe it has very seldom happened in mathematics that two men exercised such a strong and magic influence on a whole generation of students. It was a beautiful, brief time. Today there is nowhere anything even remotely comparable...."

The story of Hilbert's death came to bombed England more slowly and less accurately.

"The news has reached us that David Hilbert died during the summer (sic)," the *Journal of the London Mathematical Society* announced that fall. "The Council feels that the death of so great a mathematician should not pass, even momentarily, unnoticed; but the difficulties of obtaining an adequate account of his work and influence are at present unsurmountable."

Max Born, who was in England, had been in Göttingen when Minkowski died; and now he could not help wondering: "Hilbert had survived his friend by more than thirty years. He was still permitted to achieve important work. But who would like to say if his solitary death in the dark Nazi time was not still more tragic than Minkowski's in the fullness of his power?"

A few months after Hilbert's death, Blumenthal was swept up by the Gestapo in one of its periodic raids on the Jews in Holland. He was sent to Theresienstadt, a little Czechoslovakian village which had been converted into a ghetto for old Jews and others for whose deaths the Nazis — at least in the beginning — did not wish to take direct responsibility. It is known that at one time he was placed on a train for Auschwitz, then for some reason removed before the train left. He died in Theresienstadt at the end of 1944.

On January 17, 1945, Käthe Hilbert died. She was almost blind. There was no old friend to speak by her coffin; and since she, like her husband, had long ago left the church, no minister would perform this office. In the end, at the begging of Franz Hilbert, a woman, a good friend but not one who had known the great golden days, spoke some appropriate words.

That same year Königsberg, almost completely destroyed, fell to the Russians.

XXV

The Last Word

It might seem that the sweet sound of the Pied Piper had been stilled forever. But all over the world — in little European countries — in embattled England — Japan — Russia — the United States — there were Hilbert students, and students of Hilbert students.

Across the ocean, even during the war, one could still hear the old tune. Hermann Weyl, with his own special ardor, was attempting to create in Princeton at the Institute for Advanced Study another great center of passionate scientific life — it was his phrase — like the one he had known in his youth in Göttingen. In New York, Richard Courant was ensconced on the upper floor of a girls' dormitory in Washington Square, called cheerfully by his friends "Courant's Institute." The spirit of Hilbert lived there too.

At the time of Hilbert's death it was said in *Nature* that there was scarcely a mathematician in the world whose work did not derive in some way from that of Hilbert. Like some mathematical Alexander, he had left his name written large across the map of mathematics. There was, as *Nature* pointed out, Hilbert space, Hilbert inequality, Hilbert transform, Hilbert invariant integral, Hilbert irreducibility theorem, Hilbert base theorem, Hilbert axiom, Hilbert sub-groups, Hilbert class-field.

The ideas in his work on Gordan's Problem had extended far beyond algebraic invariants in method and significance: they had flowered into the general theories of abstract fields, rings and modules — in short, modern algebra. Much of the work of subsequent number theorists had been in the fertile fields which Hilbert had opened up with the *Zahlbericht* and his program of class-fields. The little book on the foundations of geometry — "a landmark in mathematical thought" — had set the axiomatic method securely and deeply in nearly all of mathematics. "It is difficult," a recent mathematical historian has commented, "to overestimate the influence of this little book." Since Hilbert's salvaging of the Dirichlet Principle, the theory had been simplified and extended until the Principle had become "a tool as flexible and almost as simple as that originally envisaged by Riemann." The work had also been the starting point for the development of what are known as direct methods in the variational calculus, important in both pure and applied mathematics. The theorem which Hilbert had stated and proved in his lectures on the calculus of variations at the turn of the century had been used to lay a whole new foundation for that subject. The general theory of integral equations had become one of the most powerful mathematical weapons in the arsenal of the physicists, and Hilbert Space theory had grown to such proportions that a writer complained it was impossible to treat it "in a finite number of words." And although, today, no one speaks of the axiomatization of physics, the clarifying, ordering and unifying power of the axiomatic method has penetrated that science.

But what of the last great work of Hilbert's career of formalizing mathematics and establishing its consistency by absolute proof? In spite of the blow dealt to this program by Gödel's work, Hilbert's liberating conception of mathematical existence as freedom from contradiction has unquestionably triumphed over the shackling constructive ideas of his opponents. The question of the consistency of mathematics, so seemingly simple and obvious until it was raised by Hilbert, has played a role of inestimable importance in the history of mathematical thought. "It was a good question," one modern mathematician says, "and only a very great mathematician would have thought of asking it."

Gödel (who never met nor had any correspondence with Hilbert) feels that Hilbert's scheme for the foundations of mathematics "remains highly interesting and important in spite of my negative results."

He adds:

"What has been proved is only that the *specific epistemological* objective which Hilbert had in mind cannot be obtained. This objective was to prove the consistency of the axioms of classical mathematics on the basis of evidence just as concrete and immediately convincing as elementary arithmetic.

"However, viewing the situation from a purely *mathematical* point of view, consistency proofs on the basis of suitably chosen stronger metamathematical presuppositions (as have been given by Gentzen and others) are just as interesting, and they lead to highly important insights into the proof theoretic structure of mathematics. Moreover, the question remains open whether, or to what extent, it is possible, on the basis of the formalistic approach, to prove 'constructively' the consistency of classical mathematics, i.e., to replace its axioms about abstract entities of an objective Platonic realm by insights about the given operations of our mind.

"As far as my negative results are concerned, apart from the philosophical consequences mentioned before, I would see their importance primarily in the fact that in many cases they make it possible to judge, or to guess, whether some specific part of Hilbert's program can be carried through on the basis of given metamathematical presuppositions."

Gödel also feels that "in judging the value of Hilbert's work on the Continuum Problem, it is frequently overlooked that, disregarding questions of detail, one quite important *general* idea of his has proved perfectly correct, namely that the Continuum Problem will require for its solution entirely new methods deriving from the foundations of mathematics. This, in particular, would seem to imply (although Hilbert did not say so explicitly) that the Continuum Hypothesis is undecidable from the usual axioms of set theory."

As a result of Hilbert's enthusiasm for the problems of mathematical logic and the foundations of mathematics, a whole new area of study has been added to the science – metamathematics, *beyond mathematics*.

"The future historian of science concerned with the development of mathematics in the late nineteenth and the first half of the twentieth century will undoubtedly state that several branches of mathematics are highly indebted to Hilbert's achievements for their vigorous advancement in that period," Alfred Tarski has written. "On the other hand, he will have to note, perhaps with some wonder, that the influence of this man appears equally strong and powerful in some other domains which do not owe any exceptionally important results to Hilbert's own research. An example of this kind is furnished by the foundations of geometry. I am far from underestimating the value of Hilbert's contributions . . . in his [*Foundations of Geometry*], but I think that his most essential merit was the impulse he gave to organized research in this domain. A still more striking example is presented by metamathematics. Occasional considerations in this field preceded Hilbert's Paris address; the first positive and really profound results appeared before Hilbert started his continuous work in this domain... [and] one does not immediately associate with Hilbert's name any definite and important metamathematical result. Nevertheless, Hilbert will deservedly be called the father of metamathematics. For he is the one who created metamathematics as an independent being; he fought for its right to existence, backing it with his whole authority as a great mathematician. And he was the one who mapped out its future course and entrusted it with ambitions and important tasks. It is true that the baby did not fulfill all the expectations of the father, it did not grow up to be a child prodigy. But it developed sanely and healthily, it has become a normal member of the great mathematical family, and I do not think that the father has any reason to blush for his progeny...."

In 1950, when Hermann Weyl was asked by the American Mathematical Society to summarize the history of mathematics during the first half of the twentieth century, he wrote that if the terminology of the Paris Problems had not been so technical he could have performed the required task simply in the terms of Hilbert's problems which had been solved, or partially solved — "a chart by which we mathematicians have often measured our progress" during the past fifty years. "How much better he predicted the future of mathematics than any politician foresaw the gifts of war and terror that the new century was about to lavish on mankind!"

Today, Königsberg no longer exists. Where it stood on the Pregel there now stands Kaliningrad, the most advanced naval base of the Soviet Union. The rivalry between Göttingen and Paris is a thing of the past. In both Germany and France a whole generation of mathematicians is missing. The United States finds itself immeasurably enriched, for almost all of the members of the Hilbert school and many other European scientists emigrated to that country. Among them were the following who have been mentioned at times in the course of this book: Artin, Courant, Debye, Dehn, Einstein, Ewald, Feller, Franck, Friedrichs, Gödel, Hellinger, von Kármán, Landé, Lewy, Neugebauer, von Neumann, Emmy Noether, Nordheim, Ore, Pólya, Szegö, Tarski, Olga Taussky, Weyl, Wigner.

After the war Göttingen was the first German university to re-open its doors. Eventually many of Hilbert's old friends returned to visit there, a few like Max Born to live out their lives nearby.

In 1962, on the one hundredth anniversary of Hilbert's birth, Richard Courant delivered a talk in Göttingen on Hilbert's work and its importance for mathematics.

"It is naturally impossible in a program like this to give an even approximate appreciation of a personality so many-sided as that of Hilbert," he said. "Also there is no point in attempting sentimentally to bring back the good old times. But I feel that the consciousness of Hilbert's spirit is of great actual importance for mathematics and mathematicians today.

"Although mathematics has played an important role for more than two thousand years, it is still subject to changes of fashion and, above all, to departures from tradition. In the present era of the over-active industrialization of science, propaganda, and the explosive manipulation of the social and personal basis of science, I believe that we find ourselves in such a period of danger. In our time of mass media, the call for reform, as a result of propaganda, can just as easily lead to a narrowing and choking as to a liberating of mathematical knowledge. That applies, not only to research in the universities, but also to instruction in the schools. The danger is that the combined forces so press in the direction of abstraction that only that side of the great Hilbertian tradition is carried on.

"Living mathematics rests on the fluctuation between the antithetical powers of intuition and logic, the individuality of 'grounded' problems and the generality of far-reaching abstractions. We ourselves must prevent the development being forced to only one pole of the life-giving antithesis.

"Mathematics must be cherished and strengthened as a unified, vital branch in the broad river of science; it dares not trickle away in the sand.

"Hilbert has shown us through his impressive example that such dangers are easily preventable, that there is no gap between pure and applied mathematics, and that between mathematics and science as a whole a fruitful community can be established. I am therefore convinced that Hilbert's contagious optimism even today retains its vitality for mathematics, which will succeed only through the spirit of Hilbert."

It is this optimism which will echo, as long as stone survives, from the marker that has been placed since Hilbert's death over his grave in Göttingen:

>Wir müssen wissen.
>Wir werden wissen.

Königsberg and Göttingen

An Album

Otto Hilbert, father of David Hilbert, as a university student in 1850

The Königsberg cathedral (reprinted from *Ostpreussen, Westpreussen, Danzig*, Gräfe und Unzer Verlag, München)

The Pregel river with the Königsberg castle in the background (reprinted from *Ostpreussen, Westpreussen, Danzig,* Gräfe und Unzer Verlag, München)

Hermann Minkowski when he won the prize of the Paris Academy

Adolf Hurwitz as an Extraordinarius in Königsberg

David Hilbert, 1886

David Hilbert and Käthe Jerosch, 1892

Felix Klein during his Leipzig days

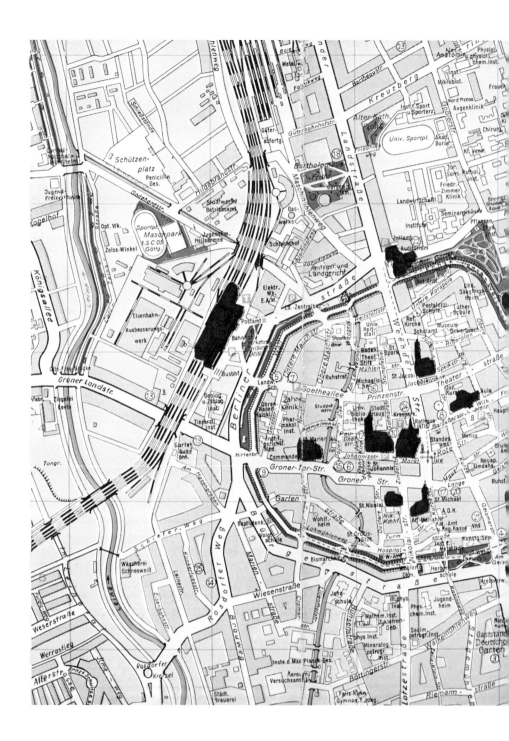

X

David Hilbert, c. 1900

Franz Hilbert, only son of David and Käthe Hilbert

The Mathematics Club of Göttingen, 1902
Left to right, front row: Abraham, Schilling, Hilbert, Klein, Schwarzschild, Mrs. Young, Diestel, Zermelo; second row: Fanla, Hansen, C. Müller, Dawney, E. Schmidt, Yoshiye, Epstein, Fleisher, F. Bernstein; third row: Blumenthal, Hamel, H. Müller (Courtesy of Kinyu-Honda)

Carl Runge

Max Born as a Privatdozent in Göttingen

Edmund Landau

A dinner party at the Kleins' with Paul Gordan (far left), Klein (center and Käthe Hilbert (far right)

Richard Courant as a student in Göttingen

Hermann Minkowski

XVIII

David Hilbert, 1912—one of a group of portraits of professors which were sold as postcards in Göttingen

Emmy Noether

The van der Waerdens and the Hopfs saying goodbye to the Courants in Zürich: Mrs. Hopf, Mrs. van der Waerden, van der Waerden, Heinz Hopf in front

Constantin Carathéodory

Hilbert's sixtieth birthday party

Front row, left to right: Richard Courant, Franz Hilbert, Mrs. Courant (Nina Runge), Hertha Sponer (later Mrs. Franck), Mrs. Grotrian; second row: Mrs. Esslen (later Mrs. Springer), Mrs. Landau, Mrs. Hilbert, David Hilbert, Mrs. Hofmann, Mrs. Minkowski; third row: Ferdinand Springer, Felix Bernstein (behind Mrs. Landau), Mrs. Prandtl, Edmund Landau, Mrs. Franck, Fanny Minkowski (at end of row); fourth row: Ernst Hellinger, Erich Hecke (behind Landau), Walter Grotrian (behind Mrs. Hofmann); fifth row: Peter Debye, Theodore von Kármán (behind Hecke), Mrs. Rüdenberg (Lily Minkowski), Paul Bernays, Leonard Nelson, "Klärchen" (second from end of row).

Hermann Weyl

Richard Courant

Max Born

The entrance to the Mathematical Institute of Göttingen

Rear view of the Mathematical Institute

David Hilbert and Hermann Weyl during the mid-twenties

David Hilbert, 1932

Courant

Korf erfindet eine Uhr,
die mit zwei Paar Zeigern kreist,
und damit nach vorn nicht nur,
sondern auch nach rückwärts weist.

Korf a kind of clock invents
where two pairs of hands go round:
one the current hour presents,
one is always backward bound.

From *Christian Morgenstern's Galgenlieder*,
translated by Max Knight,
University of California Press, Berkeley, 1963.

ALPHA AND OMEGA

IN THE SUMMER OF 1970 I received a letter from K. O. Friedrichs, a professor at the Courant Institute of Mathematical Sciences in New York City. I had sent him a complimentary copy of *Hilbert*. He wrote to thank me for it, and then added in a postscript:

"My wife and I, and many of our friends, have often urged Courant to write his reminiscences, but he has never responded to our suggestions. I wonder whether you would feel like approaching him on this matter. I know that he reacted very positively to your work on Hilbert. I can also assure you that you would find a great deal of support among all his friends."

I already knew the story of how Richard Courant, one of Hilbert's best known students, had been removed by the Nazis from his position as director of the internationally famous mathematics institute in Göttingen and how he had emigrated to the United States and built up another great mathematics institute. When I was again in New York, I went to talk to Friedrichs and his wife in his office at the Courant Institute on the corner of Fourth and Mercer streets, a few blocks from Washington Square.

Friedrichs had been Courant's student and assistant in Göttingen and his colleague at New York University for many years. It was apparent that he was fascinated by the contradictions of Courant's character as a mathematician. "Think of it—a mathematician who hates logic, who abhors abstractions, who is suspicious of 'truth,' if it is just bare truth!" He emphasized to me, however, that he did not have in mind another biography. "Courant is no Hilbert, and he would be the first to tell you that." But Courant was on the eve of his eighty-third birthday. For most of his years he had been in close personal contact with nearly all the important mathematicians and many of the physicists of the twentieth century. He had also been a part of the great migration of the 1930's which had shifted the center of the scientific world from Europe to the United States. It would be regrettable if Courant should die without communicating to a wider audience than his own colleagues and students his recollections of the events and the personalities of his time.

Friedrichs did not minimize the difficulties of getting Courant to commit himself to such a project. Courant, he told me, often seemed incapable of saying yes to anything.

Mrs. Friedrichs was more optimistic.

"Since we first came to this country in 1937, whenever Courant was by himself at home we have usually asked him to dinner. And in all that time I don't think he has ever accepted. Although he has always come."

Courant

Friedrichs had not yet spoken to Courant about the project he had suggested to me, but he had arranged with Mrs. Nina Courant that I would come out to New Rochelle the following Sunday and have lunch with the family and some of Courant's friends.

A sign on the front door of the big old-fashioned wooden house said simply "Courant." Inside, the rooms bulged with heavy German furniture—many cabinets, desks and chests, all overflowing with books and journals and papers and pictures and music. There was a grand piano in the living room and several stringed instruments stood near it.

A portrait of Hilbert hung over the fireplace in the dining room. It had been painted from photographs after his death in 1943 and was no great example of the art of portraiture; yet, better than any photograph of him I had ever seen, it managed to convey his pure and intense intelligence. At the table, counting Courant, there were four members of the American National Academy of Sciences, all of them German-born, all of them originally from Göttingen.

Courant himself had aged since I had first met him in 1965, when I was working on *Hilbert*. His small figure was even stouter through the middle. Increasing deafness had affected his sense of balance and he moved warily. Because of his deafness he also found it hard to follow the general conversation at lunch. With a dozen people around the table, he was quiet; but he listened intently, head bent forward. When someone spoke to him, his eyes became bright with interest and sheer enjoyment in what he was being told. "Ja, ja," he would say softly, nodding his head and encouraging the speaker on.

He listened to Friedrichs's proposal without comment or suggestion, but he did not reject it.

Friedrichs was encouraged. There at the table he began to plan how I would return to New Rochelle after Courant's birthday on January 8 and spend a week as a guest in the house. Courant could talk to me then about what he remembered and what he considered important, and we would simply see how the project developed.

Courant never actually agreed to Friedrichs's plan; but, saying goodbye to me, he said something almost inaudible about wishing me success with my project.

I was back in New Rochelle in the middle of January 1971, and Courant and I began to talk. He was fascinated by my small cassette recorder and commented how wonderful it would have been to have had such a thing in Göttingen during the First World War when Felix Klein was delivering his famous lectures on the history of mathematics to a small group gathered around his study table.

Looking back at the transcribed pages of our first conversation, I see that Courant always referred to our project as "this situation I find myself in."

"I have been urged several times to write down my reminiscences and collect material about my life and to talk into a machine like yours and have somebody transcribe what I said," he told me. "But I have always been much too lazy or undecided. Then Friedrichs came and said he had talked to you and maybe this would be much easier. I have wanted all the time to write something about the academic life and the sciences in Göttingen, which I was a witness of and which was a very great time. But this is not quite the right thing anymore. You have done it very well in your Hilbert book. But still, of course, there are things I would like to talk about. I have followed up some things consciously. I have no ambition to have any sort of biography of myself. But still it seems reasonable for us to talk about a few things. We do not have to start at alpha and end at omega."*

During the year 1971, Courant and I talked many times about the things he had done, the people he had known, his opinions and feelings about mathematics. I think he rather enjoyed our conversations, and he willingly gave me access to papers and letters in the house in New Rochelle and at the institute. But it very soon became clear that I had come too late for the project which Friedrichs had had in mind. Courant had neither the vigor nor the desire to go back over his life meaningfully. Maybe he never could have, for—as Nina Courant said to me—"Richard was always forward-looking."

Felix Klein, whose place he had taken in Göttingen, had spent his last years compiling his collected works—"a masterpiece," Courant told me—each important work or group of smaller works prefaced by a biographical note which set it in the framework of Klein's life and times. But, much as he admired what Klein had done, Courant could not bring himself to do something similar. He took comparatively little satisfaction in his past achievements. He was concerned about the future of mathematics and of the institute which he had created, and he was frustrated and unhappy because he could no longer help. Not only did he lack the physical and mental energy, but mathematics had passed him by.

* Omissions and editing necessitated by the nature of taped conversation have not been indicated in the text.

Courant

All during the year that I talked to him he was deeply depressed. On November 19, 1971, he suffered what at first appeared to be only a slight stroke. He was taken to the hospital in New Rochelle; and on January 27, 1972, a few weeks after his eighty-fourth birthday, never having left the hospital, he died.

The book which Friedrichs had proposed that I write—Courant's reminiscences—would never be written; yet it seemed to me that I had material for another book about Courant.

In our conversations I had been constantly struck by how important his youthful experience in Göttingen had been for him. The days as Hilbert's student and assistant had molded his view of mathematics and of science. The example of Felix Klein had guided him throughout his career as an organizer and as an administrator. Everything he had done professionally, even the writing of his books, had been influenced by these two men. But he had also put his own mark on the scientific tradition which they exemplified for him—first in Germany and then later in the United States.

I was warned that while Courant's reminiscences would have been welcomed by the scientific community, a book about him would be looked upon by many as a posthumous example of the self-aggrandizement, both of himself and of his group, for which he was so often criticized—the ultimate chutzpah. I was told that nobody who came in contact with Courant was ever able to remain neutral about him. Many people, including Friedrichs and his wife, were deeply attached to him; but there were a number of other people who disliked or even detested him. All of his life he had attracted controversy. Almost every piece of mathematical work he had ever done had brought him into a questionable position in relation to some other mathematician. Some of his best work had been in collaboration with his students and had evoked murmured references to *le droit du seigneur*. His books—the importance of which is never denied—were often reputed to have been written largely by his assistants. His way of doing things made many people uncomfortable. He was sometimes referred to behind his back as "Dirty Dick" or "Tricky Dicky."

Even some of his most loyal admirers at the Courant Institute expressed doubts about the project I proposed. "A book about Courant, sure, but it should be written fifty years from now when everybody here is dead." They were afraid that a book at this time might have of necessity to be a piece of personal and institutional puffery. And then, they pointed out, there was the impossibility of the subject, Courant himself. Almost every statement one could make about him could be matched by an opposite statement equally true. I was told by Lipman Bers, a professor at Columbia who was formerly at the Courant Institute, that he had once proved "by contradiction" the theorem: *Courant does not exist.*

In spite of the many objections, I continued to feel that an account of Courant's activities on behalf of what he conceived as "the Göttingen tradition" was the natural and necessary sequel to my book about Hilbert—the rest of the story. Although it was not the book which Friedrichs had originally proposed to me, he was willing to help; and one of the great pleasures of my life has been working with him.

The book was a long time in the writing. I think that Courant would have appreciated the fact that now, five years later, when I have finished the story of his relationship with Göttingen, I find that, if I have not written his biography, I have written the story of his life.

ONE

COURANT never volunteered much about his life before he came to Göttingen.

He told me that he had been born in Lublinitz, a small town in Upper Silesia, now Polish but then German, and that his father had been a small businessman from a large Jewish family "with a not very intense intellectual life." He also said that he had supported himself since he was fourteen. He never went back in any detail beyond that year. Children of his colleagues remembered that they had been warned by their parents: "Uncle Courant does not like to talk about his childhood."

I already knew something about the Courant family history, having read the autobiographical memoir of Edith Stein, the Catholic philosopher-martyr, who was Courant's cousin on his father's side. According to her, their grandfather, Salomon Courant, was originally a soap and candle maker. Coming to Lublinitz to hawk his wares, he saw and fell in love with twelve-year-old Adelheid Burchard, the daughter of the owner of a cotton factory, a very religious man who had earlier been a cantor. Salomon returned to Lublinitz each year until Adelheid was eighteen, then married her and settled in the town. It was a county seat, the marketing and shopping center for the largely Polish peasantry of the surrounding countryside. Working together, Salomon and Adelheid developed a prosperous "food and feed" business, produced fifteen sons and daughters who survived infancy, and—after the birth of Courant's father, the sixth child and second son—built a large house where, according to Edith Stein, Salomon "practiced unlimited hospitality." Adelheid died before Courant was born, but Salomon lived on until the boy was eight—"a small lively man always full of good humor and good ideas who could tell jokes by the dozens."

Little would be known of Courant's immediate family and the early years of his life, even by his wife and children, if it were not for another, unpublished memoir written by his father, Siegmund Courant, to justify to his sons the details of his unsuccessful business career. According to this memoir, Siegmund worked in the family business from the time he was fourteen years old. He felt that he worked hard and long but that, even when he was thirty-four and about to marry Martha Freund, he had no authority in the business. Everything was decided by his older brother, Jakob.

Martha was the daughter of a successful businessman in nearby Oels. At the age of nineteen she had come to work in the store of "S. Courant" as a clerk (a slender red-haired girl, Siegmund told his sons); but at the end of the first month, for some unstated reason, Jakob had sent her home to her father and brothers "with twenty marks, a railway ticket and a chocolate bar." Later it was suggested to Siegmund that she might be a good wife for him. He traveled to Oels and looked over the Freund business; her father came to Lublinitz and looked over the Courant business. Terms were negotiated by Jakob. Martha was to bring a dowry of 15,000 marks in cash, and her father was to take over the mortgage on the big Courant house.

After their marriage in Oels on February 17, 1887, Siegmund and Martha went to live in Lublinitz in the house of old Salomon, where several of Siegmund's brothers and sisters were already maintaining separate apartments. For some reason his sisters were very critical of Martha. He complained that she even had to cook some of the things she liked in secret.

On January 8, 1888, Martha gave birth to Richard. He was the first grandson. Everybody in the family was delighted to have a boy to carry on the Courant name.* There was a gala week-long celebration. But Martha continued to be scorned by the Courant sisters. A year later she produced another son, who was named Fritz. There began to be friction in the family about her father's rights. Siegmund, still frustrated by his inferior position in relation to Jakob, decided to leave Lublinitz. He sold his share in "S. Courant" and purchased a business in Glatz, a town in the foothills to the west of Lublinitz. That same year, when Richard was three, a third son, Ernst, was born.

In Glatz, Siegmund and Martha had a large apartment. Relatives from both sides of the family came for pleasant visits. Although somewhat undercapitalized, the new business went well. But Siegmund was burdened

---

* The derivation of the name is not known. Edith Stein suggests that the family may have lived originally near the French border. Friedrichs thinks it more likely that the name derives from the term *Thaler courant*—the current value of the *Thaler*—an expression frequently used in money exchanges.

by interest on loans from the family and assessments made, he complained, without consulting him. When his father died, he inherited only his gold signet ring with the S on it. He arranged to sell the Glatz business and contracted to buy a more modest one in another town. Jakob insisted that he abandon the purchase and come instead to Breslau, where other brothers and sisters were by then in business. Since Jakob promised to pay any judgment rendered against him, Siegmund broke his contract and went.

Richard was nine when his family made the move to the Silesian capital, now also Polish. In Glatz he had already completed one quarter of the gymnasium. In Breslau he was enrolled in the König–Wilhelms Gymnasium, a humanistic gymnasium which prepared its students for the university. His first report card ranked him twenty-first in a class of twenty-seven and bore the comment that his achievement in arithmetic was barely satisfactory, "or a little less than satisfactory."

In Breslau the Siegmund Courants found themselves in severe economic straits. They had a reputation for being "wasters," and they could not live on what Siegmund earned from a job he had obtained with an insurance company. He established a connection with a wholesale paper goods firm. The man whose business he had been going to buy sued him for breach of contract.

Richard shortly overcame what was apparently poor preparation in Glatz. By the time he was eleven, his ranking in his class oscillated between first and second, his grade in arithmetic was "very good." In the early spring of his fourteenth year—disturbed by his parents' mounting financial problems—he answered a newspaper advertisement for a tutor "qualified in all subjects." Although he got the job, he lost it very soon. The boy he had been hired to teach was lazy and slow. Richard was impatient, then angry, then abusive. The boy's mother finally slapped him and threw him out of the house. It was a cold, wet day. Courant never forgot—this he told me—how he found himself lying in the mud with a tiny gold piece in his hand, the first money he had ever earned.

In spite of this initial pedagogical failure, he soon got other tutoring jobs, becoming quickly more patient and more effective. He tried to follow the example of one of his gymnasium teachers, a man named Erich

Maschke, who was an exponent of the "Socratic method" of leading students to discovery on their own. Courant told me that he was no relation to Heinrich Maschke, who was for many years a professor of mathematics at the University of Chicago and who was also from Breslau. Maschke, the gymnasium teacher, was not a scientist and not even too well informed in mathematics. Courant recalled him as not knowing more than the elements of the differential calculus and as being hazy on the integral calculus. But he loved science and had the ability to inspire scientific excitement in his pupils. Over the years these included, in addition to Courant and his classmate Wolfgang Sternberg, such scientists as Max Born, Ernst Hellinger, Heinz Hopf, and Otto Stern among others.

While Richard was beginning his teaching career, it was becoming clear that Uncle Jakob was in serious financial trouble. The details are confusing. According to Siegmund's account, Jakob had become involved in various "shady deals" and had signed notes in the names of Siegmund and Martha without their knowledge. In the chapter of her memoirs entitled "The Sorrows of the Family," Edith Stein wrote that Jakob's own activities were blameless and that his difficulties were due to the fact that he had become involved in the business ventures of Siegmund and another brother. I was told by Helen Pick, a Courant cousin who now lives in the United States, that her parents later explained to her that there had simply been a tragic misunderstanding. At any rate, in September of Richard's fourteenth year, Jakob shot and killed himself. There was a general breakdown in the Breslau business of the family. Everyone, without exception, placed the blame on Siegmund and Martha and refused to countersign the note which would save their business.

"At a family council," Siegmund reported to his sons, "we were read out of the family."

A month later he declared himself a bankrupt.

While Siegmund was having these financial difficulties, Richard was earning a good amount of money for a boy of his age. One of his early pupils had been Wolfgang Sternberg's sister Grete, who was a student at a recently established private gymnasium for girls. The mathematics instruction there was exceptionally bad; and before long a number of Grete's classmates followed her example and came for tutoring to little Richard Courant, who was two years younger than most of them and a year behind them in school. In 1972, in Berkeley, I had an opportunity to talk to Magda Frankfurter Frei, who had been one of these girls. She remembered Richard Courant as "a very good teacher, who didn't tell you right away what you had to do but led you to understand it by figuring it out for yourself."

"What kind of a young man was he?" his cousin Helen Pick mused, repeating my question. "Very nice. Very. A little bit peculiar, you might say. You'd ask him something and he'd always circumvent it somehow, making jokes and so on. I remember my brother and me sitting together and listening to him tell stories. He always spoke in a very soft voice. So you had to listen hard or you would miss the joke. It was our greatest pleasure. I think he invented the stories because he saw what pleasure they gave us."

Among the girls whom Richard tutored, there was one who was especially good in mathematics. This was Nelly Neumann, the daughter of Justizrat Neumann, a leading member of the Jewish community in Breslau. Nelly was encouraged in her scholarly interest and ambition by her father. He was "an exceptionally kind and noble man," according to Edith Stein, and, although a Jew, looked like "a Germanic aristocrat." Since his wife's death, when Nelly was two, he had had a very close relationship with his daughter. "The happiness of their life together was disturbed only by his mother-in-law, who had remained in the house after his wife's death and continuously tortured him and his child with her bad humor."

Nelly's classmates never visited her because of her termagant grandmother. Richard, coming to give her mathematics lessons in the evening, was probably the only young person to enter the Neumann house; and in time a special friendship developed between him and Nelly.

After the bankruptcy Siegmund's account of his business activities is difficult to follow. He constantly assures his sons that "everything was in order." But the next two years were miserable. He was charged with distributing obscene postcards and fined fifty marks. Although he appealed the conviction, it was later upheld. Martha gave birth to twin sons, who survived only a few hours. In a period of six months the family lived in four different places.

During Richard's sixteenth year, his parents decided to leave Breslau and go to Berlin. He was earning enough money that he could elect not to accompany them. He rented a room and—except for a few months when his brother Fritz stayed with him—lived alone, attending classes during the day and going from house to house in the late afternoon and evening to give mathematics lessons.

Siegmund proudly emphasized in his memoir that throughout all the trouble with the Courant brothers and sisters Richard remained staunchly on his parents' side. After the rest of his family went to Berlin, the only relative he saw was Aunt Auguste Stein. A widow with eleven children, she had successfully taken over the management of her dead husband's lumberyard. Although she had sided with the others against Siegmund, she had acted as the intermediary between him and the rest of the family.

"To know of a shadow on her father's name and the conflict between her brothers was very hard on her," wrote Edith Stein, her youngest child. "And if she did not see her brother for many years after that, she showed his children a ready interest and sympathy and was happy that all of them became hardworking people who made up with industry what their parents had neglected in their education."

Aunt Auguste's favorite among the Courant boys was Ernst, the youngest; but she also admired the precocious Richard. For a time after his parents left Breslau, he came once a week to her house for lunch.

The Stein children looked forward to his coming. He always had "a wealth of witty ideas" to amuse them. But he also liked to talk to their mother "in his dry and humorous tone" about how he could help his parents. Listening to these conversations, the children did not know "whether to laugh at his funny and extremely exaggerated expressions, or to weep over the content."

In 1904–05, the year after his parents left Breslau, while he was still only in the next to the last year at the gymnasium, Richard began to prepare his girls for the *Abitur*. If they passed this examination, they would be permitted to enter the university. To some of his own teachers, it seemed a reflection upon the importance of the examination that a student who had not yet passed it would presume to prepare others. They tried to discourage his activity and finally told him that he would have to stop tutoring if he wished to continue his studies. He was becoming bored anyway. In the spring of 1905, two weeks after the beginning of his final year, he stopped going to the gymnasium classes and began to attend the lectures in mathematics and physics at the local university.

By this time he had already developed a strong interest in music. He was never able to recall for me the specific origins of it. "I heard music," he said, "and I was impressed." But he did recall vividly an occasion when Julius Stenzel, an older student at the gymnasium and a violinist, substituted for the regular teacher in the music class. At the university he became better acquainted with Stenzel, who was by then an advanced philosophy student; and Stenzel and his younger brother encouraged him to accompany them to chamber music concerts. At some point during this period, he rented a piano and taught himself to play.

(In talking to me, he always objected to the statement that he had taught himself to play the piano. "I never *learned*," he said.)

Stenzel also introduced Richard into a lively circle of young artists, musicians, writers, and students who gathered around Käthe Mugdan, a well-to-do widow.* Stenzel was tutoring her daughter Bertha in the classics and Richard soon became Bertha's tutor in mathematics and physics, and also a close friend of her mother. Many years later, when Käthe Mugdan was over eighty, she committed suicide rather than permit herself to be sent to a concentration camp.

At the end of 1905, Richard applied for permission to take the *Abitur* in the spring with his old class at the König–Wilhelms Gymnasium. He was assigned instead to the class at the Gymnasium zu St. Elisabeth. The difficulty in taking the examination as an outside student can be gauged by the fact that in the group of which he was a member he was the only one to pass.

The following year, 1906–07, Richard continued to attend lectures in mathematics and physics at the university. The difference was that now he had official status as a student: after writing an acceptable dissertation and passing an oral examination, he would be able to obtain a doctor's degree.

He originally wanted to study physics, he told me. It was a time, half a decade after Planck's discovery of the quantum, when the old physics was changing; however, the quantum was never mentioned in lectures in Breslau. Ernst Pringsheim was "a very nice but extremely boring man," and Otto Lummer was "a showman" who amused his students "but didn't let them see what it was all about." To Courant the researches of his

---

* She was a second cousin of Max Born, and it was in her home that Courant first met Born.

professors seemed "rather divorced" from their teaching. Frustrated by the poor quality of the physics instruction, he turned to mathematics.

Among the mathematicians he found Adolf Kneser "a very original man, one of the moving forces of the time," but an uninspiring lecturer. Rudolf Sturm, from whom he felt that he really learned old-fashioned projective geometry, was simply a drill master. He also heard lectures by Georg Landsberg and Clemens Schaefer, who were not yet professors.

He sat in the classes of these men and thought how much they knew and how little they were able to convey to their students. Of the professors the most successful as a teacher—in his opinion—was by conventional standards the worst lecturer. This was Jacob Rosanes, an algebraist now forgotten in his field. Rosanes came to the platform with a piece of chalk in one hand and a damp sponge in the other. As he lectured, he scribbled equations which his students never quite saw because as he wrote he hid them with his body and as he moved along he rubbed them out with his sponge. In desperation, Richard and other students, including Erich Hecke, stayed after class trying to put together something coherent out of what they had only half heard and scarcely glimpsed. They discovered that in the process they learned a great deal of algebra.

Two older mathematics students, Otto Toeplitz and Ernst Hellinger, with whom Courant had become acquainted the year before at the university, had left Breslau and gone to Göttingen to study. They wrote with enthusiasm of the mathematics at that university and of a mathematician there named Hilbert.

Breslau began to seem unbearably dull to Courant. He felt that it would be good for him to see some mountains, mathematical as well as geographical. His friend Nelly Neumann, who was also attending the university, felt the same way. In the spring of 1907, the two young people went to Zurich to study for a semester.

It was in Zurich that Courant heard the lectures of Adolf Hurwitz, one of the outstanding analysts of the day, the close friend and the real teacher of the mathematician Hilbert, who was so highly praised by Toeplitz and Hellinger. Although to the end of his life Courant always described Hurwitz's lectures as perfect, he was not stirred by them. He returned, still dissatisfied, to Breslau. That fall he took what money he had, bought a new suit, and set out for Göttingen.

TWO

THE EXACT YEAR that he arrived in Göttingen did not remain in Courant's mind. All his life, even in official documents for security clearance, he put it down variously—from as early as 1905 to as late as 1909. He could have ascertained that it was 1907 from the impressive *Diplom* which he brought with him to the United States, but he did not consider the date important. What he never forgot was how he came very soon under the spell of Göttingen and how very young he felt.

In those days, he told me, unless you were met by a horse-drawn carriage, you walked to the town, which was some distance from the railway station. It was little more than a village between gentle hills crowned by the ruins of ancient watch-towers. The newcomer saw first its red-tiled roofs, then came to an old wall and proceeded through narrow, crooked streets to the medieval town hall. In the town square a recently erected statue of a little goose girl, eternally tending her geese at the fountain, was a reminder that the Brothers Grimm had written many of their fairy tales in Göttingen.

Courant arrived at the university in the middle of October, although the semester did not officially begin until November 1. He came early on the advice of Toeplitz and Hellinger, who were very knowledgeable about the scientific situation in Göttingen. During the past year Toeplitz had been the personal assistant of Felix Klein, the organizing force behind mathematics at the university. He was currently a *Privatdozent*, or private lecturer. His mathematical interest, which in Breslau had been algebraic geometry, was now—under the influence of Hilbert—integral equations and the theory of functions of infinitely many variables. Hellinger, who had been Hilbert's assistant during the past year, had just taken his doctor's degree with a dissertation that was to make possible a great advance in this same field. What interested Hilbert—Courant quickly saw—interested almost everybody in Göttingen.

In the opinion of Toeplitz and Hellinger, their little friend from Breslau was going to need all the time he could get to prepare himself for the higher scientific standards of Göttingen. They introduced him to the *Lesezimmer*, the reading room which was the heart of the mathematical life, and put him to work studying electromagnetic theory. If he really tried, he *might* be accepted in Hilbert and Minkowski's joint seminar on mathematical physics.

1907–1908

David Hilbert was by that time the most important mathematician in Germany. He had already produced his classic works on invariants, algebraic number theory, the calculus of variations, and the foundations of geometry. For the past few years he had been working rather exclusively in the field of analysis. Hermann Minkowski, his closest friend since their university days in Königsberg, had made his reputation in a theory of his own creation called the geometry of numbers. When, however, he had joined Hilbert on the faculty in Göttingen, he had suggested that they undertake a study of classical physics for their own education. A result had been the joint seminar on mathematical physics.

Before the semester began, Courant approached Hilbert about being accepted as a member of the seminar and found him agreeable. The topics posted for individual hour-long reports began at a comparatively low level in stationary electrodynamics and proceeded rapidly to a level which Courant recognized as being far beyond his preparation—in spite of the two weeks spent in the *Lesezimmer*. He decided to sign up for the second report, which was scheduled for the third Monday of the semester.

After the semester began, he was very lonely. Toeplitz and Hellinger were busy. The other mathematics students seemed one-sided in their interests, and he saw no potential friends among them. He attended two courses given by Hilbert and one by Minkowski, and found both of the famous mathematicians very bad lecturers. He also took courses in physics and philosophy. But none of the courses interested him very much—only the preparation of his seminar report. He went for long walks by himself in the surrounding hills and decided that in spite of the many charming views he felt "closed in" by the landscape of Göttingen, so different from the plain that surrounded Breslau. It seemed to rain an inordinate amount of the time. When it was raining and he could not walk, he tried to relax from his studies by reading in the cheap attic room he had rented. One of the books he read was *Don Quixote*. His room was already quite cold, and he did not feel he could afford to heat it. Sometimes he thought that perhaps it had been a mistake for him to have come to Göttingen.

He wrote all of these thoughts in long letters to Nelly Neumann, who was still at the university in Breslau. He addressed her as *Liebe Freundin* and signed himself *Ihr Richard Courant*, using throughout the formal *Sie* rather than the intimate *du*. From his letters—I have never seen any of hers—it is apparent that he admired her self-discipline but worried about what he called her "pedantry." She admired his intelligence but warned him against "ambition."

He had hoped that he would find some way to make money in Göttingen; and he was shortly approached by Hilbert himself with the request that he tutor Franz, Hilbert's only son, a strange boy who did not seem to be able to manage the gymnasium course on his own. The job was only temporary—until the boy's regular tutor returned—but in closer contact with Hilbert and observing the social mechanics of the sciences in Göttingen, Courant conceived within the first two weeks of the semester the ambition to become Hilbert's assistant.

The position of assistant to a mathematics professor was unique to Göttingen at that time, although at all German universities it was customary for professors in the experimental sciences to have assistants who cared for laboratory equipment and set up experiments. The salary paid Hilbert's assistant was fifty marks a month and, as was jokingly said in Göttingen, *Familienverkehr*—conversation with Hilbert and his wife, Käthe. That the *Familienverkehr* was more of an attraction than the fifty marks is indicated by the fact that when Courant came to Göttingen, Hilbert's assistant was Alfred Haar, a young man from a wealthy Hungarian family.

The way to the assistantship was through the seminar.

"You are not to take me wrong and think that ambition plays a part in the fact that I am so determined to be in the seminar," Courant hastened to explain to Nelly. "What you write to me about personality and what you consider its most important ingredient—that is also true for me. And referring that to scientific relationships, I have to add that for me personal contact is more valuable than any specifically scientific stimulation. And right now, here in Göttingen—which is like a factory—there is no way to come in contact with the instructors and the professors except by doing something which makes you stand out from the great impersonal mass of students. For inner satisfaction of course there is another way—to go on studying on your own and not bothering about the professors—but that is a very hazardous thing to do. One has to think about the examination, and in my case also the assistantship."

Hilbert had noticed him, been friendly to him. Perhaps he could confirm this apparently favorable opinion when he presented his report in the seminar. But, as he had to report in a long letter to Nelly, "it turned out otherwise."

"To explain what happened, I must first tell you a few things about Hilbert and the seminar," Courant wrote in this letter, dated December 1,

1907. "Hilbert wants, so he says, to acquaint himself with the modern questions in mathematical physics. The individual students are to work through their subjects with the help of the literature so that everything fits into a system of axioms which Hilbert and Minkowski have set up. In doing this, the students are wholly dependent upon themselves, since Hilbert is little informed on the subjects and therefore can hardly give any advice which is useful.

"My report was especially difficult because it was on a beautiful paper by Hertz for which I had to work out a rigorous derivation of the electrostatic and the related electromagnetic phenomena—something which you can't find in books. My predecessor—in fact, the only person who had given a report before me—was a Dr. Helly from Vienna. (Many of the members of the seminar are already doctors, and I am probably the youngest.) He had given his talk on electrostatics to the satisfaction of Hilbert. I had worked so much on my report and had it so well in order that I was fairly certain I would show up well, too. . . . But when I stood in front of the blackboard on Monday, the unexpected occurred.

"I wrote down the fundamental equations from which I had to proceed; and immediately Dr. Helly called out loudly, 'Those equations are wrong!' There was a great commotion, everybody talking very excitedly with one another. Although Hilbert didn't say anything specifically, he looked very surprised, as did Minkowski, who seemed to me rather contemptuous. Although I was completely in the right—and knew it—I was from that moment so confused and intimidated that my report almost slipped out of my mind. I can scarcely remember now what I said. Anyway, I yielded to the objections of Dr. Helly and erased the equations. Then, hardly understanding myself, I again began to talk. But almost immediately Minkowski interrupted me. It seems that in connection with the derivation of the boundary conditions I had failed to take into consideration the surface current—the existence of which I had not been aware of up until that moment. Since the other people in the seminar were not quite clear about it either, it was discussed at length; and my report did not get beyond this point. Hilbert kept getting more and more impatient—although it's surprising that he was not more rude to me—and I became more and more upset and confused. He turned my whole outline upside down, and everybody smiled ironically. The other members of the seminar, especially Helly, talked more than I did. I was happy when the end of the hour came around."

When Courant got back to his attic room, he found a letter from Nelly waiting for him. In it she warned him that it was not good to direct himself singlemindedly toward a goal like the assistantship, and thus make himself dependent upon unknown men and their judgments of him. He agreed eagerly.

"What is important is what a man *is* and not what he appears to be. . . . I have now accustomed myself to the idea that I will not get the assistantship, which will probably become available at Easter. And if it turns out that way, I will begin to think seriously of a career as a teacher, since that would be much better than an unsuccessful life as a not so good scholar!"

The results of the disastrous day in the seminar were not all bad. Suddenly important people began to take notice of him and be very nice, especially Alfred Haar, Hilbert's assistant. Haar personally reproached Dr. Helly for his behavior "and made sure the same thing would not happen again." Toeplitz also came to Courant's support, and Hellinger gave him some good advice on how to improve his report. At the end of the week, Hilbert himself invited him and Dr. Helly and some other members of the seminar to his home on Wilhelm-Weber-Strasse—"which gave me again some courage, although it really doesn't mean anything."

The following Monday Courant presented his report again. This time Dr. Helly did not interrupt at all, and the other members of the seminar held their questions and comments until the end. Hilbert seemed pleased, "especially with some points which I myself had put in order." In fact, everything went so smoothly that Courant could assure Nelly in his letter, written the following day, "the failure of last Monday has been fairly well canceled out."

When he had the seminar report behind him, he found suddenly that in reality all of his courses were very interesting. He got to know another student named Ernst Meissner, who was getting his doctor's degree with Minkowski. "I like him very much. He is indeed—with the exception of Otto Toeplitz—the only mathematician here who is not *only* a mathematician but has also some humanistic interest and understanding." The countryside around Göttingen no longer seemed at all narrow. It had many charms. "The flowing calm line of the hills is very soothing. Everything breathes of peace. Think of it—there is not a single factory here!" He finished *Don Quixote*. "It is marvelous how one never tires of his adventures, although they all have something in common." He found a piano and began to play again, "practicing regularly and trying to concentrate on fundamentals." When Toeplitz told him that he must permit himself heat—it was by now December and really cold—he meekly agreed. He wrote enthusiastically to Nelly, "If only there in Breslau you had someone like Otto Toeplitz, from whose ability to work and reflect one can learn so much!"

Hilbert shortly suggested to Courant that he come to the weekly meetings of the *mathematische Gesellschaft* when a talk that interested him was being given. "This is somewhat unusual, since I am still very young and in general—at the wish of Klein—only students who have the doctor's degree are invited; but Hilbert now rebels everywhere against Klein's assumed dictatorship—and this explains everything."

Minkowski, who by nature was much less outgoing than Hilbert, became friendly too. A few days after the successful seminar report, meeting Courant on the steps of the lecture hall, he stopped and apologized for the rough treatment he had given him. Shortly after, he invited Courant to his house for a special evening with some other students.

Such affairs, I have been told by Minkowski's older daughter, Lily Rüdenberg, were much dreaded by the family. Minkowski was a very shy man, the students were invariably tongue-tied in the presence of a professor, and the evenings dragged painfully. This time, though, things were different. "There was this new, nondescript-looking little fellow from Breslau, and he began asking questions of my father right at the beginning. Pretty soon everybody was talking." It was a marvel in the family and to be remembered all her life by Minkowski's daughter, although she was only a little girl at the time.

My account of Courant's first days in Göttingen is taken for the most part from the letters he wrote to Nelly Neumann at the time. From them it is obvious that in the beginning the most important people in his life were the professors, especially Hilbert. When, however, during his eighty-third year, he talked to me about this period, it was not his success in the seminar that came first to his mind but the fact that he had been almost immediately accepted into what he called the "in group." No one else I talked to about Göttingen ever used this expression, or even seemed particularly impressed by the existence of such a group; but for Courant there was an "in group," and its leader was Alfred Haar.

Haar was a small, delicately built youth with the charming quality of seeming at home anywhere in the world. He possessed the kind of extremely quick, precise talent which was later to be seen at Göttingen in John von Neumann, and was even more knowledgeable than Toeplitz. The nineteen-year-old Courant was very impressed by Haar's savoir faire and his brilliant and witty conversation, also by the material success of his father, who was the owner of great vineyards in Hungary. Even at the very end of his life, whenever he mentioned Haar, he always mentioned the great vineyards of Haar's father. He told me that all the students of his day were convinced that Alfred Haar was the one of their number who would leave the deepest mark on mathematics.

After the success in the seminar, it became Courant's ambition to associate with Haar and his friends. In spite of their comparative wealth, they were not inaccessible to him. German students, even wealthy ones, lived modestly. The chief gathering of the group was the sharing of the hot midday meal at a local boarding house. Such a meal cost about twenty cents in the American money of the time. Although Courant did not feel that he could afford such an expenditure every day, he managed to arrange his budget so that he could dine with Haar and his friends twice a week.

I asked him once how it happened that he had been so quickly accepted. He replied that the "in group" was really very democratic. "When I came, I was friendly with everybody. They needed young people, who were always welcome until they became obnoxious. I guess I must not have become obnoxious."

If he did not offend the members of the "in group," he soon learned that there were others who did. One day, when he and Toeplitz were waiting in the lecture hall for the arrival of Hilbert, Toeplitz began to point out various students and tell Courant something about each of them. At length he came to a big blond young man with glasses, who sat a little apart from the others.

"You see that fellow over there," he said. "That is Mr. Weyl."

From Toeplitz's tone Courant sensed that the big blond young man was for some reason not "in." Toeplitz's next sentence dismissed Hermann Weyl with finality:

"He is someone who also thinks about mathematics."

Weyl was to become, as everyone now knows, the outstanding mathematician of Courant's generation in Göttingen—the true son of Hilbert—a man whose breadth of interest, ranging from the foundations of his subject to its applications in physics, was to equal if not surpass that of his teacher. At the time when Toeplitz made the remark quoted above, Weyl was just completing his doctor's dissertation. He was a country boy, shy and unsophisticated, from Elmshorn, a little town near Hamburg, where his father was an executive clerk in a bank. His offense in his student days, according to Courant and to others who knew him then, was a certain youthful "silliness." At times he and his intimates conversed in a language in which every syllable was preceded by the sound of the letter p (*peh* in German) ; and once at a party he lay under a chair all evening and answered questions only in barks.

When Courant told me the story about Toeplitz and Weyl, I could see that it still gave him pleasure that in the old student days in Göttingen he had been acceptable to Haar and his friends and Hermann Weyl had not. At the same time it was clear that he appreciated and enjoyed the twist that this fact gave to the whole concept of who was "in" and who was "out."

Alfred Haar became an important mathematician, particularly famous for his very original contributions to the theory of measure; but, as Courant said to me, there was to be no comparison with Weyl.

"No comparison at all."

During that first semester in Göttingen, Courant heard lectures on mathematics, physics, and philosophy.

His interest in philosophy was surprisingly strong, although—as he told me—"it didn't penetrate very lastingly." He took more courses with the phenomenologist Edmund Husserl than with any professor other than Hilbert. The philosophical objective of Husserl was the investigation of the content of human consciousness. He was not concerned with the process of conceptions and thoughts, as the psychologists were, but with the "meaning" of the content and acts of consciousness. He was after the truth upon which all other knowledge rests—"the Archimedean point." He had first applied his phenomenological method to logic and mathematics, and he was always much closer to the mathematicians of Göttingen than to the philosophers. In addition to the lectures of Husserl, Courant also regularly attended philosophical soirees in the rooms of the electric young Leonard Nelson, whose *Habilitation* was being delayed by the opposition of Husserl and the other philosophers but was strongly supported by the mathematicians, especially Hilbert.

Physics was represented in Göttingen during Courant's student days by Woldemar Voigt and Eduard Riecke, both of whom antedated Klein on the faculty. Their presentation of the subject was as classical as the presentation in Breslau had been. Neither was of the stature of Hilbert, Minkowski, and Klein. The impetus which would make Göttingen one of the great physics centers of the world would come from the mathematicians rather than from the physicists. Courant ignored Riecke after the first semester and took relatively few courses from Voigt, whom he nevertheless admired very much as a person and as a musician.

Mathematics, not physics, was by this time—as he wrote to Nelly—"closest to my heart." He took only two courses with Klein during his five semesters in Göttingen; and although each semester he took a course from Minkowski, he can be said to have "majored" in Hilbert.

For Courant—during his student days and for the rest of his life—there was never to be another teacher to equal Hilbert.

THREE

THE VERY AIR of Göttingen seemed to young Courant to be full of mathematical excitement. Even outside the lecture hall, the students talked intensely "with inspiration and real dedication" about mathematics. This enthusiasm radiated from Hilbert, a slightly built, active man whose independence, originality, and passion for the truth were as apparent in every aspect of his personality and daily life as they were in his great mathematical works.

Seven years before, at the International Mathematical Congress in 1900, Hilbert had listed twenty-three problems, ranging over all of mathematics, on which he had proposed that he and his fellow mathematicians should concentrate their efforts during the coming century. Some of the problems were abstract and general. Others were concrete and individual. Although Hilbert had no interest in what are commonly referred to as the applications of mathematics, he had a great interest in its application to the other sciences. One of his problems had been the mathematization of physics and neighboring subjects. Courant was never to forget the spectacle of a great mathematician "learning physics" at the age of forty-five.

Although on first contact he had found Hilbert a very poor lecturer, "frequently hesitating, varying and amending what he said, sometimes even getting stuck, so that he had to call on his assistant for help," he soon recognized that Hilbert brought to the lecture hall something which even the admired and inspiring Maschke had never been able to bring— a superior mind directly and creatively involved with its subject.

"Hilbert had a way which was much more stimulating than any formal perfection," Courant told me, "because while he lectured he always fought, and the fight was clarifying his ideas. You could follow him. You could *feel* his intellectual muscles."

Like other students of Hilbert, Courant found himself pulled into a struggle to solve a problem, to create or to extend a theory.

Also impressive to Courant was the fact that Hilbert did not isolate himself in order to work. A constant stream of visitors came to the Hilbert house and garden. Discussions of mathematics were interwoven with conversations on a variety of other subjects, especially politics. These were often continued on walks on the wall which still surrounded the town or on hikes into the nearby hills, sometimes on longer expeditions by bicycle. Hilbert ignored distinctions of age and rank. "Whoever you were, you could talk to him as an equal if you had something to say." The Hilbert house was "wide open." Käthe Hilbert was "a wonderful woman," in her own way as independent and original as her husband, full of interest in his young colleagues and students.

Courant was always to consider it his greatest good fortune that, in the fall of 1908, Hilbert—looking around for a student to take Haar's place as assistant—selected Richard Courant, twenty years old.

It was an especially exciting time in Göttingen. Hilbert was attacking the famous theorem of Waring, a problem which had defeated mathematicians for over a century. Minkowski had recently introduced his revolutionary conception of space-time and was working intensely to bring his new ideas to completion.

As Hilbert's assistant, Courant saw the great mathematician several days a week, usually in his garden, where Hilbert always preferred to work. Often Minkowski was present too. In spite of Hilbert's stumbling classroom performance, he spent a great deal of time and thought on his lectures. One of Courant's duties was to research the literature for him and to write reports on relevant points. He also attended Hilbert's lectures, took careful notes, and prepared a neat copy to be placed in the *Lesezimmer*, where the students could consult it at leisure. In addition, he read some of the papers submitted to Hilbert as one of the editors of the *Mathematische Annalen* and helped prepare them for publication. The salary of the assistant was still fifty marks a month (about $12.50 at that time) and *Familienverkehr*.

By Christmas 1908 Hilbert knew that he had proved Waring's theorem. He planned to present his work in the first joint seminar in January, startling even Minkowski, who had been away from Göttingen. But Minkowski was never to hear Hilbert's proof of Waring's theorem. A few days after his return, he suffered a severe attack of appendicitis, was rushed to the hospital and operated upon. He died on January 12, 1909, at the age of forty-five.

(Later in the same year there was another premature death in the Göttingen mathematical circle. Walther Ritz, whose name is attached to the spectral formula and the combination principle of spectral lines, was a *Privatdozent* for theoretical physics. He had been told by his doctors that he was perhaps fatally ill with tuberculosis and should be in a sanatorium under constant medical care. Ritz ignored this advice and worked feverishly on his scientific ideas until his death in July 1909 at the age of thirty-one. A paper by Ritz which made a great impression on Courant was one in which he utilized Hilbert's work on Dirichlet's principle to develop a numerical method of solving boundary-value problems of differential equations. It was fascinating to Courant that a purely theoretical work could lead to a work with practical applications!)

The death of Minkowski turned Hilbert to younger people for the companionship and creative stimulation he had lost. He spent more time with Courant, talked more with him, even gossiped about faculty affairs and the discussion taking place over the choice of a successor to Minkowski. It was from Hilbert that Courant got his first glimpse of Felix Klein as other than the *Bonze*—the important man who takes himself importantly.

The choice for Minkowski's successor, Hilbert told Courant, was between Edmund Landau and Oskar Perron. Both were outstanding mathematicians, but Landau was considered a brassy, rich "Berlin Jew" while Perron was generally well liked. When the time came for the choice, to the surprise of his colleagues, Klein spoke in favor of Landau.

"We being such a group as we are here," Klein said, "it is better if we have a man who is *not easy*."

Klein's opinion was of course decisive, and in the spring of 1909 Landau became a member of the Göttingen faculty.

By the time Landau arrived in Göttingen, Courant had already taken a number of courses dealing with the applications of mathematics, which were well represented at the university as a result of the influence of Klein. He had also attended lectures by Ernst Zermelo on mathematical logic.

(Since Friedrichs had described Courant to me as a mathematician who hated logic, I asked him if he had "hated" the subject even in his student days. "I didn't *hate* logic," he objected. "I was repelled by it. I *believe* in intuition.")

Most important for Courant's future conception of mathematics was the fact that no professor whose lectures he heard in Göttingen was a specialist in one single field. The mathematicians Hilbert, Minkowski, and Klein had a strong interest in physics; the latter two had even considered becoming physicists. Carl Runge, the professor of applied mathematics, had begun his career in pure mathematics. Karl Schwarzschild, the astronomer, made contributions to both theoretical and observational astronomy as well as to quantum mechanics and relativity theory. Ludwig Prandtl, the mechanical genius, developed the mathematical formulation of boundary-layer and airfoil theory. Zermelo, the logician, did his first mathematical work in the calculus of variations and made important contributions to kinetic gas theory.

This general interaction of theory and practice was always to be for Courant the distinguishing characteristic of the Göttingen scientific tradition. Edmund Landau did not fit into this conception of the tradition. His specialty was the analytic theory of numbers, and he loftily dismissed anything connected with the applications of mathematics as *Schmieröl*, or "grease." The new professor lectured as he wrote his many books, proof following theorem without a word of motivation for what he was doing or where he was heading. In Courant's opinion, both lectures and books were sometimes "so abstract that there was no relation at all to substance."

Socially, however, Landau and his wife, Marianne, the daughter of the Nobel laureate Paul Ehrlich, were an addition to life in Göttingen. They gave parties to which students were invited; and Mrs. Landau, like Mrs. Hilbert, took a lively interest in the young people and their problems.

That same spring that Landau assumed Minkowski's chair, Henri Poincaré came to Göttingen to deliver a week of lectures. Hilbert offered his own assistant to the Frenchman, and as a result Courant had the opportunity to observe together the two men who were unanimously acknowledged as the greatest mathematicians in the world at that time. They treated each other with a great deal of respect, he told me, but there was no spark between them like that between Hilbert and Minkowski.

"Which would you rate as the better mathematician?" I asked.

"I am a great admirer of Poincaré," Courant replied. "I think he was the greatest mathematician since Riemann, but you cannot compare him with Hilbert. He did not have the intensity which radiated from Hilbert and which was so wonderful. If he had had that—" More than sixty years later Courant's old eyes brightened at the thought. "But that is unthinkable—Hilbert was absolutely unique!"

During the four semesters that Courant served as Hilbert's assistant, Hilbert was devoting himself almost exclusively to subjects in analysis, the area of mathematics which developed in closest connection with problems of physics and mechanics. Analysis had not been emphasized as a field of advanced endeavor in Breslau. Although the students there studied some of the higher developments of the calculus, they tended to write their dissertations on subjects in algebra and geometry. Nelly confessed to Richard that analysis "frightened" her.

"Don't think that analysis is so much harder than other things," he wrote back. "Perhaps completely the opposite. I am getting into it now, and I find that it is not actually so dangerous."

In fact, Courant took to analysis as if it were his natural element. All his future mathematical work was to be done in subjects to which he was introduced during his student days by Hilbert. It is thus interesting to speculate what might have happened if he had come to Göttingen at a time when Hilbert was working in some other area of mathematics; for Hilbert had the characteristic of becoming so totally immersed in a subject that sometimes he appeared to forget even his own important results in areas of earlier interest.

"I think Courant might then have become the student of Felix Klein," Friedrichs suggested to me.

It is difficult to imagine anyone who would seem to contrast more with Courant than Felix Klein. Courant was small—five feet and five inches in height—and so unprepossessing in appearance that people I interviewed who had known him in his youth could never quite remember what he looked like then. Klein stood out in any group, handsome, dark haired and dark bearded with an impressive head and a memorable smile. He was the acknowledged scientific leader in Germany. Mathematicians, who are not as a rule inclined to hyperbole, invariably recall him as "a Jove." But mathematically Courant and Klein had a great deal in common. Especially notable was the admiration and empathy they had for the work of Bernhard Riemann.

Klein, whose most lasting service to mathematics (in Courant's later opinion) may well have been the opening up of Riemann's work to other mathematicians, arrived in Göttingen, a young Ph.D. from Bonn, just two years after Riemann's death. He was drawn quite early to Riemann's ideas. Many of these had come to that mathematician as the result of a unique geometric intuition which Klein seems to have been almost alone in understanding at the time. Other mathematicians admired the results but were not able to reconcile themselves to the methods by which they had been obtained. Riemann's early death also contributed to the lack of appreciation of his work, but—in Courant's opinion—it is unlikely that even alive he would have had the personal force to put over his ideas. It was left to Klein with his dominating personality and his intuitive comprehension of the geometrical relationships in Riemann's work to become, as Courant later wrote, "the successful apostle of the Riemannian spirit."

"The question, however, is *Why was Courant not immediately attracted to Klein?*" Friedrichs mused. "For Hilbert's way was not natural to Courant, as Klein's was. Of course Hilbert was *the* mathematician in Göttingen—at that time everyone gathered around Hilbert—but I think the fact that Courant was so very attracted to Hilbert has something to do with the way he was often drawn to people who were personally and mathematically completely different from him. He always had unbounded admiration for such people. In regard to Klein he felt, I think—'that I can do myself; but Hilbert, now there is something really different!'"

From the time of Courant's arrival in Göttingen, he expected to get his degree with Hilbert.

"But one must think of the drawbacks in working with Hilbert . . . ," he wrote judiciously to Nelly in his first weeks at the university. "He doesn't bother very much about the people who work with him, and people other than his assistants rarely have the privilege of receiving direct stimulation from him. Everybody says he hands out doctoral topics without regard to the personality and preparation of the students in question so that out of ten dissertations seven go completely wrong and of the other three, generally two are discovered after promotion to have been incorrect. However, since I always have firm mathematical support here in Toeplitz and also Hellinger, I don't think the danger is too great that I will get myself involved in an unfruitful topic."

As it happened, the topic which Hilbert did suggest to Courant for his dissertation turned out to be so remarkably suited to Courant's personality and to his mode of mathematical thought that it was to run, in the German expression, "like a red thread" through his life's work. This was *Dirichlet's principle*.

I asked Friedrichs why he thought that Dirichlet's principle had such a powerful attraction for Courant.

"It has esthetic appeal," he replied promptly. "There is a solution of a problem that is characterized by the fact that a certain integral, which is a very simple and clear-cut one, assumes a minimum for that solution. This fascinated him. He found it so convincing and so beautiful."

Already—in Courant's student days—Dirichlet's principle had more than half a century of high intellectual drama behind it; and for the rest of his life Courant was always to enjoy recounting its history in talks and papers meant for general audiences. He even devoted a section to it in his book *What Is Mathematics?*, causing the American mathematician E. T. Bell to remark that it was probably the first time in the history of mathematics that Dirichlet's principle had appeared in an elementary mathematical book.

The story of Dirichlet's principle, as Courant always told it, began with Gauss, who in his proof of the fundamental theorem of algebra made it clear that the first step in solving a mathematical equation is to establish that a solution of the equation does in fact exist. In the case of an algebraic equation, this step is relatively simple; but in the case of a differential equation, it can be a very knotty problem. It was for the solving of such a problem that the mode of reasoning later to be identified as Dirichlet's principle was first employed.

The problem concerned a differential equation, known as the Laplace equation, which is of fundamental importance in algebraic geometry and in mathematical physics. The solutions of this equation characterize the so-called harmonic surfaces, and the problem is to determine a harmonic surface that is bounded by a given contour. But not only is it difficult to determine the desired surface—it is also not at all certain that such a surface exists.

During the time of Gauss, however, it was observed that there is a certain integral which can be formed for any sufficiently smooth surface. This integral has the property that its value for a harmonic surface bounded by a given contour is smaller than for any other such surface. Conversely, any surface that has the given boundary and minimizes the integral is harmonic. It seemed clear that because of the positive character of the integral it would assume a minimum; and thus the existence of the desired surface—and hence a solution of the equation—was considered to be established.

In the lectures of Dirichlet, who succeeded Gauss in Göttingen, the young Riemann became acquainted with this reasoning. Impressed—as Courant said—"by its elegant simplicity and what seemed to him its inherent truth," Riemann utilized the same reasoning for the proof of a number of the fundamental theorems of geometric function theory and the theory of Abelian functions. And he gave it the name *Dirichlet's principle*.

Some time after the publication of Riemann's works, Weierstrass, one of the leading German mathematicians of the day, objected to Riemann's assumption that the "Dirichlet integral" necessarily assumes a minimum. Weierstrass's objection was the following. Since the positive character of the integral implies the existence of a lower bound to its values, Riemann was taking it for granted that this bound was a proper minimum—in other words, a minimum that would in fact be assumed by one of the admissible surfaces, or functions. But, Weierstrass pointed out, while continuous functions of a finite number of variables always possess a least value in a closed domain, there may be minimum problems of a more involved kind for which the relevant integral does not assume a minimum —even though a lower bound exists.

Riemann recognized the validity of Weierstrass's criticism. Still, he was intuitively convinced of the truth of Dirichlet's principle and of the truth of his theorems, the proof of which rested on the postulation of the principle. He died in 1866 at the age of forty without having answered Weierstrass's criticism. Shortly afterwards, Weierstrass was able to produce a very delicate and special example of a minimum problem for which the integral does not assume a minimum in spite of having a greatest lower bound.

By this time many mathematicians felt that they simply could not give up the results which had been gained by Riemann's methods. They tried to shore up Dirichlet's principle as best they could; and when they were unsuccessful, they invented various ad hoc ways of proving the important existence theorems which Riemann had thought to be derived so simply by means of the principle.

This was the situation when—almost half a century after the original work of Riemann—Hilbert "rehabilitated" Dirichlet's principle. He did this by proving *directly* that the minimum problem in question does in fact have a solution *by first finding the appropriate surface and then verifying that for it the integral assumes a minimum.*

For Courant this direct approach to the problem was always to be among the deepest and most powerful of Hilbert's mathematical achievements.

"To attack what seemed the absolutely inaccessible problem of salvaging the Dirichlet principle required the complete lack of bias and the freedom from the pressure of tradition which are characteristic of truly great scientists," he wrote on Hilbert's sixtieth birthday in 1922. "Hilbert possessed the courage, he tried and he succeeded. Completely novel and sophisticated methods of the most refined and penetrating insight were applied, and the reader of the work had to struggle laboriously to comprehend. But the great final goal had been achieved. A new weapon had been shaped by means of which an actual method of proof could be created out of the Dirichlet principle."

It happened that in 1909, when Courant was looking for a topic for his dissertation, there was a great deal of interest among the young people around Hilbert in the uniformization of Riemann surfaces and in the conformal mapping of such a surface onto a kind of standard domain. An important breakthrough in these problems had recently been made by Paul Koebe, a somewhat older man who was a *Privatdozent* in Göttingen at the time. By proving the uniformization theorem, Koebe had established conditions under which certain Riemann surfaces could be mapped one-to-one conformally on certain other domains. Hilbert, who had recently become interested in this general area, suggested to Courant that in his dissertation he should try to apply Dirichlet's principle to the same problems.

FOUR

COURANT'S dissertation was entitled "On the application of Dirichlet's principle to the problems of conformal mapping." In it he was able to modify and simplify Hilbert's approach to the principle and to apply it to certain fundamental problems of geometric function theory concerning, among other things, the conformal mapping of Riemann surfaces of higher genus. He was able to prove in a new and different way the uniformization theorem which Koebe had proved in 1908. He was also able to prove another important theorem, known as the slit theorem, which had been stated very generally in 1909 by Hilbert. He developed as well an estimate which later led him to the statement of a lemma that was to become one of his most powerful and most frequently used tools.

Unfortunately, even before he had finished writing this maiden paper, Courant's satisfaction in it was marred in a disagreeable way.

In the published dissertation (1910), there are a number of references to Koebe and his work in the footnotes. One of these is unusual.

"While I was writing this paper," Courant noted on page 5, "there appeared ... a paper by Mr. Koebe, the essential content of which is the proof of this same theorem, which was first formulated by Hilbert as a general theorem. Mr. Koebe's proof is also based on Hilbert's idea of employing the minimum property of the flow potential and agrees further with my proof of the convergence of these potentials. Therefore, I would like to emphasize here that I have found my proof myself and independently of Mr. Koebe and have so informed Geheimrat Hilbert of this in June 1909."

I asked Courant for the story behind the footnote.

"It was one of those unpleasant experiences," he said. "Koebe was already well established, and very knowing, and he could not stand having someone else do something in his field. One day he asked me—I remember it very well—what I had done in my thesis. He asked me on the street in Göttingen. We had quite a walk and I told him what I had done and then he said, 'Is that all?' I was somewhat taken aback. Then in a few days he came to me and said he had done the same thing in another way, which was more closely connected to what he had done before."

The following week, when Courant was scheduled to talk on his work before the *mathematische Gesellschaft*, Koebe announced at the beginning of the meeting that he would also like to talk on the same subject.

"Well, we have two gentlemen who wish to talk," said Klein, who was the chairman. "Who will talk first?"

Koebe promptly rose to his feet and said, "I will."

Courant could not object, since Koebe was senior to him. The other members of the *Gesellschaft*, however, were very disapproving of Koebe's behavior. Courant was six years younger and a beginner. He had written a very nice dissertation, and he should have had the privilege of presenting his results first.

Some of Courant's friends—one of them was Kurt Hahn, who later became internationally known as the founder of famous schools in Germany and Scotland—decided to pay Koebe back for his treatment of Courant. They rigged up a sophisticated mechanism which would set off an alarm at irregular intervals and hid this in a chamber pot under the lectern before one of Koebe's lectures. There was much laughter at Koebe's expense when he finally located the offending mechanism and drew it out in its container. Hahn then further publicized the prank with an article in a local newspaper.

I have been told by people other than Courant that Koebe was considered a conceited and disagreeable man with a reputation for picking up the ideas of younger people and, because he was so quick, being able to finalize and publish them first. He was, nevertheless, an outstanding mathematician; and at the end of the summer semester 1909 he was called to a professorship at another university and left Göttingen.

After Hilbert approved the dissertation, nothing stood between Courant and the coveted degree but the oral examination. This was set for six o'clock in the evening on February 16, 1910. The three examiners were Hilbert, for mathematics; Voigt, for physics; and Husserl, for philosophy.

Hilbert arrived on time and before the others. He was eager to get on with the examination and go home. But Voigt and Husserl did not appear.

"Well, well," Hilbert said at last to Courant, "I could of course start now and ask you some questions; but I know you so well, talking to you about mathematics two or three times a week, why should I ask you questions? Of course I could do that. Ja. But it would have no point."

So he sat and chatted with Courant about various non-mathematical subjects, looked frequently at his watch, wished aloud that Voigt or Husserl would come—"so that they could ask you some questions"—and then chatted some more about subjects of mutual interest and people they knew.

At last, forty minutes after the appointed time, Husserl arrived.

"Oh, Husserl," Hilbert said with a great show of relief, "now you are here so I can stop asking Mr. Courant questions."

He promptly excused himself and went home to dinner.

Husserl asked Courant one question. Before he could ask another, Courant ask him to explain a delicate point of phenomenology. Husserl talked until the time allotted for the examination was up. Voigt never did arrive.

After the oral examination and the granting of the Ph.D., Kurt Hahn decided to bring Courant's new status to the attention of the community. He and two other students rented a small horse-drawn carriage and equipped themselves with megaphones. Then, with the embarrassed Courant shrunk down in the back seat to make himself as nearly invisible as possible, they proceeded to drive around the town, trumpeting to the inhabitants of Göttingen through the quiet night, "Dr. Richard Courant *summa cum laude*!"

The next professional step was the *Habilitation*, which carried with it the license to deliver lectures as a *Privatdozent* and, although not paid a salary by the university, to collect fees from students. The decision to take this step did not lie with the young doctor but with the faculty. Klein or Hilbert took one aside—perhaps on the annual picnic and hike for mathematics faculty and students—and suggested that one might like to stay in Göttingen.

The number of private lecturers was carefully controlled. When Courant received his degree, there were seven for mathematics. Until some of these moved on to positions at other universities, a job as assistant to Hilbert looked very good indeed to Courant; and he decided to hold onto it.

There were beginning to be signs that Hilbert was moving into a new area of research. During the winter semester 1909–10, he had lectured exclusively on differential equations, the principal mathematical tools of physics; and during the summer semester 1910, he continued to concentrate on these in his lectures.

As assistant to Hilbert, Courant helped as in the past with the preparation of the lectures and took notes during their delivery. He produced hardly any mathematical work.

"I was very often depressed," he said to me, explaining a rather melancholy picture of himself taken at the time, and added, because he was by then many times a grandfather, "We forget that these things are not new."

During the late summer of 1910, he spent some time in Berlin at the home of his friend Kurt Hahn. His own parents were not in the capital. According to Siegmund's memoirs, they had been "exiled" by a younger brother, Eugen, and lived in another town with distant relatives who were under orders from Eugen to give them "one mark a day" to live on.

Courant had recently received orders to report for the compulsory year of military service. The friendship between him and Nelly Neumann had ripened into what both of them thought was love. He hoped that when he finished his army duty, he would become a *Privatdozent* in Göttingen. Then he and Nelly would be married.

On October 2, 1910, on the eve of his induction into the army, he decided to begin a diary, both as a record of his life "for Nelly and our children" and as a tool for self-examination and self-improvement.

"The decisive experiences of my life are behind me," he wrote in his student's notebook. "The course which it will take is in all probability determined within definite boundaries...."

When he looked back over the past six or seven years, he found that he was content "with what fate has bestowed on me in love, in the friendship of people, and in opportunities. And if I am still not a happy and satisfied man—if I still think that I have not made of myself what I could have made—then the reason is not the fact that I have not been spared bitter experiences and worse circumstances than most people have ever known. It is me alone.

"During these last important years I have trotted through the world as if with blinders on. I have become a scholar who does not understand much about his chosen science—who does not possess the disdain of many scholarly souls for all that is not known to them, but who still truly knows little about a thousand things about which he should know much...."

He wasted too much time talking with people. "And I will always place the human far above the intellectual!" But he hoped that keeping a journal would help him achieve the self-discipline he felt he lacked and would also help him improve his "frighteningly bad" memory for facts and experiences, "which deeply disturbs me."

It was comforting that Nelly knew his weaknesses and that they did not affect her love for him. "But no one wishes more than she that things get better."

The moment of induction into the army seemed to him propitious for such bettering, more propitious than those times in the past when he had earlier made such resolutions. "This time from outside an actual break has been made in my life. From tomorrow on, strict discipline will be applied to me. Now I must pull myself together so that it is also thus within me."

Perhaps to his own surprise, Courant found that he was more suited to the soldier's life than a young scholar might reasonably have expected himself to be. He had no difficulty in relating to the other men, although most of them were not university people but farmers from the countryside surrounding Göttingen. During his year of service he became a non-commissioned officer. But the military life was not moving him ahead in his academic career. He was envious of fellow students who did not have to share the burden. He visited Göttingen whenever he had the opportunity. He hoped to be asked to return for *Habilitation*, but there was no suggestion to that effect from either Klein or Hilbert. Then in May 1911, while he was still on duty with the reserve, he received a letter from Kurt Hensel, who was a professor at Marburg, proposing that he come there and, while preparing the thesis required for *Habilitation*, assist Hensel with the editing of the collected works of Kronecker.

Courant put off replying to Hensel's letter until, on leave, he could communicate the Marburg proposal in Göttingen. When after three weeks he still had not given Hensel an answer, he received a letter from Hellinger, then a *Privatdozent* in Marburg, scolding him for his delay. By this time Hilbert had suggested that Courant return to Göttingen. Courant wrote promptly to Hellinger and enclosed a copy of the letter he planned to send to Hensel.

"The idea of *Habilitation* in Marburg has always been tempting to me," Courant had written in his letter to Hensel, "and, therefore, your suggestion in this connection coincided with my own wish. I had looked forward very much to the activities in Marburg and particularly to the cooperation with you in connection with the editing of Kronecker's works. But suddenly now there has been a chance to remain in Göttingen, where I can get my *Habilitation* without difficulty during the winter semester."

Hellinger firmly noted on the letter: "To be changed—it sounds as if you were going to Marburg for lack of someplace better to go."

In the next paragraph Courant had spelled out the attractions of Göttingen.

"Geheimrat Hilbert promises me that at least once a year I will get to give one of the main big four-hour lectures. Because of the greater number of students in Göttingen, which will presumably not decline in the coming years, the financial side of the matter is very favorable. Even though I will have to do with considerably less income than in the old Göttingen three-lecturer arrangement, my annual income will nevertheless be considerably higher than what I could expect it to be in Marburg. Apart from my many scientific and personal connections with Göttingen—particularly with my highly esteemed teacher Hilbert—the financial consideration has been the deciding factor. I would regret it very much if this decison of mine would cause an unwelcome disturbance in your disposition."

Hellinger wrote disgustedly: "Express yourself less patronizingly."

"In any case," Courant had continued, "I ask you not to hold my decision against me. After considerable consultation with my friends and with Hilbert, I have come to the firm conclusion that I should not turn down the offer of *Habilitation* in Göttingen."

Hellinger crossed out "my friends" and wrote, "Leave them out of this."

When Hensel received Courant's letter, he replied cordially, "I can understand why you would wish to remain close to Mr. Hilbert."

In the fall of 1911, his year of military service behind him, the prospect of *Habilitation* in Göttingen ahead, Courant became engaged to Nelly Neumann.

Nelly had received her doctor's degree in Breslau the year before Courant took his in Göttingen. Her subject was a topic in geometry. Afterwards she had taken the examination and written the additional thesis required of prospective gymnasium teachers. Although "talkative and gay," she had grown into a woman who was "very thorough," according to Edith Stein. She was "particularly interested in ethical questions" and "never undertook to do anything without having carefully weighed all the pros and cons."

Courant felt that by bringing himself up to Nelly's high standards he would become the person he wanted to be. On the other hand, although she was two years older than he, he felt very protective toward her. He worried about her living habits and her health. She was inclined to drive herself. One of his letters to her ends with the exhortation: "Don't forget to swim, do gymnastics, work in the garden, take walks!"

In the journal which he had begun on the eve of his induction into the army, he had written of their relationship:

"Nelly knows every smallest and most secret thing about me, and it is wonderful and reassuring that I need to conceal nothing from her. Even though she disapproves of much, that does not affect her love. It is completely independent of such things."

The engagement was not yet formally announced, and he told nobody in Göttingen about it except Mrs. Hilbert and Mrs. Landau.

"Mrs. Hilbert said that as soon as it is official, you must come to visit me. Let's hope that will be at Christmas!"

The time between his discharge and his *Habilitation* was again an economically difficult one for Courant. During his military service he had regularly sent 120 marks a month to his parents and this had enabled them to return to Berlin. Now, no longer able to help them himself, he negotiated an allowance for them from the family business in Lublinitz. Part of the agreement was that the money was to go through him and that "every penny" was to be accounted for.

The position of Hilbert's assistant was currently being held by Erich Hecke, whom Courant had recommended for the job before leaving for the army; but Hilbert needed some additional help with his book on the foundations of a general theory of integral equations. Courant read and corrected proofs. He also found a couple of mathematics students to tutor.

"My dear, dear girl," he wrote to Nelly on the eve of the new semester, ". . . you can forgive and forget my transgressions against you, for I love you far better than it seems at times. . . . I am going to really try to conquer all the faults which you reprove me for, and with your help I will be successful."

This letter to Nelly is the only one of the period that Courant kept. The other letters I have seen are all from the first few months when he came to Göttingen. He may have kept these particular letters because they contained descriptions of people and events he considered of historical interest. As he told me in our first conversation, he had always intended to write something about the scientific life in Göttingen. This letter, dated October 27, 1911, in addition to describing a busy, bustling trip to Berlin, reports on events in the philosophical circles of Göttingen.

At this time Courant was dining once a week with Leonard Nelson, who was at last a *Privatdozent*. Nelson attracted many young people and was once described by an admirer as coming nearer to Socrates than any other modern. He wanted to establish by rigorous reasoning a system of philosophy in which everything had its determined place and every essential question was answered. He spoke of a "philosophy of right" and tried to lead his followers to a clarification and a critical examination of their own convictions. He then required them to carry out in their actions what they had recognized as just and good.

"Ethics," he told them, "is there to be applied."

Philosophically, Courant found Nelson more of "a medieval dogmatist" than a Socrates.

"Were you ever a follower of his?"

"No," he replied. "I could not stand his philosophy. I was a friend. You could not be a friend *and* a follower."

Courant's letter to Nelly also describes the problems attendant upon his proposed *Habilitation*. Now that he was back in Göttingen, the arrangements for that step did not seem to be nearly so certain as Hilbert had implied.

"Hilbert has talked with other people here, and there may be difficulties because some of them think I should perhaps have written more papers. Still, according to Hilbert, everybody wants me to be ready to start to lecture at Easter. Hilbert put it this way: 'We hope that you will have produced by then a beautiful *Habilitation* thesis.' Well. We shall see."

The winter of 1911–12, when Courant was preparing his *Habilitation* thesis, was marked by a high-spirited celebration of Hilbert's fiftieth birthday on January 23. Assistants and advanced students composed a number of verses for the occasion. One set of these described the "creation" by Klein of the position of Hilbert's assistant and the many duties which were performed for 50 marks a month and *Familienverkehr*. It concluded with the creation by Hilbert's current assistant, Hecke—"a practical man"—of the position of "assistant to the assistant." This luckless fellow did all the work but did not receive the 50 marks: "His pay for this is—ach, how nice—/ *Familienverkehr*/ And, besides, Hilbert as his boss—/ O heart, what wouldst thou more!"

The highlight of the celebration was a tour de force which devoted a couplet to each of Hilbert's romantic attachments—usually the young wife of some colleague. Beginning with Ada, Bertha and Clara, it proceeded triumphantly through the alphabet, stumbling only over Q, U, X, Y, and Z. Mrs. Hilbert was not forgotten: "God be thanked that Käthe, his wife/ Takes not too seriously his life."

(When I was writing *Hilbert*, I questioned Courant about the nature of Hilbert's relationships with his many "flames," as they were called. One mathematician had insisted to me that they were indeed affairs. "That man!" Courant said disgustedly. "He is no gentleman. He doesn't *understand* such things!")

For his *Habilitation* thesis, Courant returned to Dirichlet's principle. In a paper entitled "On the method of the Dirichlet principle," he gave a proof of the validity of the principle by means of a method which was close to Hilbert's second method but avoided various complications which were due to the great generality of Hilbert's work.

The next step in *Habilitation* was the delivery of an inaugural lecture before the Philosophical Faculty. According to custom, Courant submitted three topics of his own choice. By the time I talked to him, he had forgotten two of the topics; but he remembered the one approved by the faculty— "On existence proofs in mathematics." This was the subject on which he had personally most wanted to talk.

The interest in existence proofs was to run, with Dirichlet's principle, all through his work.

"Even the book I am trying to do now," he told me, referring to a third volume of Courant-Hilbert which was never to be completed, "is in a way about existence proofs."

Courant's inaugural lecture was delivered (without notes, as was the custom) on February 23, 1912. The dean of the faculty wrote to him on the same day, congratulating him and announcing that he had been granted the license to lecture in Göttingen.

At Easter, when the new semester began, he delivered his first lectures; and he and Nelly Neumann set a date for their marriage.

FIVE

WHEN, after the long summer vacation of 1912, Dr. Courant returned to Göttingen with a wife, it was an occasion for some surprise, especially since she was two years older than he, and rich.

"We always had the impression that their only common interest was mathematics," explained his former pupil Magda Frei, who was by that time also a young wife in Göttingen. "Nelly was the best in our class in mathematics, but not the best in everything. She never seemed to me like the marrying kind. It was a little—well, I thought at the time he probably felt 'obliged.' There was the question, could he live on what he made tutoring. I always thought her father must have helped him."

The newly married couple rented an apartment on Schillerstrasse, and Courant began to try to establish there the kind of "wide open" academic home to which he had been introduced in Göttingen. He had always admired the way in which Mrs. Husserl entertained, her housekeeping skills and cooking ability as well as the conversation at her table. The philosopher was still not a full professor, and his family was less well off than most. All was done frugally but with charm. One of the first people Courant invited to meet his new wife was the Husserls' daughter Elli, a student of art history at the university.

"I still remember it," she told me over tea at her home near Cambridge. "They lived a little outside the town. That was a time when, I think, there wasn't a single bus in Göttingen, so I walked there. And the tea—it was just very nice, but it wasn't the kind of lively atmosphere to which I was accustomed. He obviously tried his best. I remember only that one visit. It was not a roaring success. It must have been very hard for her."

In spite of Richard's efforts, Nelly was lonely. She suggested to Aunt Auguste Stein that she send her daughter Edith, who was already interested in philosophy, to study with Husserl in Göttingen. "Richard has many friends whom he has brought into our marriage, but few women friends. This would be some compensation for me." She repeated the suggestion when she was in Breslau a few months after the wedding. She also described her first adventures in housekeeping. "Things become increasingly more complicated the farther they are removed from mathematics," she observed, "and the household is the farthest removed."

As a result of Nelly's urging, Aunt Auguste permitted Edith and her older sister Rose to go to Göttingen in the spring of 1913. They were immediately made welcome by the young Courants. "Without wanting it to be true, Richard was still greatly attached to the family," Edith noted, "and always asked me about every member."

Unlike the more sophisticated Elli Husserl, Edith found the apartment in Schillerstrasse a place of easy hospitality: "Richard loved to bring unannounced guests. He had a great circle of friends, other instructors and older students; but he also liked to bring home some of his own students, young men and women with whom he wanted to discuss matters."

To Edith it seemed that "his wife's fortune" had provided her cousin Richard for the first time with "an existence free from worry and full of the youthful unstinted joy of life."

During the years from 1912 to 1914 when Courant was a young teacher in Göttingen, the famous faculty of his student days was beginning to change.

For most of 1912, Felix Klein was in a sanatorium in the nearby Harz Mountains. He continued to keep in touch with events at the university and to direct them; but at the end of the year, although he was only sixty-three, he wrote to the ministry and asked to be made emeritus.

In 1913 Constantin Carathéodory succeeded to the chair of Klein. Carathéodory was one of the many young men destined to become important scientists who received their degrees in Göttingen between 1900 and 1910. Among the others, in order of time, were Max Dehn, Felix Bernstein, Georg Hamel, Rudolf Fueter, Oliver Kellogg, Erhard Schmidt, Max Born, Ernst Hellinger, Alfred Haar, Hermann Weyl, Richard Courant, and Erich Hecke. If Klein, passing his sixtieth birthday in 1909—at the time of Poincaré's visit—had begun to look among these younger men for the one who would bring to reality his own dream of a great mathematics institute in Göttingen, it is unlikely that he would have perceived him in Courant. Most probably Klein's selection would have been Carathéodory, the scion of a powerful and wealthy Greek family, a man of unusual culture and sophistication, somewhat older than his fellow students since he had been an engineer before he decided to become a mathematician; one who in his doctoral dissertation in the calculus of variations had shown exceptional power and originality; a tall, impressive, aristocratic-looking man, although unfortunately cross-eyed. As a further recommendation for Klein, Carathéodory had an interest in educational problems. He had attended the École Polytechnique in Brussels, founded in the spirit of the École Polytechnique in Paris—a school which had revolutionized scientific education in a way which Klein much admired. In short, Carathéodory was a natural to occupy Klein's chair and his return to Göttingen was universally approved.

The scientific atmosphere of the university was also changing during Courant's days as a young teacher.

In 1912 Hilbert turned his attention from mathematics to physics with his well-known pronouncement that physics was much too hard for physicists. He asked his friend Arnold Sommerfeld in Munich to send an advanced student to Göttingen to be his physics assistant. The first such young man sent by Sommerfeld—there was to be a series that extended through the 1920's—was Paul Ewald.

Ewald and his wife, Ella, were among the few people I was able to interview who had known Courant in his youth. They remembered him as being "serious" but, as Ewald said, "having the wonderful gift of making fun of himself—this typically Jewish attitude of considering oneself as a little poor boy up against the big people on top." There was also a certain "eastern" Jewishness about him, although he was not really an eastern Jew, which—in Ewald's opinion—charmed the western "integrated" Jews of Göttingen.

After Ewald became "Hilbert's tutor for physics" (as he was promptly nicknamed), Hilbert began to look for an outstanding physicist for the faculty. He found such a person in a young Dutchman named Peter Debye, who had been Sommerfeld's first assistant in Munich; and he personally negotiated Debye's appointment to a professorship in Göttingen.

With Hilbert's new interest in physics, important physicists began to be invited to report on the latest discoveries by themselves and their students. One of the first of these was Sommerfeld himself, who came in 1912 to tell about the work of von Laue and others on X-ray diffraction in crystals, a work made possible by mathematical formulations in Ewald's not yet completed doctoral dissertation.

It may have been on this occasion that Sommerfeld talked to Courant about the quality of books on mathematical physics. He said that the older generation had produced some good books but that now something new was needed. Since his own student days, Courant had been aware of the deficiencies of most of the textbooks available, both in mathematics and mathematical physics. He had thought seriously that some day "when he grew up" he might do something about them. He did not forget Sommerfeld's remarks.

During this period Courant's fellow instructors in Göttingen included Max Born, for theoretical physics; Toeplitz, Weyl and Hecke, for mathematics; and Theodor Kármán, for mechanics (the *von*, "the sign of nobility," as Courant always liked to explain, came later).

Born was from Breslau. Like Courant, he had come to "the mecca of mathematics" on the recommendation of Toeplitz and Hellinger, but even before they themselves had come. He had been Hilbert's assistant in 1905–06 and had taken his degree in 1907. After Minkowski's death he had been asked by Hilbert to edit Minkowski's physics works.

Kármán had been introduced into the Göttingen circle by his countryman Alfred Haar, who had since become a professor in Hungary. With his vitality and charm, he had soon succeeded to Haar's position as the leader of what Courant still saw as the "in group." One of the most memorable non-scientific events of the period, at least for Courant, occurred when Kármán and the "outsider" Hermann Weyl, having both fallen in love with the beautiful and gifted Hella Joseph, got into a water fight over her. Hella had to make up her mind on the spot whose face she would wipe and, much to the chagrin of the "in group," she chose to wipe Weyl's.

Another foreign student who was an integral part of the Göttingen scientific community was Harald Bohr. At that time he was the better known of the Bohr brothers, having been a member of the Danish soccer team which had been the runner-up in the Olympics of 1908. He was also a promising mathematician and had come to Göttingen at the invitation of Landau. Everyone was attracted by his intelligence, worldliness, wide interests, and "radiance" (Courant's word) ; but Harald Bohr insisted that he was nothing in comparison with his older brother, "who was made of pure gold and would soon be recognized as one of the great scientists of our time." This excessive admiration for his brother was looked upon by most people in Göttingen as a foolish but endearingly modest trait in the otherwise wholly admirable Bohr. Courant, accepting Harald almost worshipfully, also accepted his faith in Niels Bohr.

All these young men were beginning to do work that attracted notice in the scientific world. Weyl published his lectures on Riemann surfaces. Hecke's dissertation on number theory was a significant piece of work and was followed by other important papers. Born and Kármán, rooming together in the legendary ElBoKaReBo (its name derived from the first syllables of the names of its five inhabitants), collaborated on a work in which they stated, independently of Einstein, the laws of specific heat. What Courant recalled sixty years later, however, was the impression made in Göttingen by two works from the "outside" world.

The first of these was Niels Bohr's revolutionary interpretation of the atom, which appeared in 1913—what has been called (by James Franck) "the birth hour of atomic physics." On a visit to Cambridge that same year, Courant had met "Harald's brother" and had heard him set forth his ideas in the quadrangle of Trinity.

"Thanks to prior suggestion by Harald, who had so often told me wonderful things about his brother, I was at that point immediately ready to believe that you must be right," he wrote later to Niels Bohr. "But when I then reported these things here in Göttingen, they laughed at me that I should take such fantasies seriously. Thus I became, so to speak, a martyr to the Bohr model. . . ."

The other work from the outside was that of a young American named George Birkhoff. According to Courant, when Poincaré had lectured in Göttingen, he had mentioned a certain theorem which he had been trying without success to establish. Later, although he still had not been able to prove it, he had published his work on the theorem in the hope that his results might at least be useful to others. After Poincaré's death in the summer of 1912, the theorem became "Poincaré's last theorem"—a challenge to mathematicians all over the world. In Göttingen the young men said, "That doesn't seem too hard—some rainy afternoon when we have nothing else to do we shall think about it and prove it." But the rains came and went, and even in Göttingen the last theorem of Henri Poincaré turned out to be as hard a nut to crack as it had been in Paris. Then, in the fall of 1913, a paper proving the theorem was reviewed in the *Fortschritte der Mathematik*. It had been published in what Courant always referred to as "some obscure American journal." It seems that all American journals were considered obscure in Göttingen. Overnight everyone was talking excitedly about the author, *der Amerikaner Birkhoff*.

Courant

Many Americans had studied in Göttingen; and although they were well liked and sometimes did mathematical things which people admired, they were not considered quite "top drawer" as mathematicians. In 1913, for the first time in Courant's memory, the Göttingers looked with admiration across the ocean.

In Göttingen during this period, the general direction of the mathematics and physics activities was still in the hands of Klein. As an emeritus professor he was relieved of "duties" but not of "rights and privileges." He continued to be—as he had been for more than a quarter of a century—the strongest directing force in the faculty, the most influential academic voice at the ministry in Berlin, and the representative of the university in relation to industry.

In a few years the *Vereinigung zur Förderung der Angewandten Physik und Mathematik*, the association of businessmen and industrialists which Klein had organized to support science at the university, was to celebrate its twentieth anniversary. In 1913, in preparation for a history of the association which was to be published at that time, Klein went back over his professional life, year by year. His notes, based on letters he had written at the time, are among his papers in Göttingen. In them one can glimpse the exertions, the developing strains, the illnesses which led to the breakdown that effectively ended his career as a creative mathematician at the age of thirty-three. It is clear from these notes that he recognized immediately the effect that his illness would have. "From now on," he wrote, "social effectiveness will have to substitute for lost genius." The notes also show how early the outlines for the development in Göttingen began to form in Klein's mind.

In 1914 the idea for an institute of mathematics in close physical and intellectual contact with the applications—which was to be the climax of Klein's endeavors—was on the verge of being realized. Land next to the physics institute had been given to the university, plans for the building had been drawn up, construction was ready to begin.

In later years Courant was always to recall the pre-war period in Göttingen as idyllic. The whole mathematics group—faculty and students—was "somewhat like a family." The university was "an ideal place for young scholars searching for eternal values without taking themselves too seriously"—the struggle "tempered by healthy enjoyment of an unpretentious but meaningful social life, based on real human contacts and the beautiful surroundings of the town." But in spite of the generally happy situation, Courant could not help seeing that with the appointment of Carathéodory to Klein's chair, opportunities for advancement to a professorship in Göttingen were effectively closed off. The chair of Runge, the oldest of the other three mathematics professors, would not be vacant until 1925. Hilbert would not retire until 1930, and Landau not until 1945. It was clear to Courant that he and his contemporaries would have to look to other universities for their professorships.

Courant had produced relatively little mathematics between 1912 and 1914. For the most part he had continued to be fascinated by Dirichlet's principle. There had been two more papers connected with it. One of these was especially significant because in it he stated the important lemma which had been implicit in his dissertation. In 1912 he and Born had collaborated on a paper concerning the result of Eötvös; and in 1913 he and Harald Bohr had written a paper together on the application of the theory of diophantine approximation to the zeta function of Riemann.

These mathematical works were not nearly so impressive in quality or quantity as those of some of Courant's contemporaries, most notably Weyl, who was already a professor in Zurich. However, Courant was considered for a position as an associate professor at Frankfurt; and his name appeared on the list of three sent by the Frankfurt faculty to the ministry. He did not receive the appointment (which went to Hellinger), but just having one's name on such a list was a first step up the academic ladder. In the summer of 1914, in spite of the increasing number of military maneuvers in which he had to participate as a member of the reserve, he had reason to be optimistic about his academic future.

All this changed on July 30, 1914.

At four o'clock in the afternoon of that day, Edith Stein sat in her room reading Schopenhauer's *Die Welt als Wille and Vorstellung* in preparation for a lecture by Adolf Reinach, a popular and gifted member of the Husserl circle. (In his own room Reinach studied the atlas.) A knock on the door announced a friend, who told Edith that a declaration of war was momentarily expected and that all lectures were being suspended until further notice. Almost immediately there was a second knock. It was Nelly Courant with the news that Richard had received his orders. She was leaving that night for Breslau, where she would wait with her father for the end of the war. Everyone, including Richard, was sure it would be over soon.

It was by now five o'clock, the hour for which Reinach's lecture had been scheduled. Edith closed *Die Welt als Wille und Vorstellung*, "which I never opened again," and began to pack. At seven-thirty she was at the Courant apartment. The train was to leave at eight. The carriage was already in front of the door. Nelly and Richard, who was going part of the way with them, were saying goodbye to each other in his study. "And they were not too quick," Edith reported. She was not yet aware that her cousin and his rigid, highly organized wife were finding life together increasingly difficult. Still, it was astonishing to her that Nelly would go back to Breslau before Richard had left for the war.

"In her place I would not have done so. But it was probably due to her concern about her father. And then also she was not in general like other people."

After accompanying his wife and his cousin as far as Kassel and placing them on the Breslau train, Courant returned to Göttingen to await mobilization. The next day the Kaiser's proclamation of war was posted on every street corner. Parents sent their children to look at it. Boys not out of the gymnasium stood in line to enlist. They had but one fear: that before they could see action, the war would be over. Although Courant was twenty-six, he was as eager to go as any boy—patriotically convinced of the right of Germany's cause and confident that the Fatherland would be victorious "even over a world of enemies."

War fever raged in the university town as virulently as it raged everywhere in Germany. Winthrop Bell, a Canadian student of philosophy whom Courant considered "the nicest person in Göttingen," was incarcerated as an enemy alien in the university "jail" on the third floor of the auditorium building. In the face of Klein's disapproval, his youngest and favorite daughter, Elisabeth, popularly known as "Putti," married Robert Staiger, a doctor of musicology and the conductor of the academic orchestra, just as he was about to leave for military service. The two Runge sons, neither of whom had yet completed the gymnasium, rushed to enlist and waited impatiently for the moment when they would don their field-gray uniforms. The two sons of the philosopher Husserl also enlisted.

There were, of course, some in Göttingen who were not infected by the fever. Hilbert saw no reason to swerve from his program of straightening out the physicists, and he continued to plan with Debye what was to become the famous seminar on the structure of matter. Runge, as the rector of the university, objected strenuously to the treatment accorded Winthrop Bell and insisted on transferring him to his own home, announcing that he would be personally responsible for him.

It was at this time, during the first days of war, that a neighbor woman was surprised to see Runge pacing up and down in her sheltered garden, obviously much upset. He told her that he could accept the fact that his sons, who were still only boys, should be swept up into the army; but it seemed to him wickedly foolish that the lives of young men of proved ability, like Richard Courant, should be risked.

To Courant himself the day of his mobilization and the week that followed were like "a marvelous dream."

"I shall never forget," he wrote in a diary which he shortly began to keep, "how many great and beautiful qualities appeared in people at this time, how everybody was especially kind to me, how pleased I was to find acquaintances—particularly Staiger—among my new comrades, how we were outfitted, how we left Göttingen on Saturday, August 8 (in circumstances dramatic for me), how we traveled endlessly on trains, how we marched in the full moonlight...."

The diary was among the papers which Courant brought with him to America. It is contained in a lined notebook with a cardboard cover. The first entry is dated August 13, 1914.

After a long hard march, the regiment was spending a day near the Belgian border, resting, eating, swimming, sunbathing, and conversing. Courant, who was a non-commissioned officer with a rank equivalent to that of a staff sergeant, and his friend Walter Lohse, a tall dark-haired young man who played the violin in the academic orchestra, decided to shoot a few rounds from their pistols to remind themselves of "warlike activity."

Neither Adolf Reinach, nor Putti Klein's husband, nor the youngest sons of the Runges and the Husserls, nor Walter Lohse would survive the war; but Courant's diary begins:

"It seems to me that I am on a beautiful summer vacation."

SIX

AFTER AUGUST 14, Courant and his regiment were in Belgium, moving down the valley of the Meuse with the Second Army. Everywhere they saw dreadful devastation and found as a rule an intimidated population, willing and often eager to please them. But there were also instances of "stabs in the back" by the Belgians. Courant, noting these, added, "There are probably things on our side, too."

As he and his comrades proceeded farther into Belgium, sometimes quartered in villages from which all the inhabitants had been evacuated for fear that they might give away the German position, he found himself shocked by the destructiveness of his fellow soldiers. "Yet they are otherwise good people who are now playing the harmonium [they were quartered then in a deserted school] and singing chorales."

All day and all night he heard the sound of distant artillery fire. He wrote again and again, "If only I could be there!"

Quite soon, however, he and his company were detailed to another regiment and given the job of protecting a railway station at the rear.

"We must lie around this place and fill our stomachs while others bleed," he wrote disgustedly. "Perhaps our 91st Regiment has already been in battle!"

In the coming weeks, from time to time, his company was in positions where he expected to be called up for attack at any moment; but except for one rather unusual engagement, "in which not a shot was fired or a man lost," he saw no action. There was still more railway duty. At one point, when he was quartered in a customs house, he found a piano and began to play. He took innumerable photographs and developed them at leisure. Sometimes he found himself wondering if things were going as well as had been hoped. On September 20, 1914, he wrote: "Reports of victories which are passed on every few days remain without confirmation, but I have *every confidence* in our cause. It must succeed in the end. But I am afraid it will take much longer than anticipated."

News regarding promotions began to come through. Lohse's name had been proposed, but not Courant's. "The main thing against me is that I am considered too restless. . . . I must console myself with the knowledge that I have done my duty to the best of my ability."

In Maubeuge, which his company entered shortly after its surrender, he was shocked to see that many of the officers were shipping carloads of booty back to Germany. Some, selling cattle and coal to the inhabitants of the town, were distributing the money, "which belongs to the German State," among their men. His own company commander—"Mr. von H."— had sent for his wife. "[That] is humanly understandable, but should not occur. When she spoke to me, I felt real discomfort at her illegal presence."

Finally, on October 12, 1914, the order came for Courant's company to rejoin its regiment. The first night on the march they were quartered near Reims. Everything seemed especially beautiful to Courant. The landscape, the autumnal mood, the pretty little village. He and Lohse took a walk in the evening, and Lohse said thoughtfully, "One would really like to stay alive." Courant, recording the remark later in his diary, added in French, "We shall see."

The regiment had suffered heavy losses. They were alternately in the trenches and in the cellars of the village in front of Fort Nogent, which was under constant artillery fire. Courant found himself embarrassed by the contrast between what the regiment had done and the activity of his company in protecting railway stations, guarding prisoners, and collecting weapons.

At the reunion there was no action, however. He was reminded of children playing Indians. He was almost cozy as he sat in the dugout at a table before a fire and wrote in his diary, "I would be quite content if Mr. von H. did not systematically make life and work difficult for me." He felt constantly "pestered" by the commanding officer. "Unfortunately I am very easily disturbed in my mental peace by such antipathies," he wrote at another time. "The people with whom I live and work must be in harmony with me. Otherwise I am most uncomfortable."

In a few days, much to his surprise, his youngest brother, Ernst, suddenly appeared at the front. He had been working in Rumania but had rushed back to Germany to enlist. By chance he had been assigned to the same regiment.

Now that Courant was at last in the trenches, expecting action momentarily, he hoped that fate would give him an opportunity to distinguish himself. Instead, on November 1, 1914, less than two weeks after rejoining his regiment, he was taken to the nearby village with a high temperature. Two days later Lohse was brought from the trenches with similar symptoms. Both were transported to the field hospital, where it was determined that they were suffering from typhoid fever. Courant was delirious most of the time during the next three weeks; and Lohse, who was much less ill, stayed constantly with him. "Who knows whether I would still be alive without him?"

During his convalescence Courant "burned to get back to the front." There were two reasons for this desire, he confessed to himself: first, he found sitting in the rear "unbearable" and, second, "because of a little ambition on my part."

Finally, moustached and bearded, he was permitted to return to active duty on February 5, 1915. He felt that he came back "more serious, but gladly and eagerly.... Who knows what is still ahead?"

In a few days he and the regiment were on their way to the Argonne Forest.

While Courant's battalion was setting up its position to a constant "hellish concert," his brother Ernst's battalion lost more than 220 men in a single day. Courant went to look for his brother and to his relief found him well and unharmed.

"His company chief told me that he had deported himself 'quite heroically'.... I was very proud of the dear modest youngster, who did not wish to talk about his achievement."

Everyone seemed to be sick, at least with a cold.

"But in spite of my bad cold and torn boots, I am quite well. Also my nerves have suffered less than those of many others."

A week later, after spending two days and nights building a new position under heavy fire, without sleep and without cover, he was running a fever and not able to speak. Since Lohse had arrived to take his place, he allowed himself to be sent to the rear.

"What a strange feeling it was when I left the forest by field train and saw again houses and other people!"

He was in the hospital this time for ten days and had leisure to bring his diary up to date: "In the Argonne there is no soldier that is not fed up with the war.... But that doesn't mean that we are disheartened.... No German soldier would allow the thought of defeat to enter his heart."

A few days after he returned to his company, orders came to move on to Vauquois.

Vauquois was a beautiful little village on a promontory to the east of the Argonne Forest. It had been in the possession of the Germans for some time, but the French had made a series of violent attempts to regain it. Both sides currently shared the ridge. Losses were running to more than fifty per cent.

Courant had not forgotten how, leaving the forest for the hospital, he had looked up at Vauquois—"an ocean of smoke, lightning . . . the most violent artillery fire concentrated on a small height"—and had thought, "Thank God, I am not up there." Now his regiment waited below Vauquois for the order to attack.

"I am dulled and stupefied, look neither ahead or behind, do nothing except take a few photographs. . . . What seemed interesting and important and unusual to me only a few months ago is of absolutely no consequence now. The word 'culture' sounds almost like a sneer."

When Courant and his comrades arrived at Vauquois, they were deployed on the slope and given a few moments to inspect the situation. The demarcation line between the Germans and the French ran along the highest point of the ridge, where the church stood.

Three weeks later—after almost constant fighting—Courant described in his diary what he had seen in those first moments:

"The interior of the church was filled with rubble and bodies and parts of bodies, one corpse naked, burned by a flame-thrower. At the right of the church, the position curved around the famous linden tree toward the back trench, which . . . utilized the wall of the church cemetery as its bulwark. The connecting saphead led across two coffins. In the trench there were several corpses. The heels of one of ours were sticking out, the heads of others. . . . The most dreadful sight was an arm in field grey with a white hand which protruded from the right wall of the right saphead. Everywhere we dug, we dug into corpses. More of ours than of the French. Whenever a mine exploded, pieces of flesh fell into our position. Shreds of a dead Frenchman hung from the branches of a tree. . . ."

Courant did not again take up the diary which he had started at the beginning of the war. Instead, in August 1917, he completed the record of the rest of his active duty in summary form. The following details are taken from that account.

Courant

After almost three weeks' rest in the Alsace, the regiment was called up again and the company installed in an advanced position, surrounded on three sides by the English, west of La Bassée. They had no connection with the rear, since telephone lines were immediately put out of commission by artillery fire. During six days of fighting they lost more than half their men.

Describing these days—"None of us believed we would ever emerge from this hell!"—Courant remembered a night which he had shared with Lohse. Bayonetted rifles in hand, they had crept back and forth, stumbling over the dead and wounded, trying to encourage "and console" their men. In the course of this activity they had collided. Across the grisly scene they had smiled at each other "like augurers" and "without saying a word about the situation we were in, we sat for a while there in the trench and talked about Göttingen . . . until a bursting grenade dislodged a stone which knocked my helmet from my head and my gun from my hand."

It was during this period of active fighting that one of Courant's ideas for relieving the technical difficulties of trench warfare began to be put into operation. Some months earlier in Laon, looking at the cathedral against the horizon, he had had the thought that mirrors could be used to enable the men in the trenches to see what was going on above ground without exposing themselves to enemy fire. In La Bassée, as a result of his suggestion, trench mirrors in great quantities were being produced from requisitioned mirror glass.

Trench warfare had also made him realize how valuable it would be to have a means of communication that would not be vulnerable to artillery fire and could be operated under full cover. Telephone lines, he knew from experience, were the first casualties. The light signal system, which was also sometimes used, required a straight-line connection and also required that a man leave his cover. He began to think about resonance synchronization with electromagnetic signals which would utilize the earth as a conduit.

He talked about his idea to Lohse and his fellow officers. (He had finally been made a lieutenant after the hard fighting of May and June 1915.) Later he spoke to his battalion commander and found him also receptive. He was surprised how unusual even his small amount of technical knowledge seemed to be in the army. He suggested that he should be sent to Göttingen to consult with the scientists there about the feasibility of earth telegraphy, and he shortly received a leave.

An overnight journey took him to Hannover, where he would catch the train for Göttingen. Then he would see Nelly, who had returned to their apartment. He went for a walk to prepare himself "a little bit" for what awaited him and almost missed his train.

In Göttingen, he was relieved to find Nelly not in the apartment. Leaving his things, he went to the Runges' home. They had lost their youngest son in the first months of war, and they greeted the returning Courant like a son. As Courant had expected, Runge was able to make many helpful suggestions.

Runge's daughter Nina went to look for Nelly.

"They both returned together [and] I received a rather frosty greeting. The scene ended when Nelly and I took a carriage and returned to our apartment, where she had invited three maiden ladies to live with her during my absence."

He felt that a break with Nelly was imminent, but he nevertheless enjoyed his leave. After discussing his ideas for earth telegraphy with Runge, he went to Berlin to inquire at Telefunken, the German telephone company, about the possibility of using small wireless stations at the front. Count Arco, the head of the firm, was of the opinion that the necessary distances could be achieved only by antennas three to six feet high. He showed Courant some experimental models for earth telegraphy but pointed out that they could achieve a distance of 500 meters only with a great expenditure of energy.

With one day of leave remaining, Courant returned to Göttingen, very discouraged. To his delight he found that in his absence Runge had brought Peter Debye and his assistant, Paul Scherrer, into the project. The next day the four men worked together. By evening they had succeeded in sending a message from the physics institute to the Leine River, a distance of some 1500 meters. They had as yet no usable result, but they felt they had a sure expectation of a satisfactory solution to the problem. Courant returned to the front with a statement to that effect signed by Runge under the official seal of the university. He asked immediately for another leave so that he could carry on further experiments.

When he returned to Göttingen in August 1915, he learned from Debye—the first person he met on the street—that during his absence successful experiments had been performed with the pendulum transformer of the German telephone works. For the next few weeks he worked diligently with Runge, Debye, and Scherrer in the fields around Göttingen.

Nelly had left for Berlin, and he spent most of his free time at the Runge house, playing the piano with Nina Runge, who had studied the violin in Hamburg for the past few years.

"I was just amazed at the chutzpah with which he tackled anything that was attractive to him," Nina Runge, now Courant, told me. "You know what chutzpah is? Nerve! He never suffered from any inhibitions because something was too difficult for him, but tried everything. He somehow succeeded in making a piece intelligible, getting the spirit of it, no matter how imperfect his playing was. On the other hand, really practicing, finding a method of improving something that was too difficult for him—that he couldn't do. But I learned something from him. I made up my mind that pieces imperfectly played are not spoiled by this fact as long as one understands them. I learned not to be afraid of them. Why think of the listener? We played for ourselves alone!"

On the last day of Courant's leave, Scherrer presented him with a box he had built to hold the earth telegraphy apparatus, the result of three weeks' work. It could transmit messages over two kilometers. It seemed to Courant "a magic box," which would produce good prospects for his future military career.

After his return to his unit, he drafted a letter about the instrument to the General Command and, on his own, trained a few men to use it. He hoped to be able to arrange to have Runge and Debye come to the front to conduct further experiments. But days passed without a response to his letter. Then one morning while Courant was demonstrating his apparatus for some friends, General von Etzel, the brigade commander, passed by. He was so excited by Courant's idea that he placed him immediately in contact with the communications division. That day Courant wrote Runge an exuberant letter.

"Concerning the organization, I have a daring plan . . . ," he confided. "I want to set up a little factory behind the front to produce such magic boxes. This is not actually as crazy an idea as it may at first appear to be. First, we can easily obtain workmen here, mechanics, carpenters, electricians, materials, tools. Third [sic], we can avoid all bureaucratic difficulties. Fourth, we will not have to depend upon the goodwill of industry. Fifth, it will be cheaper. Sixth, the other factories will have time to do something else. The question is only whether *I* can manage to establish such a factory. . . . The military still seems to be less impressed by the earth telegraphy apparatus than by the fact that a young lieutenant can get a *Geheimrat* and a *Professor* to interest themselves in something which this lieutenant considers important!"

While Courant waited impatiently for his transfer to the communications division to arrive, artillery fire became more and more intense. It was clear that during the anticipated offensive his regiment was going to find itself in a "particularly uncomfortable" position.

He finally received his transfer on the same day that his regiment received orders to march. There was no question where his responsibility lay. All dreams of developing the earth telegraph seemed to have been wiped out; but he decided to take his magic box to the front, just in case it could be used.

The regiment marched to Douai, where it was held in readiness for the attack. Courant heard that Ernst had been taken to the hospital with what was thought to be appendicitis. He hastened to see him, found him not so ill as he had feared, and rushed back. In the hour he had been gone, the regiment had received its orders, each battalion marching off to a different railway station.

Courant took his magic box and went as fast as he could to the nearest station. The train left in the dark in a thunderstorm. When the troops disembarked, his battalion was not to be found. Taking up his pack, he set out in a soft rain. After awhile the rain stopped and the moon came out. An ambulance picked him up and took him to Sallaumines, where his battalion shortly appeared.

All night he stood watch. He had the impression that in the section where he was the fighting was over. But suddenly there came a call to march. Soon he and his comrades arrived in a region of utter confusion. They were given ammunition and told to leave behind any dispensable baggage, which must have included the magic box.

As they proceeded, they met almost no one except, in a dugout, a few men who told them that the English were quite close. It was so dark that they could see only a little distance ahead. Sudden machine-gun fire announced that they had made contact with the English front line. There was a brief exchange, but they were able to take cover without much loss.

In the daylight Courant made a tour of inspection. Corpses were everywhere, in some places so close together that he had to step on them to get through. In front of the position there were moaning English wounded who could not extricate themselves from the barbed-wire entanglements. Toward evening he was able to have some of them pulled out and sent to the rear. Among these were two Scotchmen, with whom he "became quite friendly."

He had had no sleep for thirty-six hours, and he curled up in a hole and slept, as deeply—he later recalled—as he had ever slept in his life. Early the following morning he was awakened from "the pleasantest dreams" by Lohse, who told him the battalion was to attack Fosse 8.

The advance proceeded at a fast clip across a wide plain; but there were still losses, and screams to right and left from the injured.

All during the advance Courant had the distinct feeling that something was going to happen to him that morning. It seemed to him that he almost waited for a bullet. He found to his surprise that this premonition, instead of deterring him, made him more calm. Now one more advance would bring him and Lohse and the few other men still with them to the comparative safety of the Zeppelin Dugout. Courant, lying somewhat in front of the actual line of his men, prepared to give the command. Suddenly, out of the early morning mist and smoke, a line of English infantry appeared. He jumped up to rejoin his men. Almost instantly he felt a violent blow to his right side and fell to his knees. He was sure he had been shot in the stomach and would be dead in a short time.

". . . following some instinct of survival, I tried after a few minutes to crawl toward the rear," he wrote two years later, recalling the event. "I could not do it by myself, and two men from the neighboring battalion came to my aid. One of them was shortly hit in the leg, and immediately afterwards I was hit by a second rifle shot, which took great pieces of flesh from my left lower arm. I was beginning to lose consciousness because of loss of blood, although my other companion tightened a tourniquet about my arm in a shell crater. The man who had been hit had his leg dressed while the other man rolled me into a coat, put me on my back, and dragged me along the ground with a piece of cord from a bread sack. In my dazed state I tried to help by making rowing movements in the air with my hands and feet. We advanced only by inches. The noise, bullets and shrapnel, was infernal. The English advance was imminent. I remember that during the hours of being pulled I was constantly concerned with whether we were actually going in the right direction, since no definite line of demarcation separated us from the English. I was overjoyed when

I finally saw the familiar pile of straw in front of our barricade, and I I was taken by some strong arms and placed on a stretcher and carried across the trench to the dugout."

In one of those eerie coincidences which sometimes occur, I was translating the preceding paragraph of Courant's wartime recollections when I received a telephone call that Richard Courant had died in New Rochelle on January 27, 1972, at the age of eighty-four.

SEVEN

THE BULLET on September 27, 1915, put an end to Courant's career as an infantryman; but it had not penetrated his stomach, as had at first been feared. He spent the month of October in the military hospital at Essen. During this time Nelly came to see him and told him that she wanted a divorce.

At the beginning of November he received a convalescent leave. Nelly had already let the apartment in Göttingen, but the Runges invited him to live with them. He had the strong feeling "that this stay in Göttingen was going to be a test to see if I was worthy of having my life spared for the second time in this war."

A month and a half later, when his leave was almost up, he had to admit that he had failed the test. He had sought out company—"like a drinker seeking out alcohol." He had wasted time—"There was rarely an evening when I did not have to say 'a lost day.'" He had become involved with a young woman, a student of mathematics and physics—"I have plunged myself even deeper into guilt." On December 16, 1915, the day after she left Göttingen, he began still another journal for self-examination and self-improvement. Although the refrain of "wasting time" and "being always with people" is the same as in his earlier journals, there is a much greater urgency. He seemed to himself "a bankrupt" in life. "If I don't put an end to this carelessness and weakness, it will be the end of me."

During his convalescence he had begun to think again about his academic future. There had been a hint that he might be called to Bonn as a professor; but, going for a walk in the snow one evening with Landau, he learned that he had in fact no chance of getting the appointment. He did not feel that he deserved to be a professor "either through diligence or devotion to science," but it made him angry that people of his own age and gifts who were not soldiers were being preferred over him. When he returned to the Runges' house, there was a telegram informing him that he was to report for duty in Berlin at the first of the year. Nina Runge was still up, and he sat for a long time talking with her.

"I spoke a great deal about myself, so much in fact that I feel uncomfortable about it now. It is very difficult for me at this time to be really open with other people, especially face to face with people whose respect I want as much as I want Nina's.... Besides, it is perhaps pointless now, or even harmful, to aspire to close friendship with Nina. At the moment I am fundamentally unfit for human intercourse. . . . Perhaps people like Nina, through their way of life and their actual existence, can be of some encouragement to me now, but more than that I cannot expect from them."

At the beginning of 1916 he established himself in his parents' apartment in Berlin. Siegmund Courant was currently doing quite well with the sale of concentrates which were being sent to the front. Ernst was at home on leave. Courant found that the more he got to know his youngest brother, the more fond he became of him. But he continued to complain that he was *never alone*. He felt that he must have solitude to find himself and that the keeping of his journal should force him into that situation.

After a few entries, however, there is a gap of almost three months.

Outwardly he carried on. He established contact with people at high levels and had some success in convincing them of the value of his proposal in regard to earth telegraphy.

He and Nelly obtained a divorce. The final decree, dated February 16, 1916, stated that the marriage had been breaking up before the war began and that Courant was the party at fault. Nelly resumed her maiden name and became a teacher of mathematics. During the Second World War she was picked up by the Gestapo in Essen, where she was active in a Christian community dedicated to peace. At the beginning of 1943 her friends were told by a returning soldier, "She was led to her death in Minsk."

When Courant took up his journal again, he was in Heiligblasien to conduct experiments with his telegraphic device at the front. He no longer had the enthusiasm he had had in the past for the project. In many cases earth telegraphy was surpassed by high frequency. He had tried, so far without success, to get control of that too. Objectively, he felt, there was no doubt in his mind that it would be good if he were able to take the whole business in hand and thus counteract "the somewhat shady and uncooperative maneuvers" of business firms and ambitious officials behind the front.

"But I must confess to myself that I have still another motive in my heart," he wrote in March 1916. "I would like to survive uninjured and *on this account* to be active in other ways than in the trenches and in assaults as a platoon commander. That is perhaps too crassly expressed, and it might be different if I knew that on my return to the regiment I would have a relatively congenial situation in relation to my comrades. But what I said is so—I mustn't lie to myself."

The fact was, he conceded, that he would seize "with both hands" any opportunity—"which in my opinion is not unworthy or contrary to duty"—in order not to endanger himself beyond "acceptable" measure. He felt that his conscience would be clear if the earth telegraph project would just begin to move.

"I have no martial spirit, but I do have a very distinct social consciousness and on this account despise every sort of shirking. This is what made me a very passable officer—to share the responsibility for the actions and destinies of others, not to leave others in the lurch nor to expect more of them than I would of myself."

The next day's entry is in a completely different spirit. He had received orders to leave for Stenay in the Ardennes.

"And who knows," he wrote, "if I shall not soon have the opportunity to test earth telegraphy in action at Verdun . . . ? This turn of of events has electrified me. All the silly reflections are thrown out. Let's get on with it. That is the right way."

At the front, in the course of making some tests to obtain an idea of sounds coming through the earth against which the telegraphic signals would have to compete, Courant was startled to hear signals similar to his coming from another source. Since there was relatively little action at the time, a patrol was organized which, under the protection of "a little artillery fire," crossed over no-man's-land to the unoccupied front position of the French. Jumping down into the dugout and looking around, Courant discovered exactly what he had expected—an instrument similar to the one he had developed, but superior in some ways, and an instruction book for its use.

After this discovery there was greatly increased interest on the part of his superiors in Courant's apparatus. In his journal he begins to show concern with the effect of interpersonal relations on the project and with his own position in relation to it.

He was attached to the *Funkerkommando*, or wireless command; and as an infantryman (which he still considered himself to be), he found that he had a quite different attitude toward tactical questions from that of "the gentlemen of the *Funkerei*." He knew what it was like in the trenches. With approval from general headquarters for the experimental installation of some devices at the front, he had another concern. Now he must see that he was not forced out by the *Funkerei*, "for that would not be good for the project." He found it "idiotic" that the project should be tied up with the *Funkerei* at all. The captain was a man of "staggering narrow-mindedness." "A big blow-up" was coming. "But I shall not let the project suffer because of it."

Two weeks later what he had feared actually took place. The captain tried to force him out and replace him with an officer of his own.

"I was very upset—up to the borders of the militarily permissible in expression of my anger—but then, in the interest of the really important project, I swallowed the insult and went along to Haraumont. [Here the heaviest fighting at Verdun was taking place.] I oversaw the installation of the apparatus and spent several days in great danger."

By the middle of May favorable reports on earth telegraphy began to come in from Verdun. Plans were made to manufacture a large number of the devices. Courses were to be given so that officers would be trained in the techniques required. The chief of all field telegraphy was coming for an inspection.

Courant felt that at this point he could hardly be pushed out.

"On the other hand, recognition will essentially be steered to others. Everywhere people are now ready to take credit for the project themselves. . . . That this so easily happens is partly the result of my politics in launching the matter—namely, I have suggested to people of rank or influence that it go out as if from them. Thus I have been able to mobilize many forces where I myself would have been powerless."

The inspection, a few days later, came off successfully. The entire front seemed to be demanding the apparatus. The chief of telegraphy sent out a circular to all staff officers in which he singled out Courant for praise. There was the possibility of an Iron Cross, *First Class*.

"I must confess that now in this moment when everything appears to be secure my personal vanity and ambition awaken."

In spite of the successful outer circumstances of Courant's life, the inner situation had not improved.

"Often I observe in myself an almost pathological indecision, especially in the small things of daily life. I can, for example, vacillate over whether I should go to the theatre until it is finally too late to go and the evening is lost anyway. Most things concern me so little . . . that there is not much difference between *yes* and *no*."

In his journal he continued to castigate himself.

"[Nelly] was completely right when she said once that at the bottom of all my difficulties is the fact that I have no true relationship to my subject—and that is because I too much pursue every aspect of success rather than the subject itself."

He felt that he had been a completely different, purer, much younger person three years ago.

"I can hardly remember that person, but I shudder when I think of the difference."

In July 1916 he received news that his brother Ernst had been killed in close combat in the fighting around Verdun. During the latter part of the year, in the course of his extensive travel for the purpose of supervising the installation of earth telegraphy devices at various parts of the front, he stopped to visit Ernst's grave in Ablain as often as he could.

"But," he reported nine months later when he finally took up his journal again, "it has not helped me."

In contrast to the inner agony, the outer life which Courant led at the beginning of 1917 was one of unheard-of freedom for a young lieutenant.

"For about three weeks I have been so intensely occupied with military matters that I have hardly had a moment to myself—just as you would expect of an officer of the General Staff," he reported to Hilbert. "Of course I cannot tell you more about these matters. Only this much. I was at the Somme until January 10, where I saw a great deal. Then I was called to Headquarters in Silesia to give reports and make suggestions, with some considerable success. From here I went as the personal emissary of Headquarters, via Berlin, to the Western Front, where I am waiting for a telegram from an important personage mentioned daily in the press."

He reveled in the importance he had and, even a half century later, described with pride how at Siemens, a firm which is sometimes called the General Electric of Germany, he was taken to the dining room of the highest echelon, where the directors of the company entertained their guests. He discovered that Crown Prince Rupprecht of Bavaria, to whose division he was attached, was "very nice"—that Count Arco, the head of Telefunken, was "a very intense and very imaginative and active person" —that Otto Arendt, the director of the postal service, was "a really very unusual type, a very good engineer and a very intelligent man." He also learned that as an independent scientist he was able to talk to these people at the top "as an equal."

When he stopped in Berlin on his way to the front again, he heard that there was a possibility that Hilbert might leave Göttingen and come to the capital.

"You can imagine how closely the news touched me," he wrote to his former professor. "My first thought was—if you accept, it will be the beginning of the end of the Göttingen era, but possibly also the beginning of a new era in the life of the mathematical and physical 'world'. . . . But it seemed to me that it would be better for you personally—and for this reason better for science also—if you were to remain in Göttingen, to which you are bound by so many ties. . . . Although I have not been a regular member of this circle for quite a long time, my close ties and great attachment to Göttingen really depend upon your presence there. Your departure would leave a painful gap for me. Many others will feel likewise."

A week after Courant wrote this letter to Hilbert, Germany began to carry out its previously threatened policy of unrestricted submarine warfare.

At the beginning of February 1917, Courant received orders to go to the front and establish facilities for training a number of men in the use of the earth telegraphy apparatus.

He had no idea how to set up such a project within the framework of the military, but he put himself in the hands of a sergeant who seemed to know everything. The sergeant gave him a few samples of orders and let him use a typewriter to compose something similar. Arriving at the last point in the order, Courant had the idea that it would be useful for him to have "an assistant." He put down the name of Wilhelm Runge, who was currently at the front and whose military address he knew from Göttingen.

At this time young Runge was not someone whom most people would have selected. He had almost nothing to recommend him other than the fact that he was Nina's brother and Professor and Mrs. Runge's only surviving son. Bernhard Runge, killed in action in Belgium at seventeen, had been precociously gifted. The family could only hope that Wilhelm would turn out to be "a late bloomer." He had been unsuccessful in the gymnasium in Göttingen; and his father ("who was very patient with me") had sent him to a school in another town, where he had done only a little better. He had been in the trenches for two years and, although exposed to officers' training, had failed to make the grade.

Courant had an intuitive feeling that Wilhelm had potential. At the

Courant

same time he saw an opportunity to repay the Runges for their kindness to him by removing their son from almost certain death.

The commanding officer objected to the last line of Courant's order—it was highly irregular, he said, to request a specific person, and particularly one from another division. But Courant insisted that he needed someone he knew and had confidence in. Sergeant Runge was "a very gifted young engineer."

Wilhelm Runge later became chief director of research at Telefunken. During the summer of 1971 I talked to him in Ulm, where, although retired, he continued to oversee Telefunken's institute for basic research. At that time he described to me how in the early spring of 1917 he helped Courant establish facilities for training men in the use of earth telegraphy at Cambrai in northern France. He also told me something of their life in the occupied city.

"We had an apartment, Courant and I, in a private house. There was a grand piano there, a very beautiful thing, and Courant found some music for quartets and trios. There was one piece I liked very much. [It was Beethoven's "Serenade for Flute, Violin and Viola, Opus 25," Nina Courant told me immediately when she heard her brother hum the opening bars on my recording of our conversation.] Afterwards I always used to use it as a whistle call when I visited Courant or when we passed on the street. In my memory of that time, I always see Courant sitting at this grand piano and bringing some piece into the form that you can play on the piano." Runge apologized for talking so long. "But I wanted to bring you this picture of Courant, in this occupied land, playing the piano."

When Courant received orders to go to other parts of the front, he left the installation and instruction connected with earth telegraphy in charge of young Runge.

"Once again I am on my way," he wrote to Hilbert on March 4, 1917. In spite of the success of his efforts, there were things about his present activity which were nerve-racking. "For the simplest, most obvious things it is a constant battle against stupidity and ignorance. In most cases it is I, a low-ranking young lieutenant, opposing all sorts of high officers." There was "a mountain of work and such enormous responsibility as cannot even be imagined by someone in civilian life." He could barely remember what it was like to have had enough sleep. "Before long," he concluded, "things will probably be even worse."

Yet it was at this time that Courant apparently began to think about mathematics again. In the letter to Hilbert he mentions that he is sending along a manuscript. It is not specifically identified, however.

"I am increasingly convinced that in business and in the academic field those who give their health and strength as soldiers are being short-changed while others at home are advancing their careers without much effort."

Several weeks later he once again took up his journal, this time after a lapse of ten months. His project of earth telegraphy had had much greater success than he could ever have dreamed. Millions had been spent on it ("although, to be sure, it is almost too late"). For the past months he had been traveling in the field under orders from the chief of telegraphy "with rather great authority." Yet he found little satisfaction in his triumph.

"What do I care? Inwardly I have not been able to pull myself together."

Recently, however, he confided to his journal, he had begun to think that there might be a way out for him—a wife.

"But of all those I know it could be only Nina. Yet how could I dare now seriously think of binding Nina to me? I would then have to be so frank with her that she could not preserve the necessary respect for me. In spite of that, I have still thought that at the next opportunity I would at least ask her if later there would be a chance of her becoming my wife."

Now to his surprise a letter came to him from Nina.

While he had been thinking of her as a potential wife, she had been thinking of him as a potential husband. She knew he had been in love with the other girl in Göttingen, but she thought that he was over that. Of all the young men she knew—she told me—he was the only one to whom she could imagine herself as married. She sensed the turmoil within him.

In a brief, straightforward letter she wrote that if he felt she could help him, she was ready to become his wife.

Courant, receiving her letter on March 23, 1917, made what was to be the next to the last entry in his journal:

"In spite of doubts and difficulties, all is now basically determined. For myself I know that everything depends on whether I can succeed in grasping this hand that has been stretched out to me from a better and purer world."

A few days later, just when the deep personal dissatisfactions which had tormented him for so long seemed on the verge of being resolved, he experienced a significant setback in his military career.

Very early in his activity he had gained access to the chief of telegraphy and won his personal support. In the introduction of earth telegraphy at the front, however, he had felt constantly hindered by the jealous opposition of the entrenched staff of the regular army. The struggle is reflected in over a hundred official orders covering the period between January 1916 and March 1917.

At the end of March he suddenly found himself replaced on an important mission by another officer. He sputtered angrily in his journal:

"This man—without any knowledge of the project, without the slightest general or special understanding of it—is being sent only because he is a captain in the regular army and I am only a little lieutenant in the reserves. And this happens just at a time when it will spoil everything! My position in relation to the chief of telegraphy has evidently been undermined. I have too frankly expressed myself. Probably I will soon be completely pushed out."

The entry, dated March 26, 1917, is the final one in the journal for self-examination which he had begun during his convalescence in Göttingen. Eleven days later the United States declared war on Germany.

During that summer of 1917, Courant's old comrade in arms, Walter Lohse, was killed in action.

EIGHT

DURING the remainder of the war Courant no longer played an active role in the earth telegraphy project. After a period in Berlin, he was sent to Ilsenburg, a little town in the Harz Mountains, not far from Göttingen, where a school for training men in the use of the apparatus had been established.

The new situation was not without advantages. He could make frequent trips to see Nina, to whom he was now engaged; and he had enough time to write her long letters, often two or three a day.

Hundreds of men attended the school—"some of them quite intelligent, but most of them not." It occurred to him that a booklet of written instruction in regard to earth telegraphy might be helpful—"some little abbreviated exposition of what it was all about and how it should be used"—and he proceeded to produce a manual. It was his first attempt at expository writing and revived earlier ideas about the need for improving mathematical books.

In a letter to Nina in September 1917 he described how, having in mind "a publishing project," he had approached Arnold Berliner, the editor of *Naturwissenschaften,* for an introduction to that journal's publisher, Ferdinand Springer. On September 28, 1917, he wrote to Nina, "Today I must report to you before everything else that I have finally spoken with Springer and that everything goes well—still, of course, in the preliminary stages but completely in accord with my ideas."

Two days later he added, "The thing with Springer is something! I am afraid only that it is going to rest quite heavily on my shoulders. It will have to be very well thought out before it is made public. Therefore, please, the strictest silence! With everyone!"

Ferdinand Springer, who with his cousin Julius directed the Springer firm, had studied at Oxford. He had planned originally to become a diplomat, but he had later become convinced that an independent and free man could achieve much more than one who labored for someone else, even a great nation. He and his cousin now divided responsibilities in the family publishing business. Julius handled the engineering sciences; Ferdinand, medicine and the natural sciences as well as some areas of the arts.

Springer and Courant shared a common energy and optimism; and, as an additional bond, both had seen action in the war, Springer having served as an officer in the artillery until he was severely wounded in the foot. They took to each other immediately.

The publishing project which Courant had in mind was a series of up-to-date mathematical monographs in fields of mathematics which were particularly relevant to physics. They should make it easier for the physicist reader to comprehend mathematical ideas and methods by sparing him tiresome detours. They should also impart to the mathematician reader—along with the mathematical theorems—an awareness of the connections between mathematics and its applications.

Courant proposed to act as chief editor of the new series and, as such, to select or approve the topics and the authors. He also expected to do some writing himself. He had in mind a book which, he was certain, would have much to recommend it to Springer. He planned to approach Hilbert about the idea of a collaboration—a book based on the notes of Hilbert's lectures on partial differential equations and other areas of mathematics related to physics. As his share of the collaboration, Courant would do all the work of preparing the notes for publication. He was convinced of a need on the part of physicists for such a book; and then the combination—*Hilbert mit Courant*—that would be nice too.

Springer immediately saw in Courant a useful ally in entering a comparatively new field of scientific publishing. Although Courant was not yet professionally established, he had contacts with an impressive number of people in his subject.

Surprisingly, there was a large amount of important scientific work being done in Germany during the war. Courant insisted that it was vital to get this work into print as soon as possible. At the end of 1917, the Springer firm announced the establishment of a new journal, the *Mathematische Zeitschrift*. The first issue appeared in January 1918. The second issue contained a paper by Courant "written while in the army."

Courant had long been attracted by Rayleigh's *Theory of Sound*, which he considered full of original insights and suggestions; and in later years, when he was teaching again, he made frequent references to Rayleigh in his lectures and always urged his students to read the *Theory of Sound*. ("Which they rarely did," Friedrichs admitted. "It was very unsystematic. Nothing streamlined. Nothing put in a general framework. I couldn't read it. But Courant admired it very much.") In his 1918 paper he proved a theorem suggested by Rayleigh to the effect that among all homogeneous membranes having a given perimeter and a given tension, the one that is circular has the lowest fundamental tone. In the course of this work he realized that similar methods could be applied to the problem of the Lorentz conjecture, and he took up that problem in another wartime paper. His real work on it, however, was not to be done until after the war.

By the beginning of 1918, when Courant was returning to mathematics, the German high command knew that the war was lost; but the German people, ignorant of the real situation, were still thinking in terms of complete victory. In the spring of 1918, Ludendorff launched a desperate offensive, which he called "the Kaiser's battle." At first there were victories, but none decisive. Then—in July—the French struck back. It was at this time that Courant went to Göttingen to talk to Hilbert about the proposed book on the methods of mathematical physics.

"The whole concept so closely conformed to the secret of your own teaching success—at least to part of it," he explained later to Hilbert, "that I felt as your student I could summon up the courage to make such a proposition to you."

By October 1918, to facilitate an approach to the Allies with a request for an armistice, Germany had become a constitutional monarchy with Prince Max of Baden as imperial chancellor.

"Much could have been less painful and less trying if it had not been for the indolence and stupidity of the intellectuals and the frivolity of 'leaders' who permitted us to go to the very brink of the abyss," Courant concluded in a letter to Hilbert, devoted for the most part to their future collaboration. "Perhaps we will still have to face a rather painful purging up to the topmost levels. . . . Nevertheless, the path now opens again for all those things which make human life valuable. That this path will be taken and that hope will not fade is the task for all of us who have up to now been so critical. . . . I am looking forward to the moment when I can take off 'the king's coat' . . . and return to the Germany of Hilbert and Einstein. . . ."

Almost immediately the "painful purging" he had predicted began to take place. There was mutiny and revolt. Soldiers and Workers Councils—in imitation of the Russian soviets—were organized all over Germany. Courant himself was elected to head the Ilsenburg council. It was extremely unusual for an officer to be chosen by his men; and there were some people who were inclined to attribute Courant's selection more to left-wing sympathies than to personal popularity.

The creation of such groups as the Soldiers and Workers Councils didn't contribute much, in Courant's opinion, to clarifying the situation. Two weeks after the armistice of November 11, he described to Hilbert how at inspection nobody was sure who should give commands—the Council after consulting with the Commander, or the Commander after consulting with the Council.

"The majority of soldiers do not show any particular enthusiasm for the new order of things. They take it as an inevitable fact of life, just as they did the old situation in relation to authority. All they want is to get home as quickly as possible and to be free from further war or disturbance. This means that the moderating influence which the soldiers would have had on political developments is shrinking, and the more active and radical minority can easily gain the upper hand."

Courant went to Berlin to try to straighten out the whole confused business of the demobilization of his men. The situation in the capital seemed to him like that in an airplane when the pilot has died in flight.

He met Einstein—"a really wonderfully noble and pure personality"—and promptly "put him in touch" with Kurt Hahn, who had become the personal secretary of Prince Max. "I believe that such people [as Einstein] can do more for us now than experienced professional diplomats," he wrote to Hilbert.

Then, in the midst of post-war revolution, he sat down with Ferdinand Springer on November 24, 1918, and signed the contract for the series of books to be known all over the mathematical world as the Yellow Series.

Courant received his discharge as a lieutenant in the reserves. Among his papers in New Rochelle there is a decoration for having been wounded in the service of the Kaiser. There is also a Distinguished Service Cross, Third Class. But no Iron Cross.

The Göttingen to which Courant returned in December 1918 was very different from the one he had left. The number of students in classes ranged from forty to eighty as compared with several hundred in pre-war days. There were few foreigners—a scattering of Swiss, Scandinavian, and Dutch—but a large number of women. The men were mostly wounded veterans. Four of these were blind. Lecture halls were insufficiently heated and lighted. Everybody—professors as well as students—was at the least a little hungry. But the university was far removed from the violence of the capital.

Hecke had succeeded Carathéodory in Klein's chair. Klein continued as the scientific leader of the university, although he was almost seventy and failing in health. Hilbert, however, was taking a more active part in administrative affairs than he had in the past.

There were some new younger people. Alexander Ostrowski was helping Klein with the editing of his collected works. Adolf Kratzer was now Hilbert's assistant for physics; Paul Bernays, his assistant for logic— Hilbert's new developing interest. Emmy Noether was delivering lectures in Hilbert's name and with his support because, since she was a woman, she had not been permitted to become a *Privatdozent*.

Courant had been awarded the title of "professor" by the university while he was still in the army. It was a purely honorary title, he explained to me—"a Red Eagle, Third Class," the lowest civilian decoration. The only academic position available to him upon his return was that of assistant to his prospective father-in-law.

Like other discharged veterans, Courant still wore his uniform but with the buttons that had borne the two-headed imperial eagle removed. To people who had known him earlier, his new assurance and knowledgeableness were impressive. Even to the students, it was apparent that Klein and Hilbert were relying heavily upon him.

In the middle of December they dispatched him to Berlin as a member of a commission of faculty and students to take up the inflammatory question of increased student participation in the affairs of the universities. (One radical student organization had already proposed that professors be elected by the vote of the students.) Heinrich Behnke, who was in Göttingen at that time, recalls Courant amusing a group of young people with his vivid description of the adroitness required for such a journey to the capital. "But after ten days Courant was again in Göttingen, and we students had the impression that he had obtained the best possible results —the machinery of the university continued to function."

After the abdication of the Kaiser, none of the many political parties which existed was powerful enough to rule. All fought passionately to do so. Although as a boy Courant had entertained the idea of "becoming a revolutionary," he felt that he had seen enough of revolution. In future years he was to say that while there must be change, there must never be *discontinuities*. The best hope for Germany seemed to him to lie with the Social Democratic Party. It supported a social organization based on Marxism, but advocated winning power by taking over control of the bourgeois state rather than by overthrowing it.

Courant had long discussions about the future of Germany with Winthrop Bell, who after being released as a prisoner of war had become the German correspondent for an English newspaper. It was Courant's contention that the intellectual middle class must involve itself in the struggle to rebuild Germany.

On a Sunday morning in January 1919, Courant gave a speech at an open meeting of the Social Democratic Party. It was entitled "Social Democracy, Revolution and National Convention" and was reported at length in a local newspaper.

At the time of the truce of October 1918, he pointed out to his audience, "if one had asked the German people which party they thought would save them from catastrophe, the overwhelming majority would have answered the Social Democratic." In a few months, though, the country's mood had changed. It was no longer so much in favor of the party, mainly —in his opinion—because of the short memory of the masses and the fact that right-wing parties were agitating, successfully it seemed, to make the Social Democratic Party responsible for the defeat of Germany and all that had followed.

Courant wanted to rebut these right-wing charges. If the debacle had been caused by any single thing, he told his listeners, it had been caused by "the egotistical blind attitude of the leading segment of the population." The revolution which had followed the armistice had not been instigated by the Social Democrats "nor by anyone at all," including the revolutionaries, "who overestimate their influence in this direction." It had come like an elemental force of nature. Once it had come, the Social Democrats had taken leadership, much less in the interest of their party than in the interest of the country as a whole, "to avoid inevitable chaos and civil war." The fact that the Reichstag was soon to meet again was a mark of the party's success.

Many interests were in conflict. He urged "a human understanding," not only of the middle class and the right-wing parties, but also of the radical elements of the working class.

"The democracies of the world, powerful in number, have thrown us to the ground," he concluded, "but under the sign of Socialism, the German spirit can reconquer the world!"

In addition to university and political affairs, there were personal matters demanding Courant's attention.

He and Nina were aware that a number of people in Göttingen looked with disapproval upon the fact that the daughter of a professor and the granddaughter of Emil DuBois-Reymond, the distinguished physiologist and philosopher of science, was planning to marry "a Semite." There were also Jews who disapproved of the match. "The Runges were such *decent* people," the Jewish wife of a Jewish professor said to me. In her view, which seems to have been the common one, a poor but ambitious young man had married a homely girl with a rich father and then, discovering in Göttingen how unsuited she was to further his academic career and seeing an opportunity to marry instead a famous professor's daughter, had dispensed with her.

Nina, who since childhood had said that she intended to marry a Jew —she liked their looks—and who knew that it was she who had proposed the marriage, was unaffected by the talk.

"Nina was always a kind of royal person," her friend Elli Husserl, now Rosenberg, told me. "Her mother and her aunts and uncles had played with the Hohenzollern children on the grounds of the palace, because old Emil DuBois-Reymond was a bigwig, one of the few professors—there were always a couple—selected to be socially acceptable at the court. So the world looked different to Nina, as it does when people are really free and independent and don't have to worry about how this or that might affect a relationship. I think the way that Nina proposed to Courant had something to do with this freedom she was born into."

It was different for Courant. He had been especially sensitive to the disapproval which seemed to him implicit in the fact that the Runges had put off announcing the engagement of their daughter for some time. He realized, he had written to Nina, that he was not the most desirable prospective son-in-law. He was a Jew "from a not especially good family," was also a divorced man, and was not yet well established in his career. Still, they *were* engaged, and her parents should tell people so.

Courant's parents were not happy about the match either. Siegmund, who in 1919 was bringing his memoirs to a conclusion, noted that Richard had not consulted him and Martha in the matter, just as he had not consulted them when he had married Nelly Neumann.

"It is not Mother's wish," Siegmund wrote of the alliance, "but Richard's will."

Since Courant and Nina were of different faiths, they could not have a religious service without one of them becoming a convert to the other's religion. "Many Jews converted in Germany," Nina explained to me. "It was often for 'political' reasons. I didn't want Richard to do that, and he didn't want to do it." They were married before a magistrate in the town hall on January 22, 1919.

"Now Richard is marrying Nerina Runge," Siegmund Courant wrote —it is the last line of his memoirs. "We shall see how it works out."

That spring, the new son-in-law of Professor Runge began to lecture as a *Privatdozent*.

The number of students at the university had increased dramatically. By governmental fiat many young men, including some who had not completed the last two gymnasium years, had been awarded the *Abitur*. Others who had obtained the degree in the normal way before August 1914 had, in five years of war, forgotten much of the mathematics they had learned. Special review courses were necessary, and the responsibility for these was delegated to Courant by Klein. Courant also gave other lectures on partial differential equations and mathematical physics.

The relation of the partial differential equations of pure mathematics to the problems which arise in physics was to fascinate and absorb him throughout his career as a mathematician. In that first post-war spring and summer, it was of special significance. The book on the methods of mathematical physics—the collaboration he had proposed to Hilbert— was much on his mind. He was also occupied with the development of a theory of the eigenvalues of partial differential equations, a subject intimately connected with the vibration problem of physics.

A student who heard Courant lecture at this time was Helmut Hasse. He had just completed the gymnasium when the war began. Serving in Kiel as a cryptographer, he had continued his education by attending a night class on the distribution of primes, which was taught by Toeplitz, a professor at the university there.

"That course was very decisive for my whole later development," Hasse told me when I talked to him in the spring of 1971 in San Diego, where he was lecturing for a semester at the state university on his now classical work in algebraic number theory. "Toeplitz made me familiar not only with the whole Hilbert spirit but also with the things he himself had seen and done when he was in Göttingen."

When Hasse could no longer attend classes because of his naval duties, Toeplitz encouraged him to continue his mathematical work and mail papers back to him for comment and suggestion. As a result, at the end of the war, the young man felt he was sufficiently prepared to present himself at Göttingen as a student of mathematics.

Hasse told me that although he liked Courant personally, he did not care for his subject matter or his way of lecturing. Unlike Erich Hecke, who "presented his ideas like an artist, taking great care for refinement and balance," Courant often made mistakes. These did not seem to embarrass him. He would simply wipe them out and say, "Oh well, let's try it this way."

After Courant found out about Hasse's interest in number theory, he gave him a manuscript by Harald Bohr.

"It was a wonderful thing on the Riemann zeta function and on diophantine approximation. I worked it through and reported on it to Courant. He invited me to tea at his home one afternoon—think of it, a young student—I was only twenty-one—getting to come into such contact with a professor!"

That first post-war spring, at the suggestion of Klein, who thought it would be advantageous for the university to have a representative in local government, Courant ran for the town council and was elected. For a while, also at Klein's suggestion, he considered becoming a candidate for the state parliament but ultimately gave up the idea.

He appeared in print once again in connection with politics. In July 1919 a certain Mr. Mühlestein, who had developed during the course of the revolution from a majority socialist to an independent socialist to a communist, had delivered an inflammatory speech on the current situation of Germany. Courant wrote a long letter to the local paper taking Mühlestein to task and concluding: "If you really mean well by Germany, if you want to put in your oar with her, why don't you carry your torch of world revolution to the Allies?"

During the summer semester of 1919, Courant did a great deal of mathematical work. By the middle of August he had finished and submitted for publication a long paper on the theory of eigenvalues of partial differential equations. He had high hopes that it would reestablish him as a mathematician.

That fall Erich Hecke left Göttingen for the newly founded university in Hamburg. The chair of Klein was again vacant.

Both Klein and Hilbert were eager to have Courant succeed Hecke. They knew, however, that the ministry would not approve the faculty's proposing to call as a professor one of its own—a man who had taken his doctor's degree and had obtained his *Habilitation* there. They conceived a stratagem to get around the obstacle. Wilhelm Killing was retiring from his chair at Münster, and it was arranged to the satisfaction of all parties that Courant would be called to Münster as Killing's successor.

NINE

IN THE SPRING of 1920 the period of the Social Democratic Party's dominance in the post-war government of Germany was essentially at an end. In the elections that year, parties to the right and to the left both made great gains. Courant was relieved to have an excuse to resign from the town council. He had found its sessions "more boring than faculty meetings" and had regularly come home with the dry announcement that he "had successfully worked for the end of the debate." Leaving Nina and their new baby in Göttingen, he went cheerfully to Münster at Easter (the beginning of the German academic year) for what he hoped would be only a very short stay at that university.

The work which he had done on the eigenvalues of partial differential equations appeared in the new *Mathematische Zeitschrift* in 1920. Like so much of his mathematics, it arose out of a physical situation, reflected his interest in and enthusiasm for the work of people he knew, and had a connection with Dirichlet's principle.

The story of the work goes back to a lecture given in Göttingen before the war by H. A. Lorentz. In this lecture Lorentz referred to an interesting conjecture about the distribution of the eigenfrequencies of vibrating homogenous media, which correspond to the fundamental tone and the succession of overtones of acoustical systems. Hermann Weyl immediately began to work on the related mathematical questions; and in a series of powerful and ingenious papers, which he began to publish even before the war, he verified the Lorentz conjecture for a number of the most important vibration problems.

Courant was tremendously impressed by Weyl's work, but he felt that it lacked esthetic appeal. The method was "roundabout" and therefore did not give a really complete insight into the relationships involved. After his own work on Rayleigh's theorem during the war, he thought he saw a way in which he could verify the Lorentz conjecture more directly than Weyl had.

To get a glimpse of Courant's method and its relation to that of Weyl, one must go back to the facts on which Lorentz based his conjecture. At the time it was already known that the eigenfrequencies under consideration depend in a rather complicated manner upon the nature of the shape of the vibrating medium but that they can be described in a simple way, which becomes increasingly accurate as the frequencies become higher. It was

Lorentz's conjecture that the distribution of very large frequencies depends, not on the shape of the vibrating medium, but only on its volume.

The behavior of the eigenvibrations is described mathematically by a partial differential equation which has a solution only for special values of a certain parameter. These are the eigenvalues of the equation, the squares of the eigenfrequencies. At the time Lorentz offered his conjecture, mathematicians considered the partial differential equations which describe the behavior of the eigenfrequencies almost impossible to handle, since there were many severe technicalities connected with them. For this reason Weyl transformed the partial differential equation into an integral equation and then employed the theory of integral equations to verify the Lorentz conjecture.

"But Courant hated the method of integral equations," Friedrichs explained to me. "He thought one should be able to work directly with the relevant partial differential equations—that there was no necessity to transform them into integral equations as Weyl had done in his work. Technicalities that Weyl avoided by his approach didn't bother Courant so much. He was always inclined to feel that such things could be managed one way or another—even if he didn't know quite how at the time. And in the end, as it happened, he *was* able to straighten them out."

Courant wanted to show—as he wrote in the long paper published in 1920—that "from another point of view one can gain direct access to the whole complex of questions posed and can attain complete command of the eigenvalues . . . , the boundary conditions, and the domain in a surprisingly simple and consistent way . . . which, in economy, essentially surpasses Weyl's method and seems to be theoretically satisfying."

The leading notion of the method which Courant proposed was to characterize the eigenvalues as the extreme values of certain integrals connected with the original partial differential equation of the vibration problem. The method was of course a close analogue of the Dirichlet principle.

Even before the work of Weyl, it was known that the eigenvalues of the related partial differential equation could be characterized as those values for which a certain equation has a solution other than zero and that this solution—the eigenfunction—gives the amplitude of the vibration. It was also known that the lowest eigenvalue could be characterized by a minimum property closely related to the Dirichlet principle and that every eigenvalue could be characterized by a minimum property once all the previous eigenvalues and eigenfunctions were known. Although theoretically satisfying, this method was impractical to use in the case of very high eigenvalues. Courant, however, observed that it is possible to characterize any eigenvalue by a modified minimum property *which does not depend on the knowledge of the lower eigenvalues and eigenfunctions.*

"You can start right in there!" Friedrichs told me with enthusiasm. "Oh, you have to satisfy some additional conditions, of course; but the important thing is that you don't have *to know* the previous eigenfunctions. The conditions, the additional conditions, are of a very general character. You see, Courant's observation was the following. If you impose certain orthogonality conditions on the appropriate number of arbitrary functions and then take the minimum, that's not good enough; but if you then *maximize the minimum*, it comes out right. That gave him the possibility of direct characterization of the eigenvalues—the squares of the eigenfrequencies—and it was this that permitted him to get the asymptotic behavior of the high frequencies of the Lorentz conjecture and of the eigenvalues of the associated partial differential equations. It was a beautiful idea. *A very beautiful idea.*"

As it happened, unknown to Courant at that time, this same maximum-minimum principle had already been stated by Ernst Fischer and used for much simpler cases having no connection with the Lorentz conjecture or with mathematical physics. But, in Friedrichs's view, the important thing was not so much to have discovered the principle as to have seen that the reformulation would allow direct approach to the problem of asymptotic distribution.

"That, I think—the twist, *that* was Courant's."

Both the method of Courant and the method of Weyl had come out of Hilbert's work at the beginning of the century. At the time of the publication of Courant's work in 1920, it still seemed to most people that while his method might be more esthetically satisfying, Weyl's method—which was in some ways more specific than the method of the Dirichlet principle—was the more powerful. In the long run, according to Friedrichs, Courant's method seems to have proved superior to that of Weyl. But that was not clear at the time when Courant was a young professor in Münster trying to reestablish himself as a mathematician.

Courant's appointment at Münster was as a professor of mathematics; but, according to his contract, he also seems to have delivered lectures on the applications of mathematics as well. This was probably the reason that during the term he was approached by a middle-aged man with the kindly, open face which he always associated with peasants. Conversation revealed that the visitor, whose name was Carl Still, had studied on his own such writers as Lagrange and Clausius—with remarkable tenacity, in Courant's opinion—but now felt that he needed more than he had been able to get for himself from these authors. He talked with Courant about certain problems—"non-trivial" to the professor's surprise. Some were purely theoretical; others were important to Still's work and had to do with such applications as the determination of flow conditions by calculation and the material and energy balances for chemical processes.

The mathematical discussions led eventually to an invitation to Courant to visit the Still home. To his amazement he discovered that his simple visitor was in fact a well-known self-made industrialist of the Ruhr area.

As far as I know, the beginning of this friendship with Still was the only significant event of Courant's stay at Münster. Even before the end of his first semester there, Klein and Hilbert had begun to push openly in the faculty for his return to Göttingen.

Obtaining a professorial appointment for Courant would have been a relatively simple thing if it had not been for the fact that, at the same time, Klein and Hilbert were proposing to add *two* new physicists to the faculty to take the place of Peter Debye. Originally, Courant told me, he himself had suggested Born to Hilbert as a replacement for Debye. Born had been happy to return to Göttingen but had felt a little overwhelmed by the assignment. Although he had had experience in experimental as well as theoretical research, he did not feel that he was capable of directing the big physics laboratory, which in his student days had been divided between Voigt and Riecke. A little research on his part revealed that, although one physics professor, Robert Pohl, had recently been appointed, there was still a place on the university books for yet another physics professor. On the basis of this fact, Born made it a condition of his acceptance that his friend James Franck, an outstanding experimental physicist, also be called to Göttingen.

"Franck + Born are the best imaginable replacement for Debye!" Hilbert wrote delightedly to Courant in Münster. "I am very happy about this arrangement. We have Born's energy to thank for it!"

Thanks to Born's energy, however, Klein and Hilbert found themselves pushing for three mathematical-physical appointments at the same time. There was some opposition in the faculty—"because of a little hostility toward me personally," Hilbert informed Courant. One professor suggested that the faculty consult outside experts before making its decision. Landau proposed as a compromise that only Courant be called at that time.

Undaunted, Hilbert wrote cheerfully, "We have much reason to be optimistic. Klein holds out bravely!"

At the same time Klein also wrote to Courant, "As you may have heard from other sources, I intend to advocate your appointment in Göttingen. It would be extremely helpful for me if you would confirm explicitly in writing that you are willing to promote with energy tasks which, in my opinion, have long been unduly neglected in our educational system and new demands which I can foresee as coming up."

He then listed three items: (1) regular introductory lectures for students not headed for abstract mathematical training, (2) the taking over of the organizational duties which he himself had handled over the years, and (3) "a positive orientation" to all important questions arising out of the revamping of the educational system at the university level as a result of the post-war situation.

He noted in a postscript: "Not one of these three points should come as a surprise...."

"Klein has read me his letter to you," Hilbert explained to Courant, "and he wants your reply only to be able to assure the minister that he has your written support for his organizational efforts."

Klein was old, ill, and tired; but the rest of the faculty was no match for him. All three appointments were approved.

Born's and Franck's assumption of their duties was delayed by the fact that they couldn't obtain housing in the crowded post-war conditions of the university town. Since Courant had not given up his apartment when he left for Münster, he was able to return to Göttingen immediately.

For an American, accustomed to the American university system of separate departments, each with its own chairman, a budget and a building of its own, it is hard to conceive of the undefined nature of Courant's role as Klein's successor.

Klein had never held a formal administrative position. His fully autonomous colleagues had simply been willing to let him handle what Hilbert dismissed as "the mathematical arrangements." It is true, as Weyl later wrote, that Klein ruled in Göttingen like a god, but his godlike power came from the force of his personality, his dedication and willingness to work, and his ability to get things done. Over the years he had concerned himself with the reform of secondary education in the sciences; the improvement of the technical schools; the development of comprehensive mathematics courses for teachers and engineers; the securing of the connections between mathematics and other sciences and between mathematics and its applications; the establishment of close relations between industry and science.

In Göttingen in 1920 there was no mathematics department. There were four mathematics professors who, like the other scientists, were members of the Philosophical Faculty, which included philosophers, historians, philologists, and classicists, among others. There were also some private lecturers in mathematics. There was no budget, except for the running of the library; and there was no building—in fact, the professors did not even have offices.

Klein had dreamed of a single building—an "institute," such as those of the experimental sciences—which would house the *Lesezimmer* and all the other mathematical activities. He had desired above all else to make Göttingen the mathematical-physical center of the world. Courant was personally very much in sympathy with Klein's ideas; and he thought—he told me—that in his own way he might be able "to move them a little bit forward."

Throughout Klein's career he had had strong allies in the non-academic world. Among these was the publisher B. G. Teubner in Leipzig. But even before the end of the war Teubner had been showing himself unwilling to provide the support to mathematics he had given in the past. Like most German publishers, he was eager to pare down his operations "until normal times returned." He had let it be known that he would welcome being relieved of the *Mathematische Annalen*. He had further scandalized everyone in Göttingen by suggesting that the paper stock on hand for the collected works of Gauss—still not yet all in print sixty-five years after the death of the prince of mathematicians—should be used for other, more profitable purposes. In these circumstances Courant wasted no time in introducing his new friend Ferdinand Springer to Göttingen.

Unlike his fellow publishers, Springer recognized "that the time of aberration from 'the normal' had begun" and it would be a long time before it ended, if it ever did. Even before the war was over, he had founded the *Mathematische Zeitschrift*; and at the end of 1918 he had contracted with Courant for the *Grundlehren* series. In 1919, while Germans were expressing their outrage at the peace terms offered at Versailles, he took over the distribution of the works of Gauss from Teubner. In 1920, as the

Weimar coalition of the Social Democrats began to fall apart, he signed a contract with Teubner to become the publisher of the *Mathematische Annalen*. In 1921, as the Allies threatened occupation of the Ruhr, he published the first volume of Courant's yellow-covered *Grundlehren* series; and in the next three years, which saw the worst of the post-war inflation, he brought out a dozen more volumes.

The friendship of publisher and scientist was mutually rewarding. Springer took Courant's advice in entering the new fields of mathematics and physics. Courant made every effort "to protect my friend Springer from risks and losses."

As part of his contract with the ministry, Courant had negotiated a generous provision for two personal assistants. His first assistant was Hellmuth Kneser, who was getting his Ph.D. with Hilbert on a topic in the foundations of quantum mechanics. Kneser, who became much better known as a topologist, was the son of Adolf Kneser, who had been one of Courant's teachers in Breslau. In later years, when his own son Martin Kneser had also become a mathematician, he was to complain humorously that he had spent the first half of his life as the son of Adolf Kneser and the last half as the father of Martin Kneser. As Courant's assistant from 1920 to 1924, Kneser did the usual things to help the professor and also helped younger students in many ways. B. L. van der Waerden remembers how, as a young doctor from Amsterdam, he regularly lunched with Kneser and then took a mathematical walk, "on which I always learned a great deal."

While settling into the academic life in the winter of 1920–21, Courant became politically active again for a brief time. A plebiscite was to be held in Upper Silesia to determine whether that area, which included his birthplace, would become Polish or remain German. The right to vote had been given to all persons born in the area who would have completed their twentieth year by January 1, 1921, regardless of their current residence. The German government mounted a massive campaign to bring back as many as possible of those who had left the area to vote in the plebiscite. Courant took an active part, writing to friends and relatives he had not seen in years. On the day of the voting he was among the more than sixty Courants who gathered in Lublinitz.

Lublinitz itself recorded a German majority in the plebiscite, but it was included in that part of Upper Silesia that went to Poland. The Courants who had remained there, running the business of "S. Courant," chose to leave their birthplace forever.

After the plebiscite of 1921 Courant continued to maintain a lively interest in political affairs, but he devoted himself personally only to those aspects which directly concerned science and the university.

By the beginning of the winter semester 1920–21, James Franck had been able to establish himself as professor of experimental physics; and at the beginning of the summer semester 1921, Max Born arrived to take over his duties as professor of theoretical physics.

Born was an exemplar of what Courant considered "the Göttingen tradition." He had come originally to the university because it was "the mecca of mathematics." Then, as the tale is told, Hilbert had given him a problem which he was not able to solve—a problem still today unsolved—and Born, doubting his mathematical abilities, had turned to physics. By the time he came to Göttingen as a professor, he had done his famous work on the derivation of all crystal properties from the assumption of a lattice whose particles could be displaced under the action of internal forces; developed with Fritz Haber a heat theory which included the first example of the determination of a chemical heat reaction from purely physical data; and tried his hand at experimental research, discovering with the help of his assistant E. Bormann a method for determining the free path of a beam of silver atoms in the air.

James Franck had already produced the work on the changes of energy occurring upon the collision of atoms with electrons, for which, in 1925, he would receive a Nobel Prize.

During the year 1921 two outstanding young mathematicians also arrived at the university.

One of these was Carl Ludwig Siegel, who had been encouraged by Landau to come to Göttingen after the war and had then gone with Hecke to Hamburg. Siegel's mathematical interest was different from Courant's, and the two men had not met during Siegel's initial stay in Göttingen. Courant, however, had heard from others of Siegel's abilities and of his current situation in Hamburg, where he was cold, hungry, and unhappy.

At the beginning of 1921, Courant wrote to the ministry suggesting that the second assistantship provided for in his contract should now be activated and offered to Siegel.

It did not take any special perspicacity on Courant's part to see that Siegel was going to be an outstanding mathematician. What Courant did see was that Siegel was a young man, "not easy," who required and should receive special treatment because he was "an absolutely unique talent." He managed to convey this to the minister.

"It would make me very happy," the minister wrote to Courant, "if this grant would suffice to support so powerful a talent as, according to your presentation, Mr. Siegel is, and enable him to continue his scientific career."

When Siegel returned to Göttingen, the little Courant paternally took charge of the tall youth, found a place for him to live in a vacant room at Klein's house, invited him to swim at the faculty bathing establishment, saw that he met Hilbert there and that Hilbert learned about his work, and invited him frequently to his home. There Siegel, much to the amusement of Nina, sat on the floor and solemnly tried to get baby Ernst to repeat after him long and complicated scientific terms.

Courant also made an effort to draw Emil Artin, the other outstanding newcomer, into the inner circle of Göttingen. Artin was an Austrian with talents and interests in art and music as well as mathematics. After serving in the war, he had gone first to Vienna to study, then to Leipzig. He had taken his degree with Gustav Herglotz, who had been an associate professor in Göttingen during Courant's student days. It was Herglotz's suggestion that Artin go to Göttingen for post-doctoral study. Although, as in the case of Siegel, the mathematical interests of Artin were quite different from his own, Courant welcomed the young visitor, invited him for musical evenings at his home (where Artin, whose music was as pure and rigorous as his mathematics, shuddered at Courant's untutored approach to the keyboard), and saw that Klein and Hilbert met him and learned about his work.

This maintaining of contact between the new young people and the great men of the past was a conscious effort on Courant's part. He felt that Göttingen's mathematical-physical tradition of teaching combined with research had been almost completely the personal creation of Klein. Gauss had had little connection with the instructional side of the university and considerable distaste for it. Dirichlet had begun to give advanced courses, but these had not drawn a large number of students and the development had been cut short by Dirichlet's death, four years after that of Gauss. The sick and shy Riemann had not had the ability to carry on what Dirichlet had begun. It was Klein who had taken these great names and their wide-ranging research activity and molded them into a tradition.

Now, in the first post-war years, Courant began to weave into the Göttingen tradition the great living figures of Hilbert and Klein.

TEN

SHORTLY AFTER Courant returned to Göttingen, he introduced his friend from Münster, the industrialist Carl Still, to the Göttingen circle. At the begining of 1922, Still contributed 100,000 marks to the funds remaining in the treasury of the *Göttinger Vereinigung*. A new organization was formed with the same purposes as Klein's old group—the cultivation of the applications of mathematics in Göttingen.

That same year the German mark, which had had a pre-war value in relation to the American dollar of 4.20 to 1, fell from 162 to 1 to 7000 to 1. It was clear that imagination was going to be as necessary as money for mathematics in post-war Göttingen.

In spite of the rapidly increasing inflation, the year 1922 was to see important milestones on the road to recovery for mathematics and physics in Göttingen and in Germany as a whole. It began with a great celebration of Hilbert's sixtieth birthday. It saw, in the summer at Leipzig, the first general meeting of German scientists since the war. In the fall, in Göttingen, there was a historic week-long series of lectures in which Niels Bohr explained for the first time in public the strange new ideas which so many had rejected before the war and which were, at the end of 1922, to earn him the Nobel Prize for physics.

When, that same year, mathematics and the natural sciences were formally separated from the other specialties of the Philosophical Faculty, Courant applied for permission to change the name on the stationery which he and the other mathematics professors used for official business from *Universität Göttingen* to *Mathematisches Institut der Universität*. This seemingly innocuous request was granted.

Supported by Klein and Hilbert, Courant tried to lure back to the university the outstanding German mathematician of his generation—the young man whom Toeplitz had once identified as someone else who also thought about mathematics.

Hermann Weyl had left Göttingen in 1913 with the beautiful Hella as his wife and, after service in the army, had become a professor at the Eidgenössische Technische Hochschule in Zurich. There, for a year, he had been in close contact with Einstein. After Einstein had presented his general theory of relativity in November 1915, Weyl had given a well attended series of lectures on Einstein's ideas, later published as *Space, Time, and Matter*. Courant was among those who had immediately recognized the significance of Weyl's mathematical-physical-philosophical orientation for the new situation in science resulting from Einstein's theory. In a letter to Hilbert in July 1918, he had suggested that the mathematician Weyl should be called to Göttingen to succeed the philosopher Husserl, who had received a long overdue professorship in Freiburg.

"The department of philosophy at the universities today is conceived as if philosophy as a science in which fixed results can be formulated does not exist . . . ," Courant had written to Hilbert. "It has come to the point that a mathematician or a physicist rather fears the investigation of his own fields by philosophers and no longer expects enlightenment or even understanding from them. . . . Now, after Einstein's colossal achievement, the deep significance of which lies particularly in the philosophical area, . . . this situation must come to an end. . . . It is definitely desirable that a number of philosophical positions be held by people who are closer to mathematical and scientific thinking. Since such people are not easily found among the regular philosophers [with the exception of Husserl and Nelson], a mathematician like Weyl would be suitable for such a position," he had suggested.

Husserl's place was filled, with Hilbert's backing, by Leonard Nelson; but Courant continued to hold the idea that if such an appointment as that of Weyl was not made, the old "unfriendly" conditions between mathematics and philosophy would continue to exist. In 1922, with the establishment of a separate faculty of mathematics and the natural sciences, Courant apparently negotiated an additional chair for mathematics and offered it to Weyl.

Weyl was torn by the opportunity to return to Göttingen. On one side were the attractions of the German university to which he was always to considered himself inextricably bound and to which he often returned for the stimulation which he had received in his youth from Hilbert and Minkowski; on the other side, the incontrovertible uncertainty of the conditions of life in Germany—the new republic battered from right and left, constant threat of revolution and continuing acts of terrorism by extreme nationalistic parties, inflation now "galloping."

After a long delay and much debate with his wife, he finally set off for the telegraph office with his acceptance written out and in his hand. By the time he arrived, he had changed his mind. He wired instead his refusal.

The economically and politically miserable year of 1922 which had deterred Weyl from returning to Göttingen saw the arrival there of a number of gifted students. One of these was Friedrichs, then an awkward, quiet, and asthmatic youth, very shy.

He found the mathematical life in Göttingen "a terrific shock." The level of the subject matter was far above that to which he had been accustomed. A seminar conducted by Courant was based on a book which he had already encountered at another university, "but there, during the whole term, we had gone through chapters one and two—in Göttingen we started at chapter three."

He also found the subject of Courant's seminar an unusual one. Since his dissertation Courant had been interested in algebraic surfaces. There were currently in Göttingen two young men he much admired—Siegel, now a *Privatdozent*, and Artin; and both were well informed in algebraic number theory. Although the two topics shared the name "algebraic," they were not as a rule combined; but Courant decided there should be a seminar taking up algebraic surfaces one week and algebraic number theory the next week.

The first report in the seminar of 1922 was given by Otto Neugebauer. He was also an Austrian and had served as an artillery officer during the war. In studies in Graz and Munich he had moved from electrical engineering to physics to mathematics. From the moment that Courant heard Neugebauer's seminar report, he was completely enchanted by the young man and was soon on friendly, almost colleagual terms with him.

Friedrichs remembers the seminar as being impressively well attended.

"Siegel and Artin came, also Kneser. There was present every assistant who was in Göttingen at that time. They all had to come and participate. Such a group of people, who knew everything about everything—it was very exciting to me."

"When you say 'they had to come and participate,' do you mean that Courant required them to come, or that it was customary for everybody to come to the seminars?"

Friedrichs laughed.

"That question cannot be answered. Such a notion did not exist for Courant. He did not ask the people as a requirement as Klein might have done. No. He would say, 'It's very important that you help us with this, we need your help'; and he would manage somehow that everybody who had something to contribute did attend. That's the way he always operated."

While Friedrichs was being awed by Siegel, Artin, Kneser, and the other young mathematicians of Göttingen, he himself was putting in awe another student. Hans Lewy was several years younger than Friedrichs. He was straight out of the gymnasium, a small, bright, extremely sensitive and impatient boy, musically as well as mathematically gifted. In Göttingen for the first time in his life, he told me, he found himself confronted by a number of people who seemed to know much more than he did. He was especially impressed by Friedrichs.

When I talked to Lewy in Berkeley, where he had recently retired as a professor at the University of California, I asked him how he had liked Courant as a teacher in those days.

"Courant was, for me, a very good teacher," he replied. "Of course, often his lectures were not sufficiently prepared, and maybe he would have to change his approach during the hour because he had noticed that something was missing. But this, in some ways, was stimulating. For one thing it allowed him to say much more, to allude to subjects which he was not fully covering, to stimulate the ambition of the students to cover the gaps, to see clearer."

To Lewy it was especially important that Courant seemed to be able to communicate a lack of fear of the technical details which often overwhelm students.

"Courant had the attitude that the technical details would take care of themselves. That inspired young men. It is especially important in analysis because, as analysis is a very old subject and a vast subject, it takes a long time to get to the point where one understands where the problems lie. That, in my opinion, is partially the reason that analysis has little success in the United States and little attraction for the young. Courant was able to make a young man feel that one could just break through...."

"Yes," he concluded thoughtfully. "That was a great help."

By the time that Friedrichs and Lewy arrived in Göttingen in 1922, three of the monographs in Courant's Yellow Series were in print or in the process of publication. But his own book, the collaboration with Hilbert on the methods of mathematical physics, was still not finished.

His original plan had been merely to edit the notes of Hilbert's lectures, which he and other assistants had written up before the war. Then gradually he had begun to add things of his own and to rearrange ideas. Certain areas which did not appeal to him were dropped. Other areas which he liked very much were given an increasing amount of space. In particular, he wanted to elaborate on the theory of eigenvalues which he had developed after the war and the related theory of eigenfunctions on which he had worked since then. He felt that he had first to lay a foundation for his ideas with an extensive presentation of a number of methods suited for many different special cases. This material was already in existence but was scattered and not in easily accessible form.

Courant obtained funds, perhaps from Still, for a *Hilfsassistent*, "an assistant to the assistant," to take notes on his own lectures on partial differential equations. Several students were eager to have the job, and he had them write up the first lecture of the semester as a sample of what they could do. He ultimately selected Pascual Jordan, a physics student who had prepared a set of notes although he had not heard the lecture. Jordan also helped Courant with the preparation of the book on the methods of mathematical physics.

As the possibility of publication of the book with Hilbert continued to recede, Courant saw an opportunity to produce rather quickly a collaboration with another famous mathematician on a subject which also had great attraction for him. Adolf Hurwitz had died at the end of 1919. Although Courant had not been inspired by Hurwitz's lectures as he had been by Hilbert's, he felt strongly that they should be made available to a wider audience than Hurwitz's own students.

The Hurwitz lectures on function theory were in the spirit of Weierstrass. They were straightforward, precise, remarkable for their clarity and their esthetic qualities. Courant admired them greatly, but he did not think that they hit "quite the center of interest." He felt a need to balance them with the approach of Riemann which was—in contrast—geometrical rather than arithmetical and intuitive rather than logical and which drew its inspiration from the problems of physics rather than from those of pure mathematics.

The Hurwitz-Courant book on function theory appeared before the end of 1922. Reactions to it were mixed.

The American mathematician Oliver Kellogg was especially critical. He found the Hurwitz lectures "lucid" and "noticed no faults in logic" in them; but as to Courant's contribution—"It gives the impression of being the work of a mind endowed with fine intuitive faculties, but lacking in the self-discipline and critical sense which beget confidence. . . . What is found may, indeed, serve as an indication of some of the directions which modern investigations have taken—in fact, a very interesting one. But the proofs offered often leave the reader unconvinced as to their validity and, at times, uncertain [even] as to whether they can be made valid."

A different reaction to Hurwitz-Courant was that of Friedrichs, who read the book as a student at about the same time that Kellogg was reviewing it. To Friedrichs the first two sections by Hurwitz seemed "very neat, very clear, to the point, you could learn from them, but they were not inspiring. But the Courant section, the third chapter—when I got hold of that chapter, I started reading one morning, I read morning and night without stopping. It was the most breathtaking book I have ever read in mathematics."

(The first edition of Hurwitz-Courant sold out within two years. Courant then asked his young friend Neugebauer to eliminate the flaws in precision and logic which Kellogg had criticized. Neugebauer, whom Courant once described as "having all the virtues of pedantry and none of the vices," did an excellent job. Kellogg in a second review noted the improvement. But for some readers a little of the magic of the first edition had been lost.)

The Courant approach to function theory, so repugnant to Kellogg and many others, was that of "the romantic type" in the sciences as opposed to "the classic type." This distinction was originally made by the chemist Wilhelm Ostwald. Courant once cited it in describing Klein, whom he saw as an example of the romantic type:

"While the classic type in the sciences carefully examines every detail and repeatedly refines and polishes his work before permitting it to leave the seclusion of his study, the romantic type throws his discoveries to the public as an immense stimulus, often before his ideas have reached full maturity. The classic type prefers to lock up in his desk three quarters of his scientific output if a minor point does not satisfy him, and he never wishes to say more than he will be able to support in years to come. The romantic type, on the other hand, does not place great value on the fully matured and completed form, and he does not feel abashed if he has once said more than he can actually prove. He is interested in the immediate, vital impact. He lives within a circle of enthusiastic disciples who are greatly enriched by their contact with him and who will one day move on to do their own work."

In 1922, in addition to publishing the book on function theory, Courant also published his theory of eigenfunctions, which had grown out of his work on eigenvalues. In the earlier work, he had simply assumed the existence of the eigenfunctions associated with the eigenvalues. Now he successfully tackled the problem of proving existence. In the course of this work he used techniques that differed in some ways from those he had used before and introduced a number of elegant geometric-algebraic concepts. Today other, though related, concepts developed in functional analysis are considered more appropriate for the purpose. "But Courant was always very slow in accepting notions that grew out of this much more abstract field," Friedrichs told me. The work on eigenfunctions was not to have as great an effect as the earlier work on eigenvalues.

The paper of 1922 on eigenfunctions did, however, initiate a series of papers on the partial differential equations of mathematical physics in which Courant's primary concern was *existence*. The significance of this concern on the part of mathematicians is sometimes questioned by even quite sophisticated physicists. They are inclined to feel that if a mathematical equation represents a physical situation, which quite obviously exists, the equation must then of necessity have a solution.

Courant, however—as Friedrichs said—"simply refused to be intimidated by the objections of such physicists." He was convinced that existence investigations would contribute to the understanding of the nature of the equations and their solutions. He also felt that a knowledge of existence considerations would be helpful in setting up schemes for numerical computations. From 1922 on, he emphasized questions of existence in all his work on partial differential equations of mathematical physics.

The year 1922, which was one of the most mathematically active of Courant's career, was also a year in which shelter and food were beginning to be more immediate problems than those of mathematics. It was natural for students to turn to Courant for help, and he often turned to Carl Still.

Still was the kind of successful man of business or industry whom Courant was always to admire extravagantly. Although he was "very hard and very sharp," the more important requisites of his financial success were for Courant "expert knowledge and experience, diligence, and uncompromising adherence to the highest ethical standards of business life." Courant also saw in both Still and his wife, Hanna, a deep-rooted devotion to that which transcends mere personal values—"a *religious* quality in the real sense of the word."

If 1922 was economically miserable, 1923 was immeasurably worse. Rents were sometimes set in terms of butter, which was virtually unobtainable; and potatoes were on occasion purchased with baskets and later with wheelbarrows of currency. Money sent from home lost in purchasing power before it could arrive and then again before it could be spent. Many students worked on the railroad, shoveling gravel back under the ties, in a futile effort to supplement their incomes; but, as Lewy told me, "it was impossible to have *enough* money." Sitting before their books in the *Lesezimmer*, they could not get their minds off food.

Yet for the Göttingen mathematicians there were also hopeful signs in 1923 of a return to contact with the world outside Germany.

In that spring two Russian mathematicians arrived, the first scientists to be "sent out" of Russia since the Bolshevik revolution. P. S. Alexandroff and P. S. Urysohn were products of the post-revolutionary school of Luzin, but already they were on their way to founding their own new school of topology.

The same year which saw the arrival of these first visitors from the east saw also an opportunity offered from the west for young German mathematicians to go abroad to study once again. In 1923 the International Education Board, newly founded by John D. Rockefeller, Jr., with Abraham Flexner as Director of Educational Studies, announced a series of fellowships to assist young scientists to pursue in other countries "studies which they cannot pursue at home with equal advantage." There were places where young German mathematicians would still not be welcome; and for the first in what was to be a long series of recommendations, Courant turned to the greatest foreign friends of Göttingen—Harald Bohr and G. H. Hardy.

By 1923 there was no fixed currency in Germany. In a talk many years later, Neugebauer vividly recalled the situation as follows. Salaries and prices were expressed in basic numerical classes which were then multiplied by a rapidly increasing coefficient $c(t)$ such that the result represented their value expressed in marks at a given time. On a certain day of the week the current value of $c(t)$ was divulged to the business office of the university, hours before it was announced in the press.

Recognizing that early knowledge of $c(t)$ would greatly increase the purchasing power of funds allotted to the *Lesezimmer*, Courant offered to lend the business office the "mathematics institute's" electric calculating machine. In return he would be informed of the value of $c(t)$ as soon as it was received. Inflation was proceeding at such a rate that on July 1, 1923, the mark stood in relation to the dollar at 160,000 to 1; and on October 1, at 242,000,000 to 1. Courant's machine had a range of 19 digits. The university business office did not hesitate even an instant in accepting his offer.

By November 20 the mark had a value 4,200,000,000,000 to the dollar. The government announced that a state of national emergency existed. A new currency was introduced in strictly limited quantities and backed by a mortgage on all the industrial and agricultural resources of the country. From one day to the next, the currency was amazingly stabilized. A few months later, the government was able to declare that the state of national emergency was officially at an end.

It was during this period that Courant completed *Methoden der mathematischen Physik*—the book which he had proposed in 1918 as a collaboration with Hilbert. Since then, Hilbert's health had progressively failed. His mathematical interest—with characteristic completeness—had changed from physics to the foundations of mathematics. He showed an interest in the book his former student was writing but did not participate in any other way.

At one time, Friedrichs recalls, Courant showed him the notes of Hilbert's lectures, on which he had originally planned to base the book. These were beautiful, but they were not what is known as Courant-Hilbert. Friedrichs took his copy of the first volume out of his bookcase and leafed through it.

"It *is* a rather unsystematic book," he conceded. "Here, also, you see, Courant wanted to *combine*, like in the seminar on algebraic surfaces and algebraic numbers and in Hurwitz-Courant. There are various approaches to a theory of partial differential equations. Then he has an introductory section on algebraic notions, then some semi-algebraic methods, then some series expansion, then integral equations, then calculus of variations—and then he comes to vibrations—his own work on eigenvalues and eigenfunctions. Certain areas completely omitted, certain areas which Courant liked overemphasized. Very unsystematic."

The influence of Lord Rayleigh's *Theory of Sound* on Courant's book was apparent; the spirit of Hilbert "hung over every page" (as Paul Ewald noted). Reviewers recognized immediately, however, that in spite of the joint authorship, Courant-Hilbert was obviously written by Courant. A considerable portion of it was based on his own investigations. "Otherwise," pointed out the American mathematician Einar Hille, "the simple choice of methods, the fondness for heuristic considerations, and a certain delicate touch of pen, sometimes a bit vague but always elegant, betray the writer if nothing else does." The book also contained various errors, a number of which were to persist into the English translation made many years later. There—according to Alexander Weinstein, a later reviewer—"being by now classical, they only add to the pleasure of the reader."

The mathematical methods treated in Courant-Hilbert, as in Rayleigh, pertained to subjects which had originated in such areas of physics as elasticity, acoustics, hydrodynamics, and other classical subjects. In the hands of mathematicians these methods—which had their roots in physical intuition—had been transformed into rigorous tools backed by general theories. It was Courant's purpose to return the improved tools to the physicists for their own work.

At the time of its initial publication, however, Courant-Hilbert seemed of more interest to mathematicians than to physicists. The physics with which it concerned itself appeared to be, in 1924, rather old-fashioned to men who were striving toward some sort of understanding of the quantum of energy. In fact, Courant-Hilbert seemed a perfect illustration of Friedrichs's definition of applied mathematics as "those areas of physics in which physicists are no longer interested." But Courant was absolutely convinced that the content of his book would be important for physicists.

"You couldn't have told him different," Friedrichs said. "He wouldn't have believed you. Because of his optimism."

With a smile Friedrichs put his copy of Courant-Hilbert back into his bookcase.

"And of course he was right."

ELEVEN

THE POST-WAR YEARS saw a number of changes in Göttingen. As Klein's successor, Courant quite soon made an instructional innovation, the *Anfängerpraktikum* or "beginners' practice period," which was to have important results for the new "institute."

The *Praktikum* paralleled the calculus lectures. The students, often numbering as many as two hundred, received a mimeographed sheet of problems, some requiring inventive thinking as well as understanding of the material of the lectures. The professor regularly held a conference with a group of older students, discussing the problems and pointing out different methods of attack and various aspects of the solutions. The older students then went over the problems in the *Praktikum* with the beginning students and at the same time became personally acquainted with them. Solutions were written up and graded. Collusion was encouraged. Attendance was purely voluntary.

It is hard to realize today what a revolutionary innovation the *Praktikum* was in a German university at that time. Up until then, problems were never handed out except in applied courses, where they were usually not corrected. Textbooks were rarely utilized. Examinations were not given. The whole system was one of lecturing on the part of the professors and listening on the side of the students. The moment of truth did not arrive for several years, when the students had to take the state examination for teachers or the oral examination for the doctor's degree. For some, only then did it become clear that *knowing mathematics* is not like knowing the plot of a work of literature or the general outline of a historical period. The shock they experienced at this revelation not infrequently resulted in a nervous breakdown.

The *Praktikum* was a way of coping with a greatly increased number of students and a much lower level of ability and preparation. It required a group of older students to supervise it and these became additional "assistants" with appropriate financial support from the government. It also required *space*. As a result Courant was able to obtain rooms in an old building on Prinzenstrasse, from which Wilhelm Weber and Gauss had strung the wires of the first telegraph to Gauss's observatory. With this increased base, the paper "institute" was on its way to physical reality.

Klein never tried to interfere, Courant told me, but yielded the reins gracefully. There were, however, still signs of the old imperiousness. A favorite story of Courant's was how Klein, convinced at one point that death was actually at hand, called his assistant and dictated the arrangements for his funeral "and then was very displeased that he had not died after all." But there were also signs of mellowing. When a group of teacher-training students complained that Neugebauer, the assistant in charge of the *Lesezimmer*, had relegated Klein's collection of books on elementary mathematical pedagogy to the topmost shelves and packed them in so tightly that not a single one could be removed except with great difficulty, Klein summoned Neugebauer. He made only one comment to the young man, whose growing interest in Egyptology he was aware of.

"There came a new Moses," Klein said, "who knew not Pharaoh."

Changes were also taking place in the famous faculty.

Hilbert worked with passionate intensity on his program to shore up the logical foundations of mathematics against the attacks of L. E. J. Brouwer and his followers; but he was seriously, perhaps fatally, ill with pernicious anemia. In the spring of 1925 Runge retired; and Gustav Herglotz, a remarkably wide-ranging mathematician, the teacher of Artin, succeeded to Runge's chair. In June, that same year, Felix Klein died.

Courant wrote two moving articles about Klein. One was an account of his activities as a scientific leader; the other, an account of his life, work, and personality.

In later years, it is true, Courant was inclined to sacrifice Klein to the temptation of a good story. He most often presented him in the role of the man who planned his own funeral, the Jove who brooked no opposition, who always considered that he knew what was best for everyone, who demanded servility from his assistants and sometimes so intimidated his students that on social occasions they stood up when he addressed them—a vivid contrast to the admired, independent, unconventional, and rebellious Hilbert. When, however, I first read Courant's long article on Klein, published in *Naturwissenschaften* in September 1925, I was reminded of a remark he had once made to me about Hilbert's memoir of Minkowski.

"In that," Courant had said, "Hilbert revealed more of his own soul than he ever did at any other time."

In the *Naturwissenschaften* article Courant vividly recalled Klein's life. The gifted boy rebelling against the one-sidedness of the education offered by the humanistic gymnasium of his youth. The precocious university student, seventeen years old and already the assistant of Julius Plücker at Bonn. The recently promoted young doctor falling immediately under the spell of Göttingen. The independent scholar, stubbornly maintaining his independence from any "school," even Göttingen. The geometrically-minded and intuitive mathematician dramatically confronting in Berlin the abstract and arithmetical trend of the day. The fiery patriot, breaking off studies in Paris in 1870 to rush back to Germany to enlist in the army. The 23-year-old professor, rejected for Alfred Clebsch's chair in Göttingen because he was "too dangerous," delivering the famous Erlangen Program as his inaugural address at that university. The busy administrator and teacher of the Munich days, coming in contact for the first time with the applications of mathematics. The still young professor in Leipzig, at the height of his creative powers, locked in a furious race with Poincaré toward a theory of automorphic functions. The breakdown at the age of thirty-three which effectively ended Klein's career as a productive mathematician. The offer from America to succeed Sylvester at Johns Hopkins and then the offer from Göttingen. "The wonderful turning point"—the seemingly broken man who had lived another forty-three years and displayed the most varied sides as researcher, teacher, organizer, and administrator.

"What had been the secret of this personality and its workings?" Courant asked. "He possessed great power over men, because he combined spiritual superiority with accompanying objectivity; because he was not doing something for himself alone but was always directed toward his goal; because in the majestic dignity of his being there was no trace of vanity and no overweening opinion of himself. He did not lack the true humor which is the mark of real spiritual freedom. But all these things were eclipsed by the magic of his being, the magnetic power with which he was able to produce followers and to make co-workers out of even the reluctant."

Klein had lived long enough to see Göttingen once more an international center of mathematics and physics, foreign visitors coming again from east and west, the *Lesezimmer*—replenished and renewed—again the finest library of its kind in the world, an impressive ring of technical institutes—the result of a long collaboration between science and industry—surrounding the university and flourishing again. But he had died with the possibility of an institute for mathematics, which was to have been the climax of the whole development, still to all appearances hopelessly remote.

Then, just a few months after Klein's death, the city of Göttingen proposed a plan to build a new secondary school and to vacate the old school building, which was close to the physics institute and also to many of the other technical institutes. Such physical as well as intellectual proximity to the neighboring sciences had been an integral part of Klein's dream. Courant immediately tried to seize the opportunity, not only to bring together the various mathematical activities in research and instruction, scattered about the city, but also to relieve "catastrophic crowding" in Franck's physics institute. The government was willing to buy the old school building from the city and turn it into a mathematics institute. The only difficulty was—as Courant expressed it—"the city of Göttingen can give up its old school only if it can first build a new one and it can build a new one only if it first has the money—and it has no money."

In the late fall of 1925, during a visit with the Bohr brothers, Courant brought up this problem. The city needed 800,000 marks to build a new school, and such a loan was simply not obtainable in Germany at that time. Niels Bohr suggested approaching the International Education Board, which had recently made a grant to improve Bohr's institute in Copenhagen. Apparently Courant had never thought of the Board as a source of possible funds for the project, although he had already recommended several mathematicians for fellowships and had also asked for assistance in the purchase of foreign books and periodicals for German libraries. He immediately had a "fantastic" idea. Perhaps the Rockefeller people would be willing to lend the 800,000 marks to the city of Göttingen for the new school; then the university, with the money already promised by the government, would be able to purchase the old school. Niels Bohr advised him to forget the idea of buying and remodeling the old building and instead to apply for an outright grant. This, with the money available from the government, would make possible the erection of a new building specifically designed to meet the needs of the mathematicians and physicists.

Before the three friends separated, they decided that Courant should write a description of the problem and of his proposal in the form of a personal letter to Harald Bohr, who would shortly be in Paris, where the International Education Board had its European headquarters. Harald Bohr could then either discuss the idea with Augustus Trowbridge, a former Princeton physics professor who was the director for natural sciences, or simply present Trowbridge with Courant's letter, if that seemed more appropriate.

Trowbridge had visited briefly in Göttingen the previous fall to acquaint scientists there with the objectives of the fellowship program. It was, he had explained, less the desire of the International Education Board to improve the average of research output than to facilitate the development of exceptional scientists—or, as the French mathematician Émile Picard put it, "to make the high places higher rather than to fill in the valley with peaks." Trowbridge had also told Courant and Franck that they should consider what the International Education Board could do to help in the furthering of scientific life in Göttingen. With such an invitation, Courant did not feel that he was being presumptuous in approaching Trowbridge, but still the sum involved was staggering.

"Do you think that we dare to speak to Mr. Trowbridge . . . of such an extensive request and that we are actually justified in thinking in these terms?" he wrote to Harald Bohr in Paris on December 22, 1925. "I personally believe that through the carrying out of this plan a completely unique center of mathematics and physics would be created here in Göttingen and that the Rockefeller people do not have to worry about wasting their money. But then I am not impartial."

Harald Bohr's response to Courant's letter and his report on his hour-long interview with Trowbridge could scarcely have been more encouraging:

"The chief thing is that Mr. Trowbridge has received your general proposal with great interest and warmth, and he is willing, already now—without any official steps having been taken—to discuss and pursue the idea."

By the beginning of 1926 Courant was in correspondence with Trowbridge and, at the American's suggestion, was sounding out the German government about its willingness to cooperate by contributing funds of its own to the project. Plans were soon being made for Trowbridge and George Birkhoff to come to Göttingen in the summer for further discussions.

Birkhoff, whom Courant had never met, was by then recognized both at home and abroad as the leading American mathematician. He was a professor at Harvard, the most recent president of the American Mathematical Society, and the youngest man elected up to that time to membership in the National Academy of Sciences. His role in relation to the International Education Board—as Trowbridge described it to Courant—was that of a traveling American professor retained to advise the Board in the field of his expertise.

Unfortunately Birkhoff's visit to Göttingen was to coincide with that of another, younger American mathematician who felt that he had many grievances against Birkhoff. It was thus to play a part in the development of an unforgiving enmity toward Courant on the part of Norbert Wiener.

Wiener had studied briefly in Göttingen before the war. On his return to America, in the course of delivering lectures at Harvard, he had run "into a series of logical difficulties which were clearly pointed out to me by Professor G. D. Birkhoff. . . .

"He was, as I was later to learn, intolerant of possible rivals, and even more intolerant of possible Jewish rivals. He considered that the supposed early maturity of the Jews gave them an unfair advantage at the state at which young mathematicians were looking for jobs, and he further considered that this advantage was particularly unfair, as he believed that the Jews lacked staying power. At the beginning I was too unimportant a youngster to attract much of his attention; but later on, as I developed more strength and achievement, I became his special antipathy, both as a Jew and, ultimately, as a possible rival."

Wiener had also spent the summer semester of 1925 in Göttingen and had received the impression that he had been encouraged by Courant ("an industrious, active little man who was eager to keep all the strings of mathematical administration in his hands") to apply for a Guggenheim Fellowship and return to the university the following summer. Courant had, Wiener thought, promised him "the full cooperation of my Göttingen colleagues in making my trip agreeable and in providing me with an assistant to help organize my papers and to take care of my lapses in German."

As it turned out, although Wiener was in Courant's opinion an "extraordinarily talented and powerful" mathematician, his presence in Göttingen in the summer of 1926 was an embarrassment. Wiener's father, a professor of languages who had pushed his son into college while the boy was still in knee pants, had been the most violent anti-German agitator in American academia during the war. The publicity which the young Wiener received at the time of the announcement of his fellowship ("I had been," he conceded, "a bit loquacious.") called the attention of people in the government to the fact that the son of this abhorred man was coming to study in Germany and was actually boasting about being welcomed there. Courant was requested by the *Kurator*—the representative of the government at the university—please, to try to keep Wiener inconspicuous. When Wiener arrived, recently married but alone, he found his welcome much less enthusiastic than he had expected. Courant "scolded" him for his newspaper publicity and did not give him the help and official recognition which he thought he had been promised. As a result, Wiener felt, his lectures were less successful than he wished, both as examples of mathematical research and as lectures in the German language.

Wiener, according to Friedrichs, was regarded in Göttingen as "rather uncultivated" in his mathematical writing and lecturing.
"Now perhaps 'uncultivated' is not the right word," Friedrichs said. "But I have studied some of Wiener's writing and one of his books I've read forwards and backwards. It *is* written in a clumsy way, not only in language, but also in substance. The mathematical argument is unusually inventive, original, surprising, but it is not streamlined. So to say, the judgment that it was 'uncultivated'—'clumsy' would have been the better word—was not quite unjustified; but the conclusion that the substance could not then be so hot, was wrong. And there we made a mistake in Göttingen."

Courant was disposed, he told me, "to do something for Wiener" because the American was a cousin of Leon Lichtenstein, a good friend. But it was only with great difficulty that he was able to persuade about twenty students to attend Wiener's lectures. As time went on, that number dwindled so embarrassingly that at one point he had to pay a student to attend.

Wiener had expected that the recognition he would receive in Göttingen would enable him to get out from under what he saw as "the continuous hostile pressure" of Birkhoff in the United States—and now Birkhoff was coming to Göttingen!
"[Birkhoff] represented the American whose support Courant most wanted. Courant approached me as an avenue through which he might win Birkhoff's goodwill. I told him that I had no influence whatever with Birkhoff and that Birkhoff's entire reaction to me was hostile."
By the time Wiener's bride arrived, Wiener felt that he was on the edge of a nervous breakdown. At this point his parents also descended upon the shaky new household, "partly to share in my supposed success and partly to keep a supervising eye on the newly married couple." The elder Wiener wanted to give a public lecture in Göttingen, and the son was forced

to explain the rebuffs he had received. He found his father more concerned about the rebuff to himself. It was impossible to keep him from writing a letter to the ministry denouncing Courant.

The unhappy experience of Wiener was in complete contrast to that of another foreign visitor to Göttingen that same summer.

Since 1923 Alexandroff had returned each year, either alone or accompanied by countrymen. From 1926 through 1930 Courant always arranged for him to give regular lectures. Once the Russian mathematician taught three courses in topology, each for a quite different audience of mathematicians. The summer that Courant was trying to keep up some semblance of attendance at Wiener's lectures, Alexandroff's were crowded.

"Alexandroff and the other Russian vistors were very important, very influential," I was told by Herbert Busemann, who came to Göttingen in 1925 to study mathematics after "wasting"—as he said—several years of his life in business to please his father, one of the directors of Krupp. "The Russians filled a gap because they were familiar with certain more abstract tendencies which were not well represented in Göttingen. Courant, as probably many have told you, was rather reactionary in his mathematical outlook. He didn't see the importance of many of the modern things."

The subject matter of Alexandroff's work, unlike that of Wiener, was far removed from Courant's own—very pure and very abstract. Courant, however, was extremely enthusiastic about it, as he was always to be about the work of the many other Russians whom Alexandroff introduced into Göttingen over the years. They were all "of the first rank," he wrote somewhat later. The blind Pontryagin, who had always to be accompanied, was "an absolutely leading spirit in topology." Gelfond "had astonished the world" with his proof of the transcendence of certain famous numbers; the work on prime numbers by Schnirelman, who walked barefoot through the streets of Göttingen, "likewise." Lusternik "had the most highly original thoughts" concerning topology and analysis. Kolmogoroff was "an absolute master" in a great variety of fields.

To Alexandroff this enthusiasm for the work and the success of other mathematicians was impressive.

"There is very seldom the combination of really great scientist and great human personality," he said to me when I talked to him in Moscow in 1971. "And the emotional nature of Courant, his unselfish relation to other men, his interest in other human beings, his lack of egotism in relation to the world—he is absolutely exceptional. I do not know of any other man who has this kind of personality, this unselfish position in the science, and this *joy*—do you understand me?—in the success of other, younger men around him. Courant was himself quite a young man at that time, but he was like that even then; and I think that really explains him and his success."

The Göttingen summer of 1926 had a special significance for Alexandroff, because it was then that he met Heinz Hopf, with whom he soon established a professional collaboration and a friendship that was to last until the end of Hopf's life. Although Hopf was two years older than Alexandroff, he had only recently taken his doctor's degree in Berlin. This lateness was partly because he had served on active duty as an artillery officer during the war and partly because he had developed very slowly as a mathematician. Courant was also enthusiastic about Hopf. Recommending him the following year for a professorship in Switzerland, he predicted "in a few years he is sure to be known throughout the world."

Even in those long ago days, Alexandroff wore extremely thick lenses and was already quite bald. When I visited him in Moscow, he still looked very much as he had in snapshots I had seen of him from the 1920's. On a table in his crowded, rather Germanic living room there was a photograph of Urysohn, who, swimming with Alexandroff on the coast of Brittany in 1924, had been dashed against the rocks and killed at the age of twenty-six. I had recently left Courant, tired and depressed, in America. Hopf had just died in Zurich. Alexandroff was recuperating from an almost fatal illness which had prevented him from attending Hopf's funeral. We talked for quite a time about that long ago summer of 1926, when he had first met Hopf, and the other summers which he had spent in Göttingen.

"It was a beautiful time in my life," he told me as I prepared to leave, "and I cannot forget it yet."

That summer of 1926 was also the summer when Birkhoff came to Göttingen to discuss the proposed grant to the university from the International Education Board. The unhappy and lonely Wiener tactfully kept out of the way while Courant and Franck met with Birkhoff and Trowbridge for two days.

Courant

The most extravagant and completely satisfactory plan considered at the meeting was the erection of a new building for mathematics and the improvement of the already existing physics building. Not quite so satisfactory but still very nice was the purchase and remodeling of the vacated high school building and the including in it of the necessary additional facilities for physics. If neither of these proposals was acceptable, Courant still had hopes that the International Education Board would grant funds for some improvements in the existing mathematics and physics facilities.

No final decision was reached during Birkhoff's visit, but it was agreed that detailed sketches and estimates of cost would be made for all three proposals. For this work Courant turned to Neugebauer, whose artistic gifts and knowledge of engineering, draughtsmanship, and the needs of the mathematicians made him the perfect choice. Courant and Neugebauer were eleven years apart in age and of completely different characters. "Courant's word was *maybe*," Hans Lewy said to me. "Neugebauer's was *yes yes, no no*." But the two were fast becoming an administrative team, the purposeful disorganization of Courant balanced by the efficiency of Neugebauer. By the end of October 1926, working together, they had managed to get all the necessary plans and estimates to the International Education Board.

At Christmas word came at last from Trowbridge.

"I may say informally . . . ," he wrote to Courant, "that the action of the Board was to vote into the hands of the Executive Officers of the Board a sum not to exceed $350,000 for the construction and equipment of a building for the Mathematical Institute, and for the construction and equipment of an addition to the Building for the Physics Institute of the University of Göttingen, with the understanding that the Prussian government will provide annually a sum not less than $25,000 to cover additional maintenance costs of the Institutes of Mathematics and Physics."

Trowbridge's letter was dated December 21, 1926, a year and a half after the death of Felix Klein.

"So Klein never knew?" I asked Courant as we sat in his office on the thirteenth floor of the Courant Institute of Mathematical Sciences in New York City.

He gazed out the large corner window at the wooden water towers topping the buildings of Greenwich Village.

"No," he said. "Klein never knew."

TWELVE

THE YEARS during which a mathematics institute was coming into existence in Göttingen were to have about them, in memory, a kind of glow.

In comparison to the hard post-war period, the country was enjoying a rather remarkable prosperity. Von Hindenburg, the old victor of the Battle of Tannenberg, was lending respectability to the beleaguered new republic. Foreign relations were improving. In 1926 Germany was admitted to the League of Nations.

In Göttingen there was again a scientific paradise. Gifted students flocked to the university. There was a constant procession of distinguished visitors from all over the world. Sometimes they came merely for a single talk or a series of talks to the *mathematische Gesellschaft*. Often they lectured for a full term as guest professors. The very air seemed to Courant to crackle, as it had in his youth, with scientific electricity.

Courant himself, in his late thirties, could view his situation with satisfaction. He was a German professor, a uniquely comfortable, secure, and respected person in the Germany of the time and in a small university town like Göttingen. His personal life was highly satisfactory. Two maids relieved Nina of the responsibility of house and children. She concentrated on her own playing of the viola da gamba, an old instrument of the viol family, and also directed a group that specialized in the singing of choral music. There were yearly skiing trips with family and assistants to Arosa in Switzerland.

Courant's relations with the ministry were as close and friendly as Klein's had been. His opinion was sought when university appointments were made. He had considerable influence with the International Education Board. His files are full of letters beginning "I am writing for Kurt Hahn, a friend of our Göttingen circle," "for my old teacher, Professor Adolf Kneser," "for my colleague Franck." Trowbridge relied upon him to evaluate other professors' recommendations for Rockefeller Fellowships, "since you are yourself so familiar with the type of man we are trying to help with these fellowships."

In scientific publishing Courant and Springer were an influential team. The *Mathematische Zeitschrift* was followed by a comparable publication in physics. The old established journals and the mathematical encyclopedia had fallen years behind the scientific work actually being done. Now, for a change, important work in both mathematics and physics got into print as soon as it was completed, fields where current interest lay were promptly explored by competent authorities. The Yellow Series, emphasizing the connections between mathematics and physics, grew volume by volume.

Although Carl Still was, as Courant always said, "no Krupp," he stood generously by with financial help where it was needed. His friendship with Courant extended to Courant's friends and colleagues. On the Still estate they hunted hare, discussed their scientific problems, and played with scientific toys. Franck and Pohl set up a laboratory. Runge installed a telescope in an observation tower. Prandtl investigated the irrigation problems of the estate.

In 1925, with Friedrichs as his assistant, Courant began to work on a second volume of Courant-Hilbert. He also decided to edit and publish his calculus lectures.

Although there were a number of calculus books in print, he thought it would be very difficult for the beginner to locate a book which would open up to him "the direct way into the living essence of the subject" and give him "freedom of movement in relation to the applications." In his book, as in his lectures, he wanted to show the close connection between analysis and applications and to emphasize intuition as the original source of mathematical truth. Most important, he intended to proceed as directly as possible to interesting and fruitful topics.

The calculus was the first of Courant's books published under his name alone. It might well have been called the Courant-Klein; for, as Courant acknowledged in his foreword, "What I have here attempted is completely in the direction of my great predecessor, Felix Klein."

In spite of the single authorship, the calculus, which was published in two volumes in 1927 and 1929, was a communal effort.

Notes on Courant's lectures, written up by earlier assistants, were already in existence. These were sent to Springer and put into type. Then the galley proofs arrived. The rooms in which the *Praktikum* took place became the scene of "proofreading festivals." All the assistants took part, all offering—as Neugebauer was later to recall—"simultaneously uttered and often widely diverging individual opinions about style, formulations, figures, and many other details."

The corrected galley proofs then went to Courant and then back to Springer. The changes were made. Fresh galley proofs arrived in Göttingen.

Friedrichs shook his head, remembering.

"And they were again full of mistakes, because as in the beginning

Courant did not bother about technical details. So we assistants corrected these mistakes, and then we had first page proofs and even second page proofs. The outcome—well, you could not recognize it from what had been there originally."

The mystery was that the final product of this process was unmistakable and pure Courant.

"People who have not participated sometimes say that Courant's books were written by his assistants," Friedrichs said. "But this was not true. In a way Courant *would have liked* his assistants to write them. But they couldn't do it. Essentially the books were always Courant's books."

The assistants who supervised the *Praktikum* were talent scouts as well as proofreaders and editors. At the beginning of the term in 1925, they were delighted to discover that the answers of a new student from Yugoslavia were invariably correct and that there was no longer any need for them to solve the problems themselves. They promptly alerted Courant to the presence of Willy Feller. After the third calculus lecture, to Feller's amazement, the professor—an unbelievably august personage to a European student of that day—approached. Questioning the boy about his education in his native land, Courant discovered that Feller was already doing mathematics on his own. He told him to bring his work to the next lecture. Even thus instructed, Feller was too bashful to produce his papers on the appointed day. The next morning he was awakened by a commotion on the stairs leading to his attic room. There was a knock on the door. Courant entered and left a few moments later with the desired papers.

After Feller was "discovered" in the *Praktikum,* he was an accepted member of the new "in group" which gathered around Courant. As in the past, there was no clearly defined standard of admission to this group. The year after Feller came, Franz Rellich turned up in the *Praktikum*. He was an Austrian, only twenty years old but of such charm, openmindness, and independent judgment that he was accepted into the group even before he had proved himself mathematically.

One other young man who came to Göttingen during this time also should be mentioned; for although he was never a member of Courant's group, Courant played a role in his coming.

At the beginning of 1926 Hilbert proposed to the International Education Board that a 22-year-old Hungarian, Janos (John) von Neumann, be given a fellowship to come to Göttingen and work with him. At this time von Neumann did not yet have his Ph.D. Trowbridge replied that, although the degree was not a *sine qua non* for a fellowship, "in this case I rather judge that, combined with [the] youth of the candidate, he would probably not fall into the class which we are trying to reach with these fellowships—that is, the class of a man well advanced in his career, who has already published a sufficient amount so that one can form a judgment of his technical ability."

While the ailing Hilbert was in Switzerland, Courant took up von Neumann's case.

"It appears to me as though Hilbert somewhat unclearly expressed himself in his first letter," he explained to Trowbridge. "Mr. von Neumann, in spite of his youth, is a completely exceptional personality . . . who has already done very productive work . . . and whose future development is being watched with great expectation in many places."

The work for which Hilbert wanted von Neumann was of the greatest significance in Courant's opinion—"it concerns investigations in the foundations of mathematics and a program on completely epoch-making lines initiated by Hilbert . . . which Mr. von N. has already pursued independently."

There was also another consideration. Hilbert had failed greatly since Trowbridge had last seen him. Although it was hoped that as a result of a treatment for pernicious anemia recently discovered in America he would be able to overcome the most serious aspects of his illness, there was no time to be lost in promoting the carrying through of his scientific program: "I believe that in this way one really can perform an important service to science."

Under the force of Courant's plea, Trowbridge reconsidered Hilbert's request; and in the fall of 1926 von Neumann came to Göttingen as a Rockefeller Fellow. The young mathematicians there recognized that he was obviously a prodigy, but some were suspicious of what they saw as a certain "glibness" about him. They also found his mathematics "too abstract" for their taste.

"We were wrong about that," confessed Friedrichs, part of whose later work was to be strongly influenced by the work of von Neumann.

Friedrichs told me that he and Lewy were also somewhat skeptical of the developments which were taking place in physics.

In 1925 one of Born's assistants, Werner Heisenberg, formerly a student of Sommerfeld in Munich, had created quantum mechanics. Born and Pascual Jordan (who had come to Born from Courant) had followed this with a striking mathematical formulation which rounded out the discovery and had published this work with Heisenberg.

"I was very much repelled by what the physicists were doing," Lewy admitted to me. "To my mathematical mind they were too sloppy and also their way of talking was so glib that I had the impression—which was wrong, of course, as it turned out—that they were fourflushers. Often if you would corner them, ask them to explain precisely, they would be evasive or else you would find out that they didn't quite understand what they were saying themselves. They obviously had some physical intuition which I didn't have, but their mathematics was objectionable. The type of personality that was needed at that time in physics was not to my liking. Well, what happened teaches one humility, and it shows that different situations in science require different types of personalities. So one cannot make up one's mind what is *the right way* of being for a scientist."

I asked Courant if he had had any such qualms, and he said no. Within a few weeks Niels Bohr had publicly placed his stamp of approval on Heisenberg's work.

The physics being done at this time in Göttingen and other places was to result in a dramatic and unexpected professional triumph for Courant.

Heisenberg's quantum mechanics of 1925 was followed by Erwin Schrödinger's apparently different wave mechanics in 1926. Physicists struggled to deal mathematically with strange and unfamiliar ideas.

In Schrödinger's theory (which Schrödinger himself soon showed to be mathematically equivalent to Heisenberg's theory), a stationary state of a physical object, such as a hydrogen atom, is described by the solution of a certain partial differential equation which has solutions only if the value of a certain parameter, representing the energy of the object, is one of a particular sequence of numbers. When the numbers in the sequence were recognized as the eigenvalues of the equation and the stationary state was seen to be described by the associated eigenfunction, the physicists turned to the mathematicians to find out where they could learn about eigenvalues and eigenfunctions. To their surprise they were told that exactly what they wanted was ready and waiting for them in Courant-Hilbert, the book which they had dismissed several years earlier as nice but somewhat old-fashioned—obviously irrelevant to the modern physics of the quantum.

The particular importance of Courant-Hilbert for the physicists was not, however, the new mathematical material it contained, such as Courant's own theory of eigenvalues and eigenfunctions. Rather it was the exposition of the classical theory, which Courant had laid out in detail as the foundation for the presentation of his work.

The story of Courant-Hilbert had been told many times by historians of science.

"The book, in compact, convenient form, contained practically every mathematical method, trick, device, and special detail required for the development of the Schrödinger theory, not to mention much that was applicable to the theory of Heisenberg," Banesh Hoffman wrote in his book on the story of the quantum.

"In retrospect, it seems almost uncanny how mathematics ... prepared itself for its future service to quantum mechanics," Max Jammer observed. "Published at the end of 1924, [*Methoden der Mathematischen Physik*] contained precisely those parts of algebra and analysis on which the later development of quantum mechanics had to be based: its merits for the subsequent rapid growth of our theory can hardly be exaggerated."

It was, E. T. Bell said simply, one of the most dramatic anticipations in the history of mathematics.

As it happened, during the same period that physicists were finding a need for the mathematical methods of Courant-Hilbert, Courant was engaged in another work which would also turn out to be a dramatic anticipation.

Since his student days, when the much admired Walther Ritz had utilized Hilbert's approach to Dirichlet's principle to develop an efficient numerical method for solving the boundary-value problems of partial differential equations, Courant had been intrigued by the fact that the same methods which lead to proofs of existence, as in the case of Dirichlet's principle, can also be useful in that most practical chore of mathematics—numerical calculation. Was it possible, he now wondered, to proceed in the other direction? Could a method of approximation used for numerical calculation also be employed for existence proofs?

The particular method which Courant proposed to investigate was one long used by engineers to obtain *approximate* solutions of problems the solutions of which are given *precisely* only by the solution of the corresponding partial differential equation. In this method of "finite differences," the plane is conceived as a lattice of a given mesh. The values of the function are sought only at lattice points rather than at every point of the plane. A suitable finite difference equation can then be taken as a substitute for the harder to handle, partial differential equation.

It was Courant's idea to investigate the purely mathematical significance of this method of the engineers—to take it with full mathematical seriousness. First, he suggested, it should be determined whether finite difference equations are meaningful in a mathematical sense; that is, whether solutions of them in fact exist. Second, it should be determined whether, when the mesh is taken sufficiently fine, the solutions of the finite difference equations actually do approximate the solutions of the corresponding partial differential equations.

In 1925, in his first work on these questions, Courant carried out the finite difference method for the same partial differential equation problem to which the Dirichlet principle was applied. One of his main tools was the lemma which he had employed in earlier work on Dirichlet's principle.

The partial differential equations which Courant treated were of the so-called elliptic type. Such equations govern equilibrium; for their solution conditions have to be prescribed on the boundary of a body. Equations governing motion are hyperbolic; the heat equation, parabolic. After Courant's work the question naturally arose whether the method of finite differences could be applied to hyperbolic and parabolic equations. For the solution of these equations, conditions are prescribed at an initial time.

"It is a remarkable fact," Friedrichs told me, "that on purely mathematical grounds elliptic equations would not have acceptable solutions if initial conditions were prescribed for them; and hyperbolic and parabolic equations would not have solutions under boundary conditions. This fact—which the French mathematician Jacques Hadamard pointed out in the introduction to a very long book in 1921—was emphasized over and over again to us by Courant. At first it may seem trivial. It is not really obvious that it is deep. But it is. Very deep. I think it is one of Courant's definite merits that he spread the news, as it were, about this idea of Hadamard's. It has had a great effect in mathematical analysis, but I think it might have gone unnoticed if it had not been for Courant."

As always, following the admired example of Hilbert, Courant proceeded to bring his new work before his students. In 1926 he conducted a seminar on finite difference equations and suggested to Friedrichs and

Lewy that they should see what they could do in the hyperbolic case. This was the summer of Norbert Wiener's unhappy experience in Göttingen. He attended the seminar; but, as Friedrichs remembers, he did not give a talk. He seemed, however, quite interested in the work reported on at several seminar meetings.

That fall Courant had some more results on the elliptic equations for the equilibrium problem. Friedrichs and Lewy had some results on the hyperbolic equations for wave propagation. Courant suggested that the three of them present their work in a common paper.

In future years the Courant-Friedrichs-Lewy paper, published in 1928 in the *Mathematische Annalen*, was to play a considerable role in numerical analysis. In particular, certain severe restrictions—the need for which was first recognized by Lewy—were to become of fundamental practical importance with the development of computing methods during the Second World War.

While working on the finite difference approach, in another joint work, Friedrichs and Lewy had also been trying to approach directly the related hyperbolic partial differential equations. They had recognized that they could establish the uniqueness of the solutions of these by a certain "energy inequality," but they were baffled by the problem of establishing the existence of the solutions. Then one day when they were discussing their work on a walk in Göttingen—Friedrichs still remembers the spot exactly—Lewy suddenly stopped. "I've got it! I've got it!" he shouted. "We can use the energy inequality for existence, too!"

"I was absolutely dumbfounded," Friedrichs recalled. "It was an uncanny insight. A considerable part of the existence theory of partial differential equations has since stemmed from this idea. I think it is a good example of Courant's influence as a teacher that it was his interest in the subject that drew Lewy into it."

Once I asked Courant who of all his students had most *surprised* him, meaning which one had become a greater mathematician than he had expected him to become. At first he misunderstood my question and thought I was asking which of his students had had the greatest ability to surprise him. To this question, his answer was "Lewy—he is a most original mathematician." When I repeated my question, however—understanding this time—he replied without hesitation, "Friedrichs."

"I do not think, however, that most teachers would have perceived Friedrichs's ability as early as Courant did," Lewy told me.

"But you said how greatly you were impressed by Friedrichs when you first came to Göttingen," I reminded him.

"Yes, but I was a fellow student. I think it took a keen observation as well as a really intense human interest on Courant's part to see what was there. Friedrichs was very shy. Courant forced him out into what were difficult situations for him. He sent him to work with von Kármán at the aerodynamics institute in Aachen. I went to the station with Friedrichs. He was a very unhappy young man. If it had not been for Courant, he might have just pulled into himself and become a gymnasium teacher."

The proof sheets of the final common work with Courant were shown by Friedrichs and Lewy to van der Waerden, who was an active member of the Göttingen circle at that time. When I interviewed him in the summer of 1971 in Zurich, he still spoke with enthusiasm—almost half a century later—of the vivid impression made on him by the paper on the *existence* and *uniqueness* of the solutions of partial differential equations in the elliptic, hyperbolic, and parabolic cases. Before he came to Göttingen, he told me, he had agreed to tutor an engineer who wanted to learn something about partial differential equations so that he could solve an equation relating to the conduction of heat in a cylinder.

"I had a bad conscience about him because I had taken his money and he hadn't learned anything useful. So when I came to Göttingen, my mind was much more open to these important things. All my life," he added, "I have had great advantage from having read that paper at that moment."

At the time of publication of the Courant-Friedrichs-Lewy paper, however, engineers did not pay any more attention to the mathematicians' results on finite difference equations than physicists had paid to the mathematical methods of Courant-Hilbert when it first appeared.

Courant did not consider himself an applied mathematician, but he always believed that mathematics includes the applications and cannot be separated from them. When Runge died in 1927, Courant expressed this feeling in an article about his father-in-law—which had unexpected repercussions.

In the article he pointed out that when, in 1904, Klein had established Runge in Göttingen as the first professor of applied mathematics at a

German university, Klein had taken "the decisive step to retrieve for the applications their proper place in our science." That Klein, "who always strove to preserve science as a unified whole," had felt it necessary to take such a divisive step had been due to the conditions in mathematics at the time. But, in Courant's opinion, such a separation between mathematics and applied mathematics was no longer necessary. It was Runge's great achievement that he had repaired the broken connections with the applications and restored the unity of mathematical science, "which *includes* the applications." Thanks to Runge, it was no longer necessary to designate someone as "a professor of applied mathematics."

Such heresy brought a strong rebuke from Richard von Mises, professor of applied mathematics at Berlin.

"It is a monstrous misconception of the present situation to think that one can now renounce the cultivation of applied mathematics in special institutes and with special teaching positions."

Courant held firmly to his position.

"I would go a step further than Mr. von Mises, in that I consider we should strive for the desirable goal that *no* teacher of our subject stands coolly apart from the applications. . . . But where I am absolutely in opposition to Mr. von Mises is in the idea that it is possible to strengthen the applications by separating them from the purely theoretical science."

It was during this same year (1927) that Trowbridge asked Courant to compile a "Who Is Who" in American mathematics for his own use and that of the International Education Board.

Courant didn't feel that he was really qualified to make such an evaluation, but he responded by listing American names with which he was familiar. He also added some general observations about contemporary American mathematics.

Great improvement had taken place in the last ten or twenty years. A truly native scientific culture had developed. And it was impossible for anyone to predict what might come out of it in the near future. The predominant characteristic of American mathematicians seemed to be a tendency to favor the abstract and the so-called pure areas of mathematics. Their greatest success had been in topology. Princeton, especially, and Harvard were without a doubt the best places in the world to study that subject. Applied mathematics, however, was treated like a stepchild in America. There was no real contact between mathematics and physics—as far as Courant could see—and hardly any between mathematics and technology.

"But," he told Trowbridge, "that situation too could quickly change."

For Courant post-war Göttingen was always to remain in memory a paradise where the distinction between "applied" and "pure" mathematics did not exist, a place where there were "mathematicians, abstract mathematicians and *more concrete* mathematicians and physicists talking to each other quite intensely and very frequently—and understanding each other." In the years to come, when he was no longer in Göttingen, he was to see it always as his task "to restore this easy communication" between mathematics and its applications and "to bring mathematics back into the mainstream of science."

THIRTEEN

IN SPITE OF the glow which surrounded the Göttingen of the 1920's, it should be said that there also existed in those years a certain amount of hostility toward the scientific faculty of the university both in Germany and abroad.

Many on the outside deeply resented the tendency of the people in Göttingen to consider themselves at the center of the world and to be careless in paying attention to work done by others at other places. In 1928, when the Courant-Friedrichs-Lewy paper appeared in the *Mathematische Annalen*, Norbert Wiener pointed out that in Courant's part of the paper he had introduced, among other ideas, the probability interpretation of finite difference equations without mentioning a similar interpretation which had earlier appeared in published work by Wiener and Phillips.

Friedrichs shook his head ruefully as he said to me, "*It was always happening to him.* Norbert Wiener was very offended. I have never really understood the justification. Courant didn't know about Wiener's work. I don't think Wiener mentioned it in Göttingen—although he may have. Maybe he mentioned it to Courant, and Courant forgot that he had mentioned it to him and thought it was his own idea. These things happen whenever people work together in the same field. It is rare in mathematics, but in fields like physics it happens all the time. Anyway, later, Courant was always very careful to mention Wiener and also several other authors of earlier relevant work which he hadn't known about at the time."

Much later Wiener, still unforgiving, circulated in manuscript a novel in which the character of a professor, who was unkind to gifted young people but took over their ideas, was rather obviously based on Courant.

"It was a common failing of Göttingen people that they were not very conscientious in attributing," Lewy conceded to me. "That was true for almost all of them—Hilbert included. But when you look back at papers by some of the great heroes of mathematics, you very often find that they are careless in these matters. That greater care is taken now is, I think, due to the fact that jobs depend to a more explicit degree on the credit that a person is given. It is undoubtedly true that the group in Göttingen was careless about studying what other people had done and attributing their results to them, but I think this must be seen against a background of less care."

The Göttingers had a facetious expression for the process of making someone else's idea one's own. They called it "nostrification." There were many levels of the process: "conscious nostrification"—"unconscious nostrification"—even "self-nostrification." This last occurred when one came up with a marvelous new idea which he later discovered had already appeared in earlier work of his own.

In addition to some general hostility toward the mathematicians of Göttingen, there also existed in Germany by this time a certain amount of what Courant described as "unfriendliness" toward him personally. Most of his critics objected especially to what was accepted by his friends as "Courant's way" and found "more funny than offensive."

"Ja ja," Neugebauer agreed. "Courant's way of doing things and being active in a certain direction irritates many people. I know that well. And then there is his way of saying something and taking it back. His probing. That irritates people. And his indecision causes him trouble, I am sure. He wavers, and people think he has something secretly behind it. And yet the loyalty of his own group was always very great. He held his group together better than most people. There is no doubt about that. It is certainly a funny situation."

I remarked that someone had said to me that Courant was a person you could like and yet still be completely aware of his faults.

Neugebauer nodded.

"One takes that. So it is."

In this same connection Hans Lewy pointed out that Courant's irritatingly vague, almost inaudible manner of speaking might be a reflection of the kind of things that he did in mathematics.

"In some ways analysis is more like life than certain other parts of mathematics," Lewy explained. "For instance, than number theory. There is nothing in life like number theory. It is too clear-cut, too straightforward, too black and white. By that I am referring to classical number theory. When you come to analytic number theory, things change. Analytic number theory has the same shortcomings or, if you wish, the same advantages as analysis."

In Lewy's opinion the statements of analysis—and particularly that part of analysis which deals with partial differential equations—"are kind of hesitant statements."

"There are some fields in mathematics where the statements are very clear, the hypotheses are clear, the conclusions are clear. The drift of the subject is also quite clear. But not in analysis. To present analysis in this form does violence to the subject, in my opinion. The conditions are or should be considered temporary, also the conclusions, sort of temporary. As soon as you try to lay down exact conditions, you artificially restrict the subject. Therefore, I think in a sense Courant's way of talking is—well, you don't know—cause or effect of the subject he is dealing with."

"Courant is suspicious about clear-cut and definite formulations of the truth," Friedrichs contributed. "Well, he knows there are cases where such statements can be made. Still, he is doubtful. Before he is willing to nail himself down in a definite statement—he is very reluctant to do that, because he feels that the essence, or whatever you want to call it, is what counts more. He is inhibited out of a philosophical insight, so to say."

Besides Courant's irritating indecision, there were other things about him which were criticized. Mathematicians who were not his friends spoke of his tendency "to get himself counted," "to organize," "to meddle." Even his obtaining the grant from the International Education Board was looked upon in some circles as "undignified begging abroad."

In 1928, when the new institute was almost finished, Courant and the other Göttingen mathematicians found themselves in opposition to a number of German mathematicians, especially those in Berlin.
That spring L. E. J. Brouwer (although Dutch, very pro-German) had addressed an open letter to the members of the German mathematical society in regard to the international congress to be held late in the summer in Bologna. Repeating violent words uttered by the French mathematician Painlevé during the war—"and not yet retracted"—Brouwer had questioned whether it was possible for Germans to attend the congress "without mocking the memory of Gauss and Riemann, the humanistic character of mathematical science, and the independence of the human spirit."
Harald Bohr, in what he described as his role as "ambassador from the international world," came to Germany and visited a number of universities. He reported to Hardy on the "rather excited" situation—in his English, "which I remember you were kind enough to call a 'language' even if it was not ordinary English."
"The word *Germans* in connection with such a phrase as *whether the*

Germans will go to Bologna or not is more physically than mathematically expressed," Bohr wrote. "It is not dealing with realities, because there are just in these questions all sorts of Germans and even among the internationalistic Germans there are great differences in respect to what they call clearness concerning the international character of the congress. To take two opposite persons both completely agreeing with one another and with us in the principal view that politics and science must be completely separated from one another; namely, Hilbert and Erhard Schmidt [one of Hilbert's earliest students and a professor in Berlin]. The first thinks that the international character of the congress is clear enough and will go at any rate (if there will be no earthquake), while the other thinks that he cannot go if there will not be obtained much greater clearness."

Ludwig Bieberbach, also a professor at Berlin, followed Brouwer's letter with a letter addressed to the German academies and the rectors of all advanced educational institutions, urging that they prohibit their members from attending the congress. Hilbert responded with a strongly worded letter supporting attendance at the congress. Richard von Mises, professor of applied mathematics at Berlin, urged that "no representative of applied mathematics" go to Bologna. Other German mathematicians allied themselves on one side or the other for various, often individual reasons. Hardy joined the Italian mathematicians in their insistence that the congress was to be truly international and was not a front, as claimed by some, for the organization which had vengefully excluded the Germans from a congress held in 1924 in Toronto.

Like the other Göttingen professors, Courant accepted the assurances of the Italians as to the international character of the congress. He felt that it was impossible to boycott it "without appearing to compromise our willingness for reconciliation and our desire for peace." He regretted, however, the personal tone which had developed in the debate, especially in the exchange between Hilbert and Bieberbach. He wrote a long friendly letter to Bieberbach, whom he had known when they were both students in Göttingen, and attempted to explain Hilbert's motivation and the fact that his letter had not been directed personally against Bieberbach. Hilbert had already agreed to go to Bologna before Brouwer's letter had appeared, and he had "passionately rejected this interference and judgment from the outside."

"Why is Brouwer's interest in the subject interference," Bieberbach demanded tartly, "when that of Bohr and Hardy is not?"

In the end Hilbert led a group of seventy-six Germans to Bologna. But the battle over attending the congress and a subsequent smaller scale battle, in which Hilbert forced Brouwer out of his position on the editorial board of the *Annalen*, left a residue of bitterness in Germany against the mathematicians of Göttingen and their foreign friends.

It was not long after the congress in Bologna that the building on Bunsenstrasse was ready for occupancy. The thought of the "institute" filled Courant alternately with pride and embarrassment, satisfaction and dismay. As a German professor he had a very easy, pleasant life. As the director of a large and generously budgeted facility, he would have "to march always with a heavy knapsack on my shoulders." Forty years later he could still recall vividly how he had walked around the old wall of the city with a friend and debated his decision.

"And then this friend said this strange thing: 'If in a few years it turns out to be worthwhile, then you should do it. That would be wise. But if it turns out to be a disappointment, then it would be *such* a waste of energy!"

Since I knew that less than four years after the dedication of the Göttingen institute Courant was removed from his position as its director, I started to ask—

"But it turned out to be very worthwhile," he said with satisfaction, even before I had framed my question. "Our institute in Göttingen—you have seen it?—it was unique in the beginning. Then gradually the idea spread, here and in Germany. Now there are wonderful modern university buildings for mathematics everywhere. It is no longer such a novelty. But it was the beginning, our institute in Göttingen."

The new building—a three-level T-shaped structure—provided everything the mathematicians had ever needed or wanted in their physical surroundings. Courant always gave Neugebauer the credit for its planning. The basement contained such requirements as a bicycle room, a book bindery, and a room for refreshments. The double stairs of the main entrance led to a spacious lobby, which today contains a bust of Hilbert and is known to students and faculty as "the Hilbert space." On the main floor there were two large auditoriums, *Maximum* and *Minimum*, and four other rooms of various sizes—all equipped for lectures in applications as well as in theory—a mechanical drawing room, a room for the meetings of the *Praktikum* and of the *mathematische Gesellschaft*, smaller meeting rooms, offices, individual workrooms, consulting rooms. On the top floor were the spacious and well-planned quarters of the *Lesezimmer*, still the heart of the mathematical life of Göttingen.

Among the visitors to the new building were Artin, by then a professor at Hamburg, and his young wife Natascha. I asked Natascha Artin, now Brunswick, for a "picture" of Courant at this time; but although she remembered Nina vividly—"she was a *very* striking woman"—she could not really tell me how Courant looked.

"What I remember is that he was always surrounded by a crowd of people wherever he went. Sometimes you wouldn't even see him. You would just see the crowd moving from one place to another."

The formal dedication of the Mathematics Institute of the University of Göttingen took place on December 2, 1929. The featured speakers were the long-ago rivals, Hermann Weyl and Theodor von Kármán. Courant, Born, and Franck had long wanted to bring both of them back to Göttingen. Up to that time, Weyl had chosen to remain in Zurich. The calling of von Kármán (proposed in 1925) had never been pursued. Much as they would like to have had Kármán with them in Göttingen, his friends were well aware that in the opinion of some professors there were already too many Jews on the scientific faculty.

Standing before his Göttingen colleagues and the great crowd of visitors in the auditorium of the handsome new building, Courant was reminded of the anguished feeling of a warmly clad man who walks through a poverty-stricken section of town on a cold winter day.

"Involuntarily one asks oneself why he is in such a position. . . ."

The new building had come about as the result of decades of work on the part of Felix Klein, he told his audience. It had been Klein's belief that a mathematician is not a self-sufficient creature who needs only paper and pencil to create. He must have as well a library, models and other demonstration materials, instruments; and he and his science must stand "in active and reciprocal relation" to the other sciences and to society. And yet, to justify the existence of the institute, it was not enough—in Courant's opinion—merely to cite the historical development and the interaction the institute would provide between mathematical and other interests.

"The ultimate justification of our institute rests in our belief in the indestructible vitality of mathematical scholarship. Everywhere there are signs to indicate that mathematics is on the threshold of a new breakthrough which may deepen its relationship with the other sciences and demand their mathematical penetration in a manner quite beyond our present understanding. It is not at all certain whether in the course of such a development the so-called applied sciences may not have to take second place to things now considered remote abstraction. . . ."

Courant

The old Hilbert was exultant.

"There will never be another institute like this! For to have another such institute, there would have to be another Courant—and there can never be another Courant!"

The dedication of the new institute took place a little more than a month after the American stock market crash. Signs of slump were already appearing in Germany, unemployment up, wages down, bankruptcies increasing daily. But the mathematicians, happily installed for the first time in a building of their own, were unaware of them.

With Hilbert's retirement in 1930, all the heroic figures of Courant's first student days were gone from the active faculty; but, as in the past, each of the mathematics professors of Göttingen had a highly individualized approach to his research and to his teaching which attracted a particular type of student.

Landau was the senior member of the new faculty. During the 1920's he had been the center of a fantastic flowering in the analytic theory of numbers, and he had written a number of influential books on that subject. In lectures, as in books, his ideal was absolute rigor and completeness. The assistant was instructed to interrupt if the professor omitted anything at all. Standing before the big blackboards in the lecture halls of the new building, Landau wrote with speed—theorem, proof, theorem, proof— while a menial with a sponge hastened after him to erase what he had written so that there would be room for him to write more. He never gave any explanation of where he was going, but he had organized his material so well that there was—I have been told—a quality of incredible clarity about his lectures.

Being Landau's assistant was a wearing experience. Werner Fenchel, who held the position for several years, recalls that when he arrived in Göttingen, Landau demanded to know when he got up in the morning. "Oh, about eight," Fenchel hazarded. The next morning, precisely at eight, Landau presented himself at the door of Fenchel's room. "But I learned to adjust," Fenchel smiled. "Landau took things very literally, always. In the beginning I would say, 'I have an appointment,' and he would say, 'That you can change.' But then I could tell him straight, 'I have no time.' That he would accept."

Unlike Landau, Herglotz ranged over a variety of areas, which included number theory, celestial mechanics, continuous groups, the mechanics of particles and continua, optics, to name a few. It was his custom, almost without exception, to offer a new course each semester.

"His lectures were works of art in every respect, including the lecturer himself," Charlotte John told me. "There was something of the nineteenth century about him, which he cultivated, I think. He had the elegance of that century. In fact, his dress was actually like that of Goethe. And he had really luminous eyes. He looked like someone who enjoyed himself extremely—and who saw things. And as you heard him lecture, you had the feeling that you got a glimpse of what he saw. There was a great beauty about his lectures. Unfortunately, when you came home and tried to work out the notes you had taken, you found that there was a lot to be filled in—you hadn't really seen what he had seen."

After delivering his beautiful lectures, Herglotz, who was a bachelor, would leave the hall without speaking to anyone and go alone to his house. He had almost no contact with students other than Hans Schwerdtfeger and Hanna Mäder, whom Schwerdtfeger later married. But he sometimes dropped in at Busemann's rooms, having learned that the son of the Krupp director, unlike most students, served very good wine.

Hermann Weyl was Hilbert's successor on the faculty. His coming in the fall of 1930 had been preceded, as in 1922, by extended debate with himself and with Hella. In addition to the economic and political situation in Germany, he had been concerned about his own fitness for the position. Perhaps he was too old. He had brought up the name of young Artin, but the suggestion had been rejected in Göttingen. Only Weyl should succeed Hilbert.

Weyl was the mathematical hero of many members of the younger generation, including Friedrichs. Others, like Lewy, were repelled by his literary and personal approach to mathematics and by his philosophical aspirations. They complained that one had to scrape away the words to find out what he was saying. His mathematical papers and books do convey a quality of personality rare in mathematical writing. His great influence came through them.

As a teacher, Weyl was completely different from his famous predecessor and never developed "a school" of his own. There was something tremendously impressive but remote about him. Even his most elementary courses were for only the most gifted. When it came his turn to give the calculus lectures for the beginners, he presented the subject entirely from the point of view of the Intuitionists.

"A very interesting experiment for the lecturer and three of four students," Courant later wrote, "but less useful for the others."

Emmy Noether was not a full professor, but she contributed importantly to the mathematical atmosphere of Göttingen during this period. She and her students, few in number and many of them foreign, represented the trend toward abstraction and generalization which was to become more and more dominant in mathematics during the coming years.

She was a very poor lecturer, writing on the board and wiping out almost immediately what she had written. She spoke quickly and sometimes condensed many syllables into one or two. To Friedrichs it seemed that her speaking never quite caught up with her thinking. "I have no doubt she had a very clear understanding of what she was saying," Hans Lewy told me, "but she didn't have a clear idea of what she was going to say."

She was devoted to her students, who came to her with all their problems, personal as well as mathematical. She was especially popular with the Russian visitors; and when they began to go around Göttingen in their shirtsleeves—a startling departure from proper dress for students—the style was christened "the Noether-guard uniform."

Much of the social life in Göttingen depended on the parties which the professors gave at various times during the year. These were characteristic. Landau's parties were intellectual tests, to which Hilbert never came. Games were played, there were winners and losers. "Hilbert didn't like the premises," Courant explained to me. Herglotz, naturally, gave no parties. Emmy Noether was famous for her "children's parties," to which Hilbert did come. The Weyls hosted a tea dance on a Saturday afternoon—very elegant and formal with many pretty girls present. At the Courants' house there was an unending succession of musical evenings, to which some students were always invited.

I asked Courant if in his new official role as the director of the Mathematics Institute he ever had any difficulty with the other professors.

"No—mostly they were glad to have someone else take over the work, and things were going the way they wanted them to go," he replied. "But I am afraid I was a little bit autocratic. I was always inclined to be as autocratic as the world permits. Only Weyl sometimes objected to what I did. He wanted to be consulted about things; and one time we sat down and had a talk. But otherwise—no."

The new institute was generously staffed in accordance with the government's agreement with the International Education Board. There was an *Oberassistent*, a tenured and salaried official who handled many of the administrative duties. This position was held by the indispensable Neugebauer. There were also ten full-time assistants, all actual or prospective *Privatdozenten*, and a varying number of part-time assistants according to the needs of the *Praktikum* and other activities. In this way, it seemed to Courant, the old "family life" of the mathematics group in pre-war days was being preserved in spite of the much greater number of students since the war.

"[But] it was really restricted to the comparatively small, though by no means exclusive, top group," he later wrote. "This class division could not fail to create among the less fortunate majority the psychological atmosphere of an inferiority complex, which so easily becomes dangerous for scientific situations...."

An American who came to the new institute in Göttingen in 1931 was Saunders MacLane. Up until the time of Birkhoff, MacLane told me, it had been considered impossible for an American to get a proper mathematical education without going abroad. Germany was the place to go, and Göttingen was the most frequent choice. Of 114 American mathematicians who took foreign degrees between 1862 and 1934, a total of 34 received theirs in Göttingen. The second most popular universities were Leipzig and Munich with 9 each. In the early years the attraction was Felix Klein, and five future presidents of the American Mathematical Society were his students in Leipzig or in Göttingen. Later, of course, the attraction was Hilbert. By 1931 not many Americans any longer went abroad for their degrees, but Göttingen was still a popular place for post-doctoral study.

"There was no place like it in the world," MacLane recalled. "It was a real intellectual center. There was heavy excitement going on. There was somehow the feeling that this was the real stuff. That this was the center of things. Everybody talked about mathematics all the time. It was the first real top-class mathematical center that I had ever been at. It was much better than Chicago had been when I had been a graduate student—infinitely better than Yale, where I had been an undergraduate. As I look back now, of course, I have seen many other places that are just as lively and jumping—today in this country Berkeley or Chicago or Harvard—there are many—but it was the first one. It was the real thing."

I asked MacLane how the political situation in Germany had looked to him in the fall of 1931.

"People did not seem very concerned," he said. "Things were always in disorder, but they accepted that. Different people, of course, had different views. My impression that first year was that probably Hitler shouldn't be taken too seriously. Politics in Germany seemed a great big mess. I distinctly remember buying a pamphlet that was labeled 'The 27 Parties of Germany.' There *were* 27 of them, and the NSDAP—the Nazi party—was just one."

FOURTEEN

BY THE END of the academic year 1931–32, the full impact of the worldwide depression was beginning to be felt in Germany. Stringent economy measures were adopted, and university faculties were ordered to dismiss most of the younger assistants.

The members of the mathematics and natural science faculty objected strenuously. Not only would it be unfair to struggling young scholars, who had meager incomes at best; but it would also bring to a virtual standstill the activities of the various institutes. Instead, they proposed that professors pay the salaries of a number of the people affected out of their own pockets.

"The battles this caused with the faculty [as a whole] still make me shudder," Max Born, who was the dean of the science faculty, later recalled. "In the course of an interminable meeting we won with a considerable majority. But those who were outvoted displayed an animosity we had never before experienced...."

It was at this time that Courant received an invitation to lecture in America during the spring and summer of 1932. The work of Americans and their interests had become increasingly prominent in mathematics. Several of them served on the editorial board of the *Zentralblatt für Mathematik*, the new review journal recently founded by Springer. Born and Franck as well as many of Courant's other friends had already been in the United States, and he had heard their experiences. He was "quite curious."

Since Runge's death in 1927, the growing Courant family had moved into the big Runge house on Wilhelm-Weber-Strasse. This was the avenue favored since the turn of the century by the Göttingen professors for their homes. There were now four children, all of whom were cared for and supervised by a much loved housekeeper named Martha Meyer. The household also included Hilde Pick, a cousin of Courant's, who served as his secretary at the new institute. When Courant learned he was going to America, he immediately invited Saunders MacLane to come and live at the house also in exchange for dinner-table lessons in English. Just before he left on his trip, he made still another addition—a young woman student named Charlotte Woellmer, now the wife of Fritz John, a professor at the Courant Institute.

Shortly before Courant died, I talked to Fritz and Charlotte John in their home in New Rochelle; and they told me how he had helped them to continue their educations in the desperate days of the Depression. They had already been friends when Fritz had set out for Göttingen with very little money but the optimistic hope that he would be able to obtain some sort of scholarship there.

"I guess I didn't know how to go about such things," he reflected. "Anyway I was not very successful. But Courant found out about my case and he helped me to get a scholarship. Then he continued to be interested in me. He was always inviting students to musical evenings at his house. I was not musical, but I went anyway. Then fortunately Charlotte arrived, and she became a member of Nina Courant's singing group. But Charlotte had even more financial problems than I."

Charlotte explained to me that she had been able to come to Göttingen only by exchanging room and board with a girl from Göttingen who wanted to live in Berlin.

"Unfortunately—it was very sad—before the end of the semester this girl became ill and died. On the last day of school I met Courant in the hall at the institute and he said, 'Goodbye—see you next semester.' I said, 'I doubt it.' He said, immediately, 'Why? What is the difficulty?'"

When she described her situation to him, Courant suggested that she come to lunch and they would try to see if something could be done. After lunch he said to her, "I am going to the United States soon. How would you like to come and live here at our house for a while?"

"Courant tried—he always went out of his way to help people and to draw people into mathematics," Fritz John said. "Many people—most professors, I think—did not make the effort. There were very few who really had a program of furthering mathematics. They just gave their lectures. Courant had more vision than others in this respect, and he felt a kind of moral obligation to create conditions that would help people."

During the increasingly hard times, a number of informal assistantships came into being at the institute in addition to the official ones funded by the government. Often duties were vague or non-existent. Courant once gave a student a stipend because he thought the young man was on the verge of a nervous breakdown and needed a skiing vacation. He also contrived to have some students work without pay. One of these was Busemann.

"Normally I would have become an assistant after I got my degree," Busemann explained to me, "but this was of course the depression time by then; and so, since my father had money and Courant knew it, he asked my father if he wouldn't support me and make it possible for some other youngster, who did not have money, to become a mathematician."

In the early spring of 1932, as Courant prepared for his trip to America, he was concerned about the academic futures of some of his older students. Although Friedrichs was by this time a professor at the technical institute in Braunschweig, Lewy was still a *Privatdozent*; and as times got harder, there was beginning to be evidence of anti-semitism in appointments at German universities. Neugebauer's future was another concern. That young man had chosen a field of interest for which no university in Germany maintained a professorship—the study of ancient mathematics. Since 1923 Courant had managed in various ways to support Neugebauer so that he could continue what Courant—in spite of his own quite different interests—considered important pioneering work. He was currently *Oberassistent* and editor of the *Zentralblatt* and other Springer journals, including a journal for studies in ancient mathematics which Springer had founded for him. He was also a *Privatdozent* for ancient mathematics. With these various sources of income, Neugebauer, now married, managed to live quite comfortably. When he was offered a professorship in Darmstadt, he refused it because, as Courant explained to me, "It would have defeated him." Still, Courant thought, Neugebauer should have a position appropriate to his superior abilities. Perhaps, even during a depression, there would be a place for him as well as for Lewy in the land of millionaires.

During the semester that Courant would be away, he arranged that Oswald Veblen would give lectures in Göttingen. Next to Birkhoff, Veblen was the most influential force in American mathematics. Although not of Birkhoff's stature as a mathematician, he had a personality which was much more appealing to Europeans. ("Veblen combines the best qualities of an American with the best qualities of an Englishman," Hardy used to say.) Courant thought enough of Veblen to let him use his first precious automobile, a Röhr, which he and Nina had just recently learned to drive.

Courant left Germany immediately after a presidential election which pitted the 84-year-old von Hindenburg against four other candidates, one of whom was Adolf Hitler. He arrived in America late in March 1932, when people were more concerned about the recent kidnapping of the Lindbergh baby than about the fact that Hindenburg had failed by 0.4 per cent to obtain a majority of the vote and would have to submit to a runoff election between himself and Hitler.

The American lecture tour took Courant from Columbia in New York City to the Ivy League colleges, then to Chicago and the great midwestern universities, then to the west coast, the California Institute of Technology, Stanford, and the University of California at Berkeley. Everywhere he went he met men and women with earlier connections to Göttingen. The head of the Rockefeller Foundation, with whom he had lunch shortly after his arrival, was Max Mason, who had taken his Ph.D. with Hilbert. Maria Goeppert-Mayer, a Göttingen girl who had married an American, helped him with the English of his first lecture at Columbia. At Cornell he met Virgil Snyder, one of Klein's Ph.D.'s, who had married a Göttingen girl; and at Princeton he found numerous physicists who had been in Göttingen during the 1920's as well as Solomon Lefschetz, who had given lectures there the year before. At Yale there was Einar Hille, who had also recently spent some months as a visitor in Göttingen. At Harvard he met Birkhoff again and Kellogg, who had been a student of Hilbert, and William Fogg Osgood, a student of Klein. At Brown he saw R. G. D. Richardson, who had been in Göttingen during his own student days, and R. C. Archibald, who came regularly to Göttingen each summer to purchase lecture notes and books. At Chicago the department head was G. A. Bliss, who had studied in Göttingen during the age of Hilbert and Minkowski. On the west coast Courant renewed old times with von Kármán at the California Institute of Technology; and going up to see the telescope on Mt. Wilson, he found Walter Baade, who had been Klein's assistant and the director of the observatory in Göttingen. At Berkeley the head of the mathematics department was M. W. Haskell, who had taken his Ph.D. with Klein and had made the English translation of Klein's Erlangen Program. Courant could not help feeling at home.

"The people are extraordinarily nice," he wrote to Neugebauer. "I am frightened sometimes when I am overwhelmed with attention by people here, for I have hardly paid any attention at all to those same people in Göttingen."

America did not disappoint him by failing to produce millionaires. His letters are full of "the big people" he has met. The words appear invariably in English in the German of his letters. He came over on the boat with an American banker who turned out to be a friend of the elder Busemann. At the luncheon given for him by Flexner he met the head of Macy's ("the largest American department store"). Someone else introduced him to a member of the House of Morgan ("the greatest American banking house"), and he tried—unsuccessfully—to "wangle" money out of him for the Mathematics Institute. In San Francisco, through a banker friend of Carl Still, he met A. P. Giannini, the president of the Bank of America, which until recently had been the Bank of Italy. He was very impressed by this Italian immigrant's son, who had started out as an errand boy in a produce business. He and Giannini had a long talk about the situation in Germany.

"It is simply striking how optimistic about Germany and how relatively adverse to America this man and others of the business people here are in regard to the long-term situation."

Throughout his journey Courant himself nervously followed political events in Germany. In Princeton, with Lefschetz, he listened to the results of the runoff election, in which Hindenburg polled 19,359,642 votes to Hitler's 13,417,460, and discussed with Lefschetz anti-semitism in America. When the Brüning government fell shortly after the election, he begged Nina to save all the newspaper reports for him. Three weeks later he complained, "I still do not understand politics in Germany at all!"

Of the American colleges and universities he visited, he was most impressed by the "old and high" intellectual level at Harvard. But the physical accommodations for mathematics were dreadful. An inadequate library. Bad lecture halls. Impossible blackboards. Everything separated. No applied mathematics. It was obvious that "there was never anyone like Klein to see to the improvement of these conditions."

Princeton impressed him by its luxurious accommodations.

"We try to *Americanize*," he wrote to Neugebauer, "while Princeton tries to surround itself with German *Gemütlichkeit*. In the professors' clubroom ... there are even newspapers and magazines for entertainment!"

The American university which charmed him most was Berkeley. He wrote at length to Nina about how "nice and friendly" the people were and what a great interest there was in music, "although, strange to say, Debussy and Tchaikovsky play a great role." He found the intellectual standards of the university not very exacting. Although there was a great deal of interest in his lectures, he missed the "resonance" he had felt at Harvard and Princeton.

I asked him once if on this trip he had ever considered remaining in the United States. He said yes, "fleetingly"—at Berkeley.

In the course of his travels Courant did not forget that he was looking for places for Neugebauer and Lewy. In spite of "the really fantastically increasing hardship" in America, he was certain that Neugebauer, whose work seemed to be known and admired, would have no difficulty in obtaining a suitable position. Placing young Lewy would be more difficult, although it was not excluded. The general economic situation in America, which was in the midst of the Depression, seemed to him worse than that in Germany.

"Probably the main difference is the existence of large material reserves everywhere here ... and, generally speaking, a more stable attitude on the part of the population, who have gone through long periods of economic well-being. But for the rest, there are the same phenomena. . . ."

Three hundred mathematicians and physicists—"some of them quite good"—were going to be unemployed the next year.

While he was in Chicago, Courant heard about a young Ph.D. from the University of Chicago who was planning to spend a year as a National Research Fellow in Germany. He promptly offered to support Edward J. McShane in Göttingen if McShane would translate the Courant calculus books into English. McShane had had only "half a summer's course" in German to prepare himself for a pre-doctoral examination in the language. But times were hard. In the year since he had received his degree only one job opportunity for a mathematician had come to the notice of the head of the mathematics department at Chicago.

"That was from some obscure teachers' college in the sticks, very poor paying even for those times; and Bliss said he was going to do me the favor of not recommending me," McShane recalled. "So I had the brass to accept Courant's offer."

Courant had come to America with a great curiosity about the new Institute for Advanced Study which Abraham Flexner was setting up in Princeton. One of the first appointments there had already been offered to Weyl. Courant knew that two telegrams lay on Weyl's desk in Göttingen, one accepting and one refusing Flexner's offer, and that regularly each morning Weyl debated afresh with Hella the sending of one or the other.

"My first impression was that one should advise Weyl to accept," Courant wrote to Nina. "But [since being here] I have become somewhat more doubtful, in spite of all the difficulties in Germany. . . ."

Late in the summer of 1932—in the middle of an American presidential campaign in which Franklin D. Roosevelt was opposing the incumbent, Herbert Hoover—Courant returned to Göttingen. Weyl refused Flexner's offer; and the winter semester 1932–33 began in Göttingen on November 1, 1932, with the famous faculty intact.

McShane remembers that, in the fall and winter of 1932–33, politics was a favorite conversational topic at the Bahnhof Restaurant, where he and Saunders MacLane regularly ate with the *Privatdozenten* for mathematics. He also remembers that although he and his wife spent the New Year of 1933 in Berlin, they heard nothing about the Nazi riots that were taking place in the capital at that time until they received newspaper clippings about them from their worried families in America.

The Papen government had fallen in December, and in January the Schleicher government also fell. President von Hindenburg went back to Papen and instructed him to explore the possibilities of forming a government under Hitler "within the powers of the constitution." On January 30, 1933, Hitler became chancellor of Germany.

I asked Courant and many others to tell me their reaction, how they felt and what they thought about the future of Germany at that time. Most of them said that they really do not remember any longer—that they have tried to forget and, besides, "so much has happened since then." Even the American MacLane told me he felt he had suppressed the memory of the time. He did recall that after the Reichstag fire on February 27 and the new elections people began to get very nervous. "Immediately there began to come all sorts of regulations, talk about things that would be done at the universities. And then it was unpleasant. At that time I became aware of Courant's particular position, because the Nazis, as far as I can recollect, were especially out to get Courant, as compared to other professors."

Brown shirts and swastikas suddenly began to appear in mathematics lecture halls. The wearers of these were not members of the "in group," but several of them were good mathematicians. Werner Weber, one of the most active of the pro-Nazi students, had been a *Privatdozent* since 1931. The 20-year-old Oswald Teichmüller was extremely gifted. The Nazi sympathies of these and others came as a complete shock to Courant and the young mathematicians of the "in group."

At lunch in the Bahnhof Restaurant, a box was placed in the center of the table and anyone who mentioned politics was fined.

In those nervous days in the early spring of 1933, Courant became possessed with the idea that he must finish the second volume of Courant-Hilbert. Since 1925 he had been trying to work on the book with Friedrichs when that young man was in Göttingen. But somehow he hadn't been able to bring himself to pursue the project. Constantly he had pushed it off. Then in 1930 Friedrichs had gone to Braunschweig as a professor. He continued to visit frequently in Göttingen; and on those occasions Courant would lead him up to a tiny room on the third floor of the new institute building, a retreat which he pretended that no one else knew about. He would then bring out some new section which he had written up—some idea that had caught his interest—and discuss it with Friedrichs.

"He always wanted me to work with him, to read what he had written and talk with him about it, make proposals. He could not seem to work alone. But in all those years, from 1925 on, nothing really happened."

Friedrichs smiled at the memory.

"If it was true, as people said, that Courant's assistants wrote his books, I would have written Courant-Hilbert II."

At the beginning of March 1933, Courant arranged that Friedrichs and Franz Rellich would accompany him and his family to Arosa during the spring vacation. They would ski and work, and they would *finish* the second volume of Courant-Hilbert!

Courant was aware of rumors which were being circulated in Göttingen that he was planning to "flee" to Switzerland. He considered cancelling the ski-work trip because, as he explained to Franck, "it seems unfeeling of me to sit calm and free abroad at this time." But his children needed the fresh air and sunshine of the mountains. Friedrichs and Rellich were already waiting in Arosa. He decided to go ahead with his plans.

The sequence of events, the emotions, and the reasoning from which Courant and others acted during the next few months were to fade in the coming years, what was actually known at the time confused with what became known only later, what should have happened with what did happen. For this reason I have tried to limit my account to those documents which Courant took with him to the United States in 1934 and which he turned over to me in 1971, a few months before his death. He could not bring himself to go over them in detail but merely glanced at them, saying he found it impossible to believe that he had written what he saw there.

While Courant was in Arosa, Hitler pushed through the Enabling Act, which gave the government power to issue decrees independently of the Reichstag and of the President. A nationwide boycott of Jewish business and professional men was announced for April 1.

On March 30 Courant wrote to Franck to ask his advice about whether he should return immediately to Göttingen. He was sure that if he could stay in Arosa as planned, he would be able to finish at least half of the second volume of Courant-Hilbert. And yet—he had been somewhat alarmed the day before when he had heard through the servants that people were saying in Göttingen that he would not be back.

"Incidentally, it is my impression from newspaper reports that things have been going peacefully and quietly in Germany," he wrote to Franck. "From the first I have been appalled to read reports of how such people as Einstein have made statements and what lies and 'latrine rumors' have been utilized to make Germany's internal situation a butt for general political agitation abroad. If I had known Einstein's address, I would have written to him. . . ."

Einstein, who had been in America for the past few months, had been making a number of widely publicized statements deploring "brutal acts of violence and oppression against persons of liberal opinion and Jews . . . in Germany [which] have aroused the conscience of all countries remaining faithful to ideals of humanity and political liberties." He hoped that world reaction would be sufficient "to preserve Europe from regression to the barbarism of past epochs." On March 29, the day before Courant wrote to Franck, the government in Berlin had announced that Einstein had inquired about taking steps to renounce his Prussian citizenship.

"Even though Einstein does not consider himself a German," Courant wrote, "he has received so many benefits from Germany that it is no more than his duty to help dispel the disturbance he has caused. Unfortunately, as I see from the papers, a reaction to these events has set in.... I very much hope that it will be possible to deter the intended boycott [of the Jews] at the last moment. Otherwise I see the future very black.

"What hurts me particularly is that the renewed wave of anti-semitism is ... directed indiscriminately against every person of Jewish ancestry, no matter how truly German he may feel within himself, no matter how he and his family have bled during the war and how much he himself has contributed to the general community. I can't believe that such injustice can prevail much longer—in particular, since it depends so much on the leaders, especially Hitler, whose last speech made a quite positive impression on me."

Franck's reply is undated, but it seems to have been written on March 31. He advised Courant to try to quiet the rumors about himself by returning to Göttingen immediately.

"It is also possible that any day decisions may be necessary which can only be really understood on the spot."

April began with the "non-violent" boycott of Jews, although officially it had been withdrawn by the government. Hilbert, who was still an inveterate gardener, picked a bouquet of flowers and sent them to his Jewish doctor.

"It will not be long before the German people find Hitler out," he said, "and then they will put his head in the toilet."

On April 7 Courant, who had returned to Göttingen without his family, wrote reassuringly to Nina. On that same day the government announced a series of laws—the *Reichsgesetze*—for the "restoration" of the professional civil service. The intent of the new laws was to remove from government employ all persons of "non-Aryan" descent with certain exceptions, mainly military, and also to remove persons who were considered politically unreliable. Since university professors were civil servants, the new *Reichsgesetze* applied to them as well as to judges, to officials and employees of innumerable public and semipublic agencies, and to many others.

In Göttingen, Courant and Franck could take comfort in the fact that non-Aryans who had fought for Germany or her allies were specifically exempted; unfortunately Max Born, who had done war-related research, was not so protected. How far would the government go in carrying out the *Reichsgesetze*? In the second week of April 1933, it seemed that the violent anti-semitism simply could not last.

But—demanded James Franck of his friends—didn't the very fact of the new laws, removing as they did German citizenship from many of Germany's Jews—declaring that their children were *not Germans*—require response and protest on the part of Jews?

FIFTEEN

THE University of Göttingen had a certain tradition of political protest which was as old as its mathematical-physical tradition. During the time of Gauss, seven famous professors, including the physicist Wilhelm Weber but not Gauss, had protested the repeal of the liberal Hanoverian constitution. Their subsequent dismissal had aroused a storm of indignation throughout Europe.

In 1933, almost a hundred years later, Courant, Born, and Franck considered a joint protest in the spirit of the Göttingen Seven. Courant and Born thought that such an act might precipitate action from the government which would not otherwise be forthcoming, but Franck insisted that he wanted to resign. They debated whether they were not morally obligated to follow him. Although Born was more vulnerable under the law than Franck, he drew back. He and Courant agreed that they should remain at their posts and do all in their power to save what they had helped to create in Göttingen. Courant, even more than Born, felt a personal responsibility to remain because the Mathematics Institute had been so recently built as a result of his personal negotiations between the Americans and the German government. He was also inclined to postpone irrevocable action.

For a week the discussions among the three professors and Neugebauer continued. Then, on April 13, several individuals at other universities in Germany were summarily placed on leave. On Easter Sunday, April 16, Franck came to his decision. He wrote to the minister asking to be relieved of his duties. In a letter to the rector of the university, he explained his action:

"We Germans of Jewish descent are being treated as aliens and enemies of our homeland. It is required that our children grow up with the knowledge that they will never be allowed to prove themselves as Germans. Those who fought in the war are supposed to have permission to continue to serve the State. I refuse to avail myself of this privilege, even though I understand the position of those who consider it their duty to remain at their posts."

He intended, he said, to try to continue his scientific work in Germany.

Angry response to the announcement of the resignation of Franck, a Nobel laureate, came first from forty-two members of the faculty, who issued a statement condemning him for giving the foreign press material for anti-German propaganda. Many of those who signed were the same people from agriculture and forestry who had bitterly opposed the scientists' proposal that the professors pay the younger assistants out of their own salaries. Rumors began to circulate that Born and Courant were part of a plot to "sabotage" the new government and that they had had Franck alone resign simply for "tactical" reasons.

Two days after the faculty statement, without any direct notification to those involved, an announcement appeared in the local newspaper on April 26 that, in accordance with the new *Reichsgesetze*, six Göttingen professors were being placed on leave until further notice. Four of the six were from the scientific faculty: Felix Bernstein (who was in the United States at the time), Max Born, Richard Courant, and Emmy Noether. The headline on the little story announced that more would follow.

Courant heard the news when a friend who had seen the paper telephoned him.

He immediately sent a letter to Harald Bohr. The optimism of the first weeks of April was gone. He wrote Bohr that he was convinced—as was everyone else in Göttingen, including the local authorities—that the "leaves" would shortly and irreversibly be converted into dismissals.

"I have been harder hit by the turn of events and less prepared than I should have thought," he confessed. "I feel so close to my work here, to the surrounding countryside, to so many people and to Germany as a whole that this 'elimination' hits me with an almost unbearable force."

His personal plan was to remain quietly in Göttingen—financially this would be possible for a number of months—and to do his best to complete the second volume of Courant-Hilbert. "Only in this way can I maintain the energy and the moral justification for continuing in my scientific career." As for the more distant future, positions would have to be found for himself and the other dismissed professors. The best hope seemed to him to lie in America with Flexner and his many connections, but he did not feel that he should personally approach Flexner.

"How hard it is for us to make use of your friendship in this manner, we need not assure you," he told Bohr. "I hope it will never be necessary to show you that at any time we would be prepared to do the same for you."

It may have been this letter from Courant to Bohr which Neugebauer remembers carrying to Hamburg so that he could place it directly in the hands of the Danish consul there.

"The very first moment was physically extremely dangerous," he told me. "One didn't know what the S.A. would do. From the time they took over, there was brutality."

Support for Courant came immediately from many friends and former students—in some cases even from people from whom he did not, perhaps, expect support. One of these was Hellmuth Kneser, who had been his first assistant in Göttingen. Kneser was a professor in Greifswald and was known to be impressed by the new government, which "despite everything" had—in his opinion—brought about a revolution with quite unrevolutionary peace and order and had been able to unify the *Reich*— something which neither Bismarck nor the Weimar government had been able to accomplish. He was considering joining the National Socialist Party —"not," as he said, "for the sake of influence or advantage—if I had desired that, I would have joined earlier—but so as not to be farther removed externally than internally." He wrote Courant, "very startled," the day the news of the "leaves" appeared. "I don't understand at all what has been done, unless it is intended to prevent disturbances [at the university] in the future."

Courant replied with an extensive letter. He had not yet received any official notification of his status. He was preoccupied with "the reason" for his dismissal, since he should have been exempted from the *Reichsgesetze* because of his wartime service. He mulled over the details of his short-lived political career but could not imagine that the facts of this could be sufficient to provoke a leave order. He recognized that he and the scientific faculty were "instinctively abhorrent" to various groups. There was probably a certain amount of unconscious jealousy involved. Wild stories and rumors had developed which could be summarized by the description of the Mathematics Institute as "a fortress of Marxism." Much of the hostility had centered upon his person, because it was he who had been responsible for the erection of the institute. And then there was the case of Franck—"I can imagine that it was Franck's resignation which actually provoked the ministry's action."

The situation at the Mathematics Institute became increasingly volatile. Kneser, Courant knew, had personal ties with an influential party member in the ministry, who might be able to help. The day after his first letter, he dispatched a second letter to his former assistant.

"Not only are the students apparently determined to try to prevent Landau and Bernays from lecturing, but they are also attacking Neugebauer as 'politically unreliable'—that is, communistically oriented. The dean has stood up for Neugebauer, but it does not look as if Neugebauer can endure the pressure. He now holds—as my representative—the position of director of the institute; but I am afraid that he will give it up if the students don't withdraw their threat to boycott him. Since Weyl is not a strong and stable personality and since Herglotz cannot be considered for the directorship either, I see the future of our institute as very dark....

"It is a pity to think what treasures are going to be destroyed in this way after more than ten years of work at reconstruction. It pains me most to see what senseless damage will be done to Germany. Only look at the Americans and other foreigners who are now getting ready to break off their studies and go home! In any case, the spirit of our institute has already been destroyed. Ugly signs of opportunism have become evident. I am much afraid that quite apart from what has happened to me irreversible actions have been taken. . . . Still, if the matter of Neugebauer could be straightened out, I would be very happy."

After his letters to Kneser on April 28 and 29, Courant fretted constantly about counteracting the rumors that were being circulated about him. According to one of these, he had carried a red flag during the days of the 1918 revolution and had disarmed troops returning from the front. He was convinced that such unfounded stories, or at least the hostile attitude from which they stemmed, were responsible for his being placed on leave. Although he had not yet received official notice of his status, he made up his mind to approach officialdom.

Seeking a colleague unaffected by the *Reichsgesetze* who could present his version of his activities to the administration of the university, he settled upon Prandtl. The professor of mechanics was generally considered a somewhat naive man. But during the wild week following the announcement in the paper, he had acted with courage and decision, firing one of his assistants when he had discovered that the man was an informer for the Nazi forces at the university.

It was decided that Courant should write a letter to Prandtl setting out the facts of his political activities after the war. Prandtl would then present the letter to the *Kurator*.

In this letter, dated May 1, Courant covered essentially the same points he had made in his first letter to Kneser, with the exception of the fact that he also took up a charge that he had "jewified" the Mathematics Institute. He pointed out that Lewy (who had already left Göttingen—the first to go) was the only Jew whom he personally had appointed to an assistantship.

"I don't have to say anything about my attitudes," he concluded. "Better than words, the facts testify. I have served the general interest in all my activities, from which the university has surely benefited."

Since the previous Thursday, when the announcement of his leave had appeared in the paper, Courant had not gone to his office in the Mathematics Institute. From a practical point of view there had been almost no disruption in activities, since he and Neugebauer had always worked closely together. But he continued to worry about how long Neugebauer would be able to handle his increasingly difficult role.

Over the weekend Neugebauer was asked to sign the required oath of loyalty to the new government. He refused and was promptly suspended as *untragbar*—a word which means in German, quite literally, "unbearable." Although he continued to receive his salary as *Oberassistent*, he was forbidden to enter the institute building.

As May and the new semester began, for the second time within a week, the Mathematics Institute of Göttingen was without a director.

Although initially Courant had felt that he should not personally approach Flexner for help in obtaining another position, he now wrote the American a long letter.

There was, he explained, a small and constantly diminishing number of people who still remained aloof from the new nationalistic ideas and enthusiasms. "The government possesses a power and ability to act such as probably has never existed in the world before." The picture of the German universities which Flexner had presented in his famous book was no longer a true one. "To my distress I foresee that the flowering of our scientific group here is past."

Neither he nor Neugebauer had been able to make any personal decisions about the future. He was continuing to try to finish the second volume of Courant-Hilbert, and Neugebauer was continuing his historical researches.

"Quite frankly, I am writing you to ask whether you can help us. You may be sure that I do not have in mind one of those brilliantly distinguished positions at your institute, but would be perfectly satisfied with some modest place. On the other hand, I believe quite objectively that I could again be productive and of use in new surroundings. The same naturally applies to Neugebauer."

Spring 1933

On May 5, nine days after the announcement of their leaves in the newspaper, the six affected professors received official notice of their status.

Courant renewed his efforts to set his personal record straight. In response to the official notification, he composed a lengthy description of his activities and sent this for transmission to the ministry with a personal letter to the *Kurator*: "I repeat that my letter should not give the impression that I am asking for mercy. I am interested only in not being stigmatized as a bad German, or as a negligent civil servant."

Max Born, shortly after receiving the official notice of his leave, left Göttingen for the Dolomites, where he and his family had already planned to spend their summer vacation. On May 24 he sent his resignation to the ministry, expressing his basic agreement with Franck's position as expressed in Franck's letter of resignation.

Franck himself remained in Göttingen, conducting a physics seminar in his home and trying to help his assistants and students to place themselves outside Germany.

Courant's own activities and statements during the period often appear contradictory. From the moment the leaves were announced in the newspaper, he was convinced—as he wrote Harald Bohr the same day—that they would inevitably become dismissals. At the same time, during April, May and June, he put up a determined fight to reverse his leave.

Perhaps Max Born, writing to Einstein from the peace of the Dolomites, had the clearest understanding of both Franck's and Courant's actions at this time:

"Franck is resolutely determined not to go abroad while he has the slightest prospect of finding work in Germany (though not as a civil servant). Although there is, of course, no chance of this, he remains in Göttingen and waits. I would not have the nerve to do it, nor can I see the point of it. But both he and Courant are, in spite of their Jewishness, which is far more pronounced than in my case, Germans at heart."

After Neugebauer was declared *untragbar*, Hermann Weyl became acting director of the Mathematics Institute and immediately joined in the efforts to have the leaves of Courant and Emmy Noether lifted.

In Courant's case, Weyl thought that the best approach would be to try to obtain official documentation of his services to Germany during the war. By the middle of May letters from old comrades and co-workers began to arrive. Some were obviously brush-offs, but many were sincere testimonials.

Most heartwarming was the letter from the major under whom Courant had first served in 1914 and whose life had been saved in the Argonne by the bravery of Ernst Courant. Now a lawyer in a small country town, this man wrote a sympathetic personal note and enclosed a detailed account of the wartime activities of both brothers while they were under his command. He described Ernst as "fearless and extraordinarily brave" and Richard as "an absolutely reliable soldier and a true comrade . . . helpful and tireless. . . . His quiet sense of duty made him an example to officers, non-commissioned officers, and men."

These letters, as well as typewritten copies of newspaper reports of Courant's post-war political activities and a reminder of his work in getting out the vote in the plebiscite of 1921, were sent by Weyl to the ministry on May 18.

At about this same time Hellmuth Kneser had an opportunity to approach the ministry's representative, Theodor Vahlen, about Courant's case.

Vahlen, a mathematician who had written the first German textbook on ballistics, had been a student of Klein in Göttingen. Once, on the occasion of the Day of Nationality, when he was a dean at Greifswald, he had personally ripped down the black, red, and gold flag of the Republic and hoisted in its place the old black, white, and red of the Empire. When Kneser had left Göttingen, he had gone to a professorship in Greifswald; and Courant was always inclined to attribute his later political views to Vahlen.

"Kneser was young, and he got some ideas under the influence of this dean," Courant explained to me. "I don't think he ever did anything really bad. He is a good man. A very good mathematician."

In spite of Vahlen's new position in the government, he still came regularly to Greifswald on Saturdays for a lecture. He was occupied most of the day with the local S.A. leadership, but he spent an hour with Kneser talking over the cases of Courant and other professors who had been placed on leave. Kneser reported to Courant that Vahlen had assured him that only established facts would be considered and that every affected person would be informed of the various points of the law which applied to him and would be given an opportunity to rebut them. Vahlen knew Courant only from rumors, "such as that you were a Zionist, which of course I

Spring 1933

denied immediately." He convinced Kneser of "a serious intent to be scrupulous." The most important thing in the conversation seemed to be the fact that when Kneser said of Courant, "But I know him—I was his assistant for four years," Vahlen replied, "Then you also could be asked to appear."

"So if you think of any point on which I could testify for you," Kneser instructed Courant, "call me or other suitable people."

While Kneser was approaching Vahlen on Courant's behalf, Friedrichs came to Göttingen to see Neugebauer. The two men agreed that they must do something to help Courant. They decided on a statement about his activities as professor and director of the Mathematics Institute. This would be signed by former students and other colleagues with connections to Göttingen and then sent to Bernhard Rust, the new minister for education.

By the time they finished writing, they had a long statement running to almost four double-spaced typewritten pages. Invoking revered and Aryan names, they began by pointing out how Klein and Hilbert had given the great scientific tradition which had developed in the work of Gauss, Dirichlet, and Riemann "a decidedly new turn." It had been they who had called the young Münster professor to Göttingen to carry on their work after the war. Courant, "a student of Hilbert," had devoted himself "with never tiring energy" to this task. His scientific works, particularly the great book written with Hilbert on the methods of mathematical physics, were now classical tools of analysis. In line with Klein's educational ideas, he had reorganized instruction in Göttingen to meet the needs of the unexpectedly great number of students coming to the university after the war. "As we are indebted to Klein for the mathematical *Lesezimmer,* which is available to all students, so Courant has established the great *Praktikum* for mathematical beginners and has extended it through years of work." It was thanks to Courant's initiative and energy that Klein's plan for the building of a mathematical institute had finally been realized— "offering opportunities for Germans as well as foreigners that are not offered anywhere else in the world." His "constant care" had been devoted to more than the welfare of mathematics—they cited the improvements to the physics institute, the building of the mineralogical institute, the extension to the institute for organic chemistry. In the latter part of their statement they turned to the activities of Courant with which they were most familiar—the discovery of talent, the financial aid and support of students, Courant's role as counselor and friend to the young.

"Rarely has a man tried so little to get credit for his achievements in the eyes of the outside world. He has never accepted opportunities offered him to go abroad because he is so closely connected with his work in Göttingen and because Göttingen has become so entirely his scientific and personal home. It is because of his personality that many of us also remember our time in Göttingen as a life in a scientific homeland."

In one of our conversations Friedrichs brought out the faded folder which contains a rather tattered copy of his and Neugebauer's petition, the final list of signatures, and the letters written by many who refused to sign.

On various pieces of paper in the folder, eighty-seven signers are suggested. Of these, twenty-two have been eliminated for various reasons, frequently it seems because they were teaching outside Germany and anything which could be construed as "meddling from abroad" was known to antagonize the government. Of the sixty-five remaining, sixteen did not respond at all. These included one man who had been Courant's assistant for several years. Of the other forty-nine, twenty-one refused to sign but wrote letters explaining their reasons.

There were various reasons—they were no longer close to Göttingen, or they did not know Courant well enough to judge whether the statements made about him were true. One writer explained that he and a colleague had signed immediately and had set about to obtain more signatures for the petition. They had, however, encountered so much negative response and had been so "urgently requested" by their colleagues not to participate that they had become hesitant and "after two days of consideration" had come to the conclusion that they should think the matter over further. There was an admitted feeling on their part that, much as they personally would like to support Courant, they did not want to bring themselves as individuals to the attention of the government. Was the Göttingen *Kurator* personally supporting the petition, they asked, and had any mathematicians who were "positively oriented" to the new government signed? If so—and if there were at least twelve signatures and if Friedrichs and Neugebauer thought it was necessary—they would add their signatures by wire. He was ashamed, the writer concluded, but he had to admit that he was afraid of the consequences if he signed.

Some did not consider a petition a proper vehicle for approaching the ministry. One found parts "disrespectful and offensive"; another said that he was sure "our minister of education, Mr. Rust, will do the right thing." Some thought the petition should be postponed or changed.

"I cannot sign," wrote one man who had been a student in Göttingen, "because I disagree very much with the general tenor of the petition. . . . Courant's stature as a mathematician is uncontested and well known. . . . His political status should be clarified. Proof should be found that his activities in this connection were undertaken for the sake and safety of the university."

"I remember a remark of Klein's to the effect that he did not care for Courant's political outlook and was inclined to think that he had Marxist leanings," another professor contributed. "I want to warn people against signing this petition, which will not help Courant and which may be very harmful to those who sign it."

Others, refusing to sign, said that they would nevertheless be willing to support Courant personally vis-à-vis the ministry. Among these, Carl Ludwig Siegel, by then a professor at Frankfurt, rewrote the petition in a way which seemed to him more precise and to the point and sent his version directly as a personal letter in support of Courant.

Kneser, who also did not sign, thought that a more effective way to help Courant would be to have a small number of colleagues and former students or assistants ("preferably party members before January 30, but they will hardly be found") write directly to the minister and ask to be allowed to testify as to their personal knowledge of Courant's character and his activities. The suggested letter, which was signed by Friedrichs, Kneser and Prandtl, began:

"Each of the undersigned knows Professor Courant as the result of a number of years of close collaboration. To our knowledge in all of his activities he has felt himself a German citizen and a representative of German science and has conducted himself as such. . . . The mathematical facilities of Göttingen, which since 1921 have been developed essentially through his efforts, are of great significance for the scientific culture of Germany and will not without essential damage be separated from his person."

It concluded with a request that the three be heard in person and, if this was not possible, that they be permitted to contribute their testimony regarding Courant in writing.

At the end of two weeks, twenty-eight men, more than a third of them physicists, had agreed to place their names on the Friedrichs-Neugebauer petition. These included such well-known scientists as Artin, Blaschke,

Carathéodory, Hasse, Heisenberg, Herglotz, Hilbert, von Laue, Mie, Planck, Prandtl, Schrödinger, Sommerfeld, van der Waerden, and Weyl.

Neugebauer found it "quite a nice list" as he conveyed it to Friedrichs on June 8—"although there are many not to be seen who should be seen."

"Several of the names on the list are those of people who later were considered to be Nazis or near-Nazis," Friedrichs pointed out to me, "and even at the time some of them were known to be in sympathy with the regime."

Today the judgment of motives and actions seems to him more difficult to make than it did then.

"You see, it is not clear. It is not clear at all."

One of those who refused to sign the petition in support of Courant was Hecke, who had known him since their earliest university days in Breslau and had come to Göttingen and been chosen as Hilbert's assistant on his recommendation. Although Hecke was in sympathy with the purpose of the petition, he felt that it would have either no effect or the opposite effect from the one desired.

It simply did not direct itself to the political mentality of the government.

"And of course Hecke was right," Friedrichs said as he put the old letters back into the folder.

SIXTEEN

THERE WAS no response from the ministry to either petition.
Courant sat at home and fidgeted. He tried to concentrate on the second volume of Courant-Hilbert, but the work refused to move forward. Psychologically he was in a worse state than he had been since the war. He leaned heavily on the Bohr brothers and the contact they represented with the outside world. Once they came to Germany, and he and Neugebauer met them on a North Sea beach and enjoyed the pleasure of speaking freely for a whole afternoon.

Weyl still served as director of the Mathematics Institute. Since Hella Weyl was Jewish, his tenure in that office was precarious; but he worked tirelessly, writing letters, conferring with government officials, trying to assure continuity.

Busemann, who was a quarter Jewish and thus still unaffected by the racial laws, remembers being begged by Weyl to remain at the university. "But I had already lived under a dictatorship when I was in Italy, and I had made up my mind to leave Germany and sit out Hitler in Copenhagen."

Fighting to save something in Göttingen, Weyl was still full of self-recrimination that he had refused the offer from Flexner and had doomed his family to Nazi Germany. Courant worried the question, if Weyl left, who would carry on the tradition of Klein and Hilbert? He placed his hopes on Rellich, who, although not a professor, had been Fritz John's examiner; but he feared that Rellich, while personally inoffensive, was too closely connected with the *Juden-Fakultät* of Göttingen to be acceptable to the ministry.

The future of Fritz John, who was half Jewish, worried Courant too. There had been a last minute crisis about John's getting even enough money to pay the fees required for the doctor's examination. To make matters worse, ten days after he got his degree, he had married the equally poor Charlotte Woellmer, who was not Jewish. Marriages between non-Aryans and Aryans, while not yet forbidden, were looked upon with such disapproval that Fritz and Charlotte had told no one about their plans. During the summer they survived on small sums paid by Courant from vague funds he still had. They also made some money from the sale of Fritz's lecture notes, typed by Charlotte. These were purchased by an American professor at a university called Brown which, they assumed from its address, was located on an island off the coast of the United States.

Courant was convinced that Fritz John would be an outstanding mathematician, and in the midst of his own immediate problems he fretted about finding a position for John outside Germany.

At the end of the summer semester, as MacLane and McShane and their wives prepared to leave Göttingen, Courant gave McShane an envelope of money addressed to Harald Bohr and asked him to mail it outside Germany. McShane told me that the money was used by Bohr to support young Jewish mathematicians from Germany who were already by then in Copenhagen.

It was at this same time that Courant received a letter from a former Turkish student, who had heard that he and Franck might accept appointments at a newly reorganized state university in Istanbul. Ten days later, at a meeting in Zurich, the two professors were surprised to be formally approached by a representative of the Turkish government. Courant took a very reserved position.

"I am—though on leave—a Prussian employee and would consider it impossible (either personally or professionally) to enter into negotiations regarding service in a foreign government without informing my superiors," he explained in a letter written later from Göttingen to the Turkish representative. "The most important reason to me, however, is that—if conditions force me to go abroad—I go, not as an embittered emigrant, but as a proud representative of German culture who neither internally nor externally has severed his ties with his homeland."

He was nevertheless "very interested" in the idea of creating "a scientific center of European quality" in Istanbul and eager to help "to the fullest extent of my ability." Later that summer, apparently assured that his current status would not be affected, he made a trip to Istanbul with Franck. On the way—or perhaps it was on his earlier visit in Zurich—he let Flexner know that if Weyl were again offered a position at the Institute for Advanced Study he would accept.

In Istanbul, Courant and Franck found a beautiful but alien city. Hitler's dismissal of so many outstanding scholars in every field had coincided with the desire of Ataturk (the name was not actually adopted until the following year) to staff his new state university with the most distinguished faculty in the world, At first Courant had been attracted by the idea of going to Turkey, and Nina had been very much in favor of it; but observing the political conditions on the spot, he decided that nationalistic Turkey was not for him.

"But why did you even consider going to Turkey instead of going to the United States?"

"It was closer to Germany."

Before the end of summer 1933, Courant received an invitation to spend the coming academic year in England as a visiting lecturer at Cambridge. His official status was still that of a German civil servant—although one on "forced leave"—and he had to obtain official permission to accept the English invitation. This was a delicate maneuver, since the question of the lifting of his leave was pending. Finally, ten days before the beginning of the term, he was informed that—in view of the fact he would not be in Göttingen during the coming academic year—his leave was being expressly lifted; in short, he was no longer on "forced leave" but was simply "on leave" at another university.

Courant left his family in Göttingen. The children continued in school with only a few incidents—in Courant's view, these were for the most part comical rather than tragic. He was more concerned about the indoctrination taking place in the schoolroom and on the playground and the natural desire of young people to be like their peers. A few days after he arrived in England and while he was still staying with Born, who also had a temporary appointment in Cambridge, he took the opportunity to write freely to Flexner about the importance of getting young scientists out of Germany immediately:

"The Nazis have remained consistent only in regard to the so-called Jewish question. . . . They indoctrinate a ridiculous racial theory (the basis of which is anti-semitism) through propaganda of all kinds . . . and it may easily happen that once this poisonous seed has germinated, an atmosphere much worse than that now existing will have been created."

A week later Courant wrote again to Flexner to tell him that although the government appeared to be having second thoughts about the removal of well-known professors—as in his own case and that of Schrödinger—the situation of Jewish professors in Germany was hopeless.

"The following event is characteristic of the course of things. Professor Landau, [to whom the *Reichsgesetze* do not apply in any way], went to start his lectures last week. In front of his lecture hall were some seventy students, partly in S.S. uniforms, but inside not a soul. Every student who wanted to enter was prevented from doing so by the commander of the boycott. [This was Weber, who had once been Landau's assistant.] Landau went to his office and received a call from a representative of the Nazi students, who told him that Aryan students want Aryan mathematics and not Jewish mathematics and requested him to refrain from giving lectures. . . . The speaker for the students is a very young, scientifically gifted man, but completely muddled and notoriously crazy. [From this description Friedrichs recognized Teichmüller.] It seems certain that in the background there are much more authoritative people who rather openly favor the destruction of Göttingen mathematics and science."

Courant's appointment at Cambridge was very specifically for one year only, and almost immediately he realized that in spite of the kindness of the English mathematicians he would not be able to live permanently in England.

"Mit den Engländern I get along very well," he wrote in a mixture of German and English to Franck, who was now in the United States. "All the objections you had regarding my temperamental faults have proved to be unjustified, because I have really taken advantage of the change in environments in order to change my outward attitude to people I do not know as well as I know you. I believe, therefore, that I 'pass' here. Still, if serious prospects develop in America, they will have to be given preference for the long term."

After receiving the Cambridge invitation, Courant had received an invitation to spend a semester at Berkeley. He had had to refuse it, since it was for a shorter period of time; but it had aroused in him the hope that there might eventually be a place for him at the University of California.

Flexner wrote that he was "actively exploring possibilities" for him. Veblen, who was now at the new Institute for Advanced Study, also wrote, "Your friends in America are trying to find a worthy position for you."

In Cambridge, Courant found that enthusiasm for finishing his book very soon began to evaporate. He was overwhelmed by guilt that in spite of freedom from the administrative responsibilities that had excused his failure for so long, he was still not able to write. People who had known him earlier found him "lost" in Cambridge.

At Christmas he returned to an "empty" Göttingen. The only one of the mathematics professors left at the university was Herglotz; the only physics professor, Robert Pohl. Weyl and his family had gone to America.

(Weyl's son Joachim recalls that during the preparations for the move his father went about the house intoning, as he often did, appropriate lines of poetry. "I cannot remember exactly," Joachim Weyl told me, "but there was something about 'if the beast of tyranny takes over his country, he puts the torch to his own house and leaves.' The last line was *zu dienen im Dunkeln dem fremden Mann*—'to serve in the dark the unknown man.'")

The sad state of the Hilberts—Käthe Hilbert was almost blind—added to the bleakness of the holidays for Courant. Also Martha Meyer, the housekeeper whom he had hired for Nina when their third child was born—a woman who had been like a mother to all their children—announced that she felt it would be better if she did not remain any longer in the household of a Jewish family. It was rumored that Heisenberg might take Born's place, but that Franck's professorship would be done away with. The most shocking development was that Werner Weber—the assistant who had led the demonstration against Landau—was the new director of the Mathematics Institute.

When Courant returned to Cambridge after Christmas, he was accompanied by Nina and their eldest son, Ernst, as well as by Fritz John, for whom he had managed a small fellowship. But he was still miserable.

"Day and night the thought tortures me what is supposed to become of the family in the summer," he wrote to Franck. "Please do not be angry with me . . . that I am deeply depressed. I know very well that, as in the case of Ehrenfest, all such difficulties actually come from within. . . . I know that I must cope with them."

Weyl sent happy letters from Princeton. In Fine Hall, where Flexner's group was temporarily lodged, German was spoken as much as English. He frequently saw Emmy Noether, who was nearby at Bryn Mawr. He wrote that he was "thoroughly satisfied with [the Institute for Advanced Study], the spirit and the work which is being done."

At the end of January, Courant received a long heralded offer from New York University. He would be given a two-year contract with the possibility that the position might become permanent. The salary was to be $4,000 a year, half of it to be paid by the Committee in Aid of Displaced German Scholars and the other half by the Rockefeller Foundation.

Courant was disappointed. Even in the early 1930's, $4,000 did not look like very much to a German professor who had been receiving a salary equivalent to $12,000. It was, in fact, a $1,000 less than what Felix Klein had been offered by Johns Hopkins University in 1883 to succeed Sylvester. Klein had refused *that* as inadequate and had gone instead a few years later to Göttingen.

Courant's disappointment was intensified by the fact that at the same time he received the offer from New York he also received another semester's invitation from Berkeley. It would pay him $3,000 from the middle of August to the end of December. He was torn again. Veblen wrote from Italy, Franck from Baltimore, Weyl from Princeton, all in favor of New York. "I think you simply *must* accept this offer," Weyl told him sternly. But Courant hesitated. Would it be possible, he wondered, for him to go first to Berkeley and then to New York? His friends assured him that it definitely *would not*.

Courant had no idea where—in New York—New York University might be. On his visit to the city in 1932 he had been aware only of Columbia. He wrote helplessly to Veblen, reverting to German, although by now he was customarily writing in English to the Americans: "Wer sind dort die Mathematiker?" *Who are the mathematicians there?*

Veblen replied that the only mathematician he knew at New York University was an assistant professor named Donald Flanders.

". . . he will be of special interest to you because he played a role in your history. Flanders is a Ph.D. from the University of Pennsylvania who studied in Princeton for a year or two as a National Research Fellow. His field is topology, and he is very zealous about research but has not been correspondingly successful. Early in the academic year he came to see me to get my advice about improving the mathematical situation in New York University. He was very anxious to get them to call in some real mathematicians. I was quite impressed with his unselfishness in the matter, because it was clear from the first that it would mean calling people in who outrank him in every respect, and he was under no illusions on the subject."

Courant wrote immediately to Professor Flanders: "Every kind of information about the general level of lectures, the type and preparation of the students, the facilities of mathematical studies would be very precious to me."

In spite of Courant's own troubles, all through the year at Cambridge he continued to be the person other professors who had been placed on leave turned to for help. A record of their pleas and questions and his responses is contained in a bulky folder of correspondence labeled "1933-34," which he later brought to the United States. Many of his replies were typed by Charlotte John, who came to England in March to join her husband. Courant paid her from funds he had from Göttingen, since Carl Still had insisted that he not return money that Still had already given him for the year.

From Germany that spring came rumors of a disturbing speech by Bieberbach in which the Berlin mathematics professor had applied a current theory of personality types to the practitioners of his own subject. One type was represented by "true Germans"—Hilbert and Klein were examples—and the other, by Frenchmen and Jews.

To support his thesis that "a German essence" exists in mathematical creation, Bieberbach had quoted a statement made by Klein in America in 1893: "It would seem that a strong naive space-intuition were an attribute preeminently of the Teutonic race, while the critical pure logical sense is more fully developed in the Latin and Hebrew races."

One of the great achievements of true German mathematics, according to Bieberbach, was Hilbert's work on axiomatics; and it was most regrettable that "abstract Jewish thinkers" had succeeded in turning this work into "an intellectual variety show."

"Our nature becomes conscious of itself in the malaise brought about by alien ways," Bieberbach had explained. There was an example in the "manly rejection" of Edmund Landau by the students in Göttingen. This man's un-German style in research and teaching was "intolerable to German sensibilities." The important task for "National Socialist science" was to recognize the existence of the "German essence" in works of science and then "to proceed to action"—the nurture of that essence.

At the time of Bieberbach's speech, Courant was already beginning to make preparations to go to the United States. There were various difficulties for him in leaving Germany. His first task was to obtain permission to take out of the country more than the amount of money permitted by law. The second was to obtain an exemption from the tax which was placed on all those who were emigrating—this alone would be equivalent to about a quarter of his worth. Both of these problems could be solved with comparative ease if he were to be simply on an extended leave.

To support the request for an extension of his leave, Courant argued that his presence in America would in fact be an asset to Germany. He furnished several testimonials to support this contention. Among these was a statement by Ferdinand Springer that it would be in the interest of German scientific publishing for Courant to accept the American invitation.

"When in 1932 efforts were being made in the United States to boycott German scientific literature, Courant—who was then in that country—achieved an improvement in the general situation by negotiating with certain important people," Springer explained. "Difficulties and dangers for German scientific literature in America are going to continue; and for this reason alone it would be desirable that a mediating personality like Professor Courant, who knows the situation on both sides, be in the United States."

The dean of the Göttingen faculty also pointed out that Courant had been responsible for bringing a great deal of American money into Germany and it was important not to sever the connection he represented with the source of that money.

In spite of his personal problems, Courant continued to be concerned about the future of the Mathematics Institute.

Before leaving for America, Weyl had urged Helmut Hasse, at that time a professor at Marburg, to accept a call to Göttingen when and if one came to him. Hasse was an outstanding mathematician, although he was not in the broad mathematical-physical tradition which had flourished at the university. He was known to be politically conservative. Under the circumstances he seemed to Courant, as he had to Weyl, the best possible person to be director of the institute. As soon as Courant received the New York invitation, he wrote Hasse and brought up the subject of Rellich, who was currently a *Privatdozent* in Göttingen.

"If you really go there," Courant told Hasse, "Rellich will be of enormous help to you in the administration and also otherwise."

By Easter, Hasse had received the expected call to Göttingen and was negotiating with the ministry about the terms of his acceptance. At the same time, with a little coaching from Courant, he was trying to support Courant's application for a two-year extension of his leave so that he could accept the New York position without giving up his official position in Göttingen. The minister favored the "emeritization" of Courant rather than another leave and suggested to Hasse that it would be nice if the request for emeritization would come from Courant himself.

Courant had earlier explained to Hasse that there were considerations for him other than financial ones: "My feelings of belonging to Germany and the Mathematics Institute, which after all I created, are decisive. It is clear that even permission for a leave would not make a later return to Göttingen very feasible, for both external and internal reasons; but it is my personal disposition not to turn down the slightest possibility that such a chance might exist."

All during the early summer of 1934, the situation at the Mathematics Institute was chaotic. It was announced that a Nazi party member named Tornier would succeed to the chair of Landau and would act for Hasse as director until the latter officially took over. On his own authority Tornier dismissed Franz Rellich. Courant, shocked and dismayed, communicated this action to Hasse.

"I inform you of this because I think that the apparently premeditated act of not telling you what is being done in your name creates a very bad situation. . . . I hope and wish that things here will soon change so that you can come with pleasure and without hesitation."

But in his heart he was not very hopeful, as a letter he wrote to Carl Still on June 19, 1934, indicates.

Still was trying to promote high quality appointments to the scientific faculty in an effort to rebuild what had been destroyed by the removal of the Jewish professors, and he had asked Courant to help him. Courant felt that he had no choice but to decline. He had come to the belief, he wrote to Still, that the mathematical Göttingen which he and Klein had built up had been irreparably damaged. Germany had not benefited, only countries abroad. America was now trying feverishly to create scientific centers similar to the one that had been in Göttingen.

"The appointment of a few good people cannot put an end to the current anarchy or bring back the regard which the world had for Göttingen."

The day after Courant wrote to Still, the first official extract of Bieberbach's remarks appeared in print. The speech had already become a cause célèbre. Harald Bohr had responded with an answer in a Danish paper based on what Bieberbach angrily called "a ridiculous caricature" of his remarks. Now, on the basis of the published extract by Bieberbach himself, Hardy carefully summarized in *Nature* the ideas which had been put forth and came to "the uncharitable conclusion" that the Berlin professor really believed what he had said. Bieberbach published a strongly worded "Open Letter to Harald Bohr" in the *Jahresbericht* of the German mathematical society. This was done over the opposition of the other two editors, Hasse and Konrad Knopp.

In July, Hasse came to Göttingen to attend to some details of his appointment with the dean of the science faculty. As he walked past the Mathematics Institute, he was greeted by such an unpleasant demonstration on the part of the pro-Nazi mathematics students, led by Teichmüller, that he returned disgustedly to Marburg.

The ministry's response to Courant's request for a leave remained negative. Courant recognized a veiled threat. He requested emeritization—retirement—as of April 1, 1935, but agreed that if the ministry wished for an earlier date for the sake of making new appointments at the Mathematics Institute he would consent. On July 30 Theodor Vahlen wrote to him, "I take this opportunity of sending you my appreciation and special thanks for your valuable academic activities...."

Friedrichs was surprised and impressed that Vahlen had written to Courant.

"I thought it was just a formality," I said.

"Oh no!" he replied. "The Nazis didn't bother with formalities."

From a financial point of view, Courant had come off rather well. His regular salary as a professor was paid to him from the time he returned from Cambridge to the time he left Göttingen. He was excused from the emigration tax in view of the services which it was expected he would perform in the interest of Germany in the United States. He was given permission to take a larger amount of money than was customary out of the country.

With the decision to leave Germany made, he began to delight in the largesse of what everybody called "play money"—the marks still in his bank account, which he could not take to the United States but which he could spend as he wished in Germany.

One morning, before breakfast, Ilse Benfey, a young neighbor, came rushing over to the Courants' house with the news that marriages between Aryans and non-Aryans were to be officially forbidden in Germany. The family of Ilse Benfey's Jewish father had lived in Göttingen for more than two hundred years, but that morning in the summer of 1934 Courant advised her to come with him and his family to America.

"But I do not have the money to do that," objected the young woman, who at an earlier time had given gymnastic instruction to the Courant children.

"Ja, ja. But you should not worry about that," Courant mumbled. His voice trailed off into something about how she could help Nina a little with the children and he would pay her passage.

The elder Benfeys insisted that there was no need for their daughter to leave Germany, but Ilse began to help Nina with the packing. Everything was to be taken, since the shipping expenses would be paid out of the "play money." Heavy furniture that had belonged to the Runges. Two grand pianos, one of them left behind by Hans Lewy, the other purchased by Courant when he was a young officer in Berlin. Innumerable other musical instruments, pictures, books. The collected works of Gauss, of Riemann, of Klein, of Hilbert, of Minkowski. The many volumes, now 43, of the Yellow Series. Mathematical reprints. Letters and diaries. Currency was stuffed into every possible hiding place.

When the moving men arrived, Courant was everywhere, making dry jokes with the men, fussing over the packing of the musical instruments, passing out cigars, tipping everybody generously.

From Göttingen the party traveled by train to Bremen, where they stayed overnight. "What did you do your last night in Germany?" I asked Nina. Her face lit up. "We made music with some friends."

The next morning they sailed for America.

SEVENTEEN

IT WAS on the evening of the tenth day—August 21, 1934—that Courant and his party first saw lights in the distance. Early the next morning they hurried out on deck with the other passengers, several of them also refugees. The day was already sweltering by European standards. Out of a bright morning haze the famous skyscrapers emerged like a range of great mountains. As the *Stuttgart* steamed slowly into New York harbor, a little boy—also a Jewish refugee—greeted "Tante Liberty" with a sweetness and sincerity that touched them all.

At the pier, immediately, they saw Hilde Pick, the cousin who had been Courant's secretary in Göttingen. While they were going through the formalities of customs, several other people appeared to greet them: Franz Hirschland, a German-American industrialist whom Courant had met through the Busemanns on his earlier trip to America; Donald Flanders, the young mathematics professor who had started the negotiations which had led to Courant's new position; a student whom Gregory Breit, a physics professor at NYU, had thoughtfully sent to help them with their luggage. Then suddenly Dolli Schoenberg, one of Landau's daughters, also appeared; and finally, at the barrier, they found Hans Lewy, for the past year an instructor at Brown University. In this situation, as Nina wrote back to Germany, how could they feel that they were strangers?

It had been arranged that, until they found a place to live, Ilse Benfey and the girls would stay in the city in Flanders's apartment while Courant, Nina and the boys would be guests at Hirschland's estate in Rye. They were all packed into two cars—Hirschland's equipped with a uniformed chauffeur—and were driven up through Manhattan, first to lunch and then to Flanders's apartment to leave Ilse and the girls.

Courant had come to the new world resolved to make the best of what he knew would be a modest situation, but he was not prepared for the apartment of the young professor. As he later learned, the self-effacing Flanders and his wife were intellectual individualists who placed their children's education at expensive progressive schools above material possessions and surroundings. They spent their summers on a farm in upstate New York, from which Flanders had come to welcome Courant. During the school year they lived in New York City in what is known as a railroad flat.

As the party climbed flight after flight of stairs, they were increasingly disturbed by what they saw. The flat itself was *schrecklich*—Nina reported —frightful, dreadful. Five rooms, kitchen and bath lay all in a row. Only one room had a real window, and it looked out on the street. How anyone could manage in such a dwelling with two grownups and three children, even for only part of the year, was incomprehensible to the visitors from the big Runge house on Wilhelm-Weber-Strasse. Courant searched in vain for Flanders's study. Outside, some sort of elevated train roared by. He looked at Nina in dismay.

Was this how a professor lived in New York?

The shock of Professor Flanders's home—the lack of a study—was always to be the Courant family's most vivid memory of their arrival in America. In contrast, the Hirschland estate in Rye was everything a home should be in a land of millionaires. A long curving driveway took them through a natural arca and then a park-like garden. On the way to the house they passed tennis courts, a lake with boats, stables with riding horses, a garage with yet another car. The house itself was "extremely elegant." In a single room they noted a Cranach, a Corot and an El Greco.

Since Courant was not inclined to describe anything in detail, I looked forward to finding in Nina's letters some account of his first impressions of the university which was to be his new base. But I was disappointed. She describes Flanders: "a blond, somewhat plain-looking young man but exceedingly friendly and modest—certainly no *typical* American"; but she does not devote a word to New York University. In fact, she does not mention her husband's work in her letters until February when one of her correspondents asks, "What about Richard?" From the first she was—as she told me—completely confident that Richard Courant would be as successful in New York as he had been in Göttingen. Her private name for him was always that of the clever cat who impressed the king, outwitted the ogre, and won the hand of the beautiful princess for his young master—"Puss in Boots."

When I tried to draw Courant out about his feelings in August 1934 when he first saw New York University, he said only, "It was very different from what I had been accustomed to in Göttingen."

At that time, although its buildings were scattered all over the city, there were two main locations of the university. One of these, very much like a typical small-town college campus, was situated on the heights of the Bronx overlooking the Harlem River. The other, consisting of a number of buildings without a campus, was at the foot of Fifth Avenue just off Washington Square. This was where all graduate courses met. Courant would have his office at the Heights but would deliver his lectures at the Square, commuting between uptown and downtown by subway. It was still summer vacation, and there were only a few faculty members and students around. One of these was a young man named Morris Kline, who was an instructor working toward his Ph.D. in mathematics. When I talked to Kline in 1971 in his office at the Courant Institute, where he had recently retired as director of the Electromagnetic Division, he told me he remembered very well that first day when Courant had turned up at University Heights. He wanted to know how to go about buying a car—he seemed to be convinced that a person could not exist in America without an automobile—and so Kline found a student to go out and look at cars with him.

In the next couple of weeks the pace and Brobdingnagian scale of American life did much to mitigate for Courant the dismal first impressions of the lot of a professor at NYU. A friend from Germany meeting him inquired, "Ah, Courant, and will you be playing quartets again in this country?" "In this country not quartets," Courant replied. "In this country, octets!" He found many old friends and acquaintances already established in the city and in nearby Princeton, which Nina reported as *sehr göttingisch*. In New York itself the New School of Social Research, which was being referred to as "the university in exile," seemed staffed almost entirely with refugees, as did New York University's Institute of Fine Arts. In fact, Courant and his family found themselves so surrounded by refugees and by German Jews who had come in earlier times that in her letters Nina stopped to describe in detail any "real" American whom they met. She also invariably identified new acquaintances to her correspondents as *arisch* or *judisch*, indicating—to the quarter—the degree of Jewishness. This surprised me—it was so unlike Nina—and then I realized it was the result of what she had so recently experienced in Germany, where a person's future depended on the exact amount of his Jewish blood.

Although Courant continued to follow the activities of the National Socialists in the newspapers, he felt far away from Germany and its problems. And yet—"The situation here is also full of uncertainty and tension," he wrote to Max Born. "But, in the first place, one does not immediately feel and understand it and, in the second place, it is still in seriousness and danger not comparable with that in Europe. I believe, however, that also in America a complete far-reaching transformation of the whole social organization is taking place."

While the children visited from friend to friend, Courant and Nina with the help of Ilse Benfey tackled the problems of settling a family in a strange country. They had been in the new world just two weeks when they rented a house, and Nina was able to write to her mother, "We now have a real address, 142 Calton Road, New Rochelle, USA." It was to be Courant's home address for the rest of his life, and the house where Nina still lives with Ilse Benfey, now retired after a long and active career as a social worker.

New Rochelle, located on Long Island Sound, northeast of New York City, is some twenty-five miles from Washington Square, where Courant was to teach. Friends had recommended the town for its good public schools. Nina, who had some Huguenot blood, was pleased to note that its name and its architecture in many sections memorialized the French Protestants who had fled their homes in Europe to escape persecution for their religion. The principal monument, a few blocks from the Courants' new home, was a statue of Jacob Leisler (born in Frankfurt am Main), who made it possible for the Huguenots to settle in New Rochelle.

The house which the Courants had rented was a big, comfortable, but very modest structure standing upon a knoll with other similar houses. They found it one of the curious things about America that people often lived in such wooden houses all year around, not just in the summer, and also that there were no fences between their garden and the gardens of their neighbors. A pleasant tree-lined street led down to the high school, in front of which there was a little park and a small lake. There was a garage, surprisingly "right on the street"; and within a week after they had moved in, Courant had negotiated the purchase of a secondhand Chevrolet through a cousin who was an insurance agent—one of the four Courants of his generation already in America in addition to Hilde Pick and himself.

Hilde Pick agreed to act again as his secretary and came out to New Rochelle to live with the family as she had in Göttingen. Ilse Benfey took over the management of the household and the children. Nina began immediately to "make music." The change in the family's situation seemed hardest on Ernst and Gertrud, the two older children; but every evening as 10-year-old Hans lay in bed, his mother heard him singing *Deutschland, Deutschland über Alles* to himself, "in the softest tones of which he is capable." Six-year-old Lori marveled that no one in school asked her if she was Jewish.

What was it like, I asked Courant, to leave his home and country and start over at forty-six in a situation where he was again financially and professionally at almost zero.

"It was hard," he said simply. "There was such a little bit of money. I felt myself responsible for many people. There was really nothing scientifically at NYU. And then of course"—his voice trailed off as it usually did—"I was so attached to Göttingen."

When I asked the same question of Nina, she replied, "Of course it was something we could never have brought ourselves to do on our own, but when it was forced upon us and there was nothing else we could do, it was wonderful—like being young all over again!"

I have been told that, on the occasion of their fiftieth wedding anniversary, when Courant in a little speech described Nina as "a heroine" who had courageously left family and friends to follow her husband to a strange land and a precarious position, she had stood up at the table and, rapping on her glass, had asked to say a few words herself. It had not been the way Richard had described it at all. In fact it had been rather a relief to her to leave the little university town of Göttingen where she had grown up as one of the four daughters of Professor Runge and where she and her family had lived, after her father's death, in the old house that had belonged to her parents. She had been no heroine, she insisted. She had been happy and eager to leave the old world for the new!

Watching Courant's face while Nina spoke, many of the guests had the feeling that he was displeased. For him it had been different.

After he arrived in America, he continued to seek news of the Göttingen institute from European friends outside Germany. In September, he received a firsthand report from the English mathematician Harold Davenport, who had recently visited Hasse.

In the late summer of 1934, at the urging of other German mathematicians, who were disturbed by what had happened to mathematics in Göttingen, Hasse had agreed to go back to that university; but his life was still being made miserable by Tornier and the pro-Nazi students led by Teichmüller.

Davenport's letter also brought news of the annual meeting of the *Deutsche Mathematiker Vereinigung*, at which Tornier had appeared in the company of a storm trooper in civilian dress (later asked to leave). Although the election of Blaschke as president had been considered a victory for the moderate forces, the members of the DMV had passed a resolution censuring Harald Bohr "most sharply" for his attack on Bieberbach "to the extent that it was an attack upon the new German state and upon National Socialism." They had deplored the fact that Bieberbach had published his reply to Bohr over the opposition of the other editors, but they had formally recognized that he had been motivated by his concern "for the interests of the Third Reich."

Courant was incensed by the affront to Bohr. He also felt an urge to write a few lines of sympathy to Hasse, "but I am afraid such a correspondence observed by the Tornier guards may do him more harm than give him comfort."

The students Courant faced at Washington Square were very different from those he had known in Göttingen. He found them "not ungifted but extraordinarily poorly prepared." Almost all were Jewish, the sons and daughters of immigrants from eastern Europe. As undergraduates they had usually attended the City College of New York, since it charged no tuition. After graduation they were able to continue their education with a class or two at NYU only because that university regularly offered its graduate courses in the late afternoon and evening.

Flexner described these young first-generation Americans from the nation's largest city as "a great reservoir of talent," and Courant now made this phrase of Flexner's his own. In the coming years he was to use it over and over. The metaphor touched a responsive belief in him.

"There are many outstanding people, potentially, everywhere at all times," he told me, "but of course the conditions are not always conducive to their development."

"That is something you really believe, isn't it?"

"Very much."

The role Abraham Flexner played in helping Courant during his first years at NYU is not generally known, and it was in fact so minor in relation to Flexner's many other activities that he does not mention it in his autobiography. His correspondence indicates, however, a sincere interest in the project and in Courant, whom he saw as an "idealistic, energetic, and unselfish" man. During the years from 1934 to 1939, when Flexner retired as director, the Institute for Advanced Study cooperated with Courant in various practical ways. Flexner personally also gave him introductions to wealthy New Yorkers who might be interested in the improvement of the graduate mathematics program at New York University.

"How *was* the program being offered in mathematics at NYU when Courant came?" I asked Morris Kline.

"No better than mediocre, maybe even a little worse," he replied promptly. "There were only a few people who could lead doctoral candidates, since not all members of the faculty had doctor's degrees themselves. I think that when Courant came I would have shifted over to him if it would not have meant going into an entirely new area of mathematics. I was interested in topology then—like most young American mathematicians of the time."

Kline's evaluation of NYU's graduate mathematics program in 1934 is borne out by a study which was being concluded that same year by R. G. D. Richardson in the hope that, as he wrote, "those concerned with the strategy of promoting mathematical thought and achievement in America can find in [it] several signposts for their future guidance."

As one criterion for the excellence of the program at a given university, Richardson took the number of Ph.D.'s which had been awarded. (Another was the amount of work produced, as measured by published pages.) In the course of his survey, he discovered that since 1862, when Yale had awarded the first doctor's degree in mathematics, more than one-sixth of the total of 1286 degrees had been conferred by the University of Chicago. Six universities (Chicago, Cornell, Harvard, Illinois, Johns Hopkins and Yale) had been responsible for more than half. In the decade immediately preceding (1924–34), only 29 universities had awarded five or more doctor's degrees and could thus be considered an important factor in the mathematical education of America.

New York University was not one of these.

1934-1935

Harry Woodburn Chase, the chancellor at NYU since 1933, was definitely interested in improving the graduate mathematics program at the university. Unlike Flexner, whose father had been an itinerant peddler of hats when he first came to America, Chase had a background completely different from that of most of the NYU students. He came from an old American family, had grown up in a small New England town, and had attended Dartmouth College. He had taken Courant with the assurance that Flexner, who had earlier refused the chancellorship, intended to maintain a practical interest in mathematical developments at NYU.

"The organization and the inner workings of the giant university are very complicated, and it will be a long time before I understand them completely," Courant wrote to Born after meeting Chase. "It is clear, however, that—in principle—a really rewarding project offers itself here. I am not quite sure how it should be carried out. Also it is by no means certain that I shall have the opportunity."

The ties with Göttingen which he had struggled so hard to maintain during the past year began to be severed almost immediately. Notification came from Germany that his emeritization, which had been set for April 1 of the following year, had been moved up to October 1. This formal conclusion of a career as a German professor—under normal circumstances—would not have occurred until 1956. On the first day of October 1934, Courant wrote drily to Franck, "Today I celebrated my sixty-eighth birthday."

Before the end of the year another tie was to be severed. Protesting the censure of Harald Bohr "for remarks which he had made as a private person," Courant resigned from the *Deutsche Mathematiker Vereinigung,* which had been founded in happier days by Klein and Hilbert and others.

Franck, concerned about his friend's future in a new country where people might not understand and appreciate his virtues, cautioned him:

"Please, dear Courant, prepare your lectures well and try not to organize right at first!"

Courant intended to follow this advice and, as he put it, "to regard present conditions as passively as possible and to try to give good lectures from which people can really get something."

He hired a man in the speech department to tutor him in English and carried in his pocket a small notebook in which he was constantly writing down colloquialisms he came across in conversations or in books. He dictated carefully prepared lectures and then translated them from German into what he hoped was "good English." Although he also dictated innumerable letters in German to friends still abroad and to his other German friends now in America, he tried his hand at composing letters in English whenever an opportunity offered. Even the American Cancer Society, New Rochelle branch, received a long apologetic account of his financial situation as a refugee professor along with his check for $5.00.

He had done no real mathematical work since the Courant-Friedrichs-Lewy paper. He knew he should begin to publish again, but he found himself unable to do mathematics or to work on the still unfinished second volume of Courant-Hilbert. He could not accustom himself to the quiet, the lack of people around him, the silent telephone. Sometimes, in his little office at the Heights, he would lift the receiver from its hook and listen in the hope that there had been some mistake and the phone had simply failed to ring.

In spite of his resolution to follow the admired Franck's advice, he was eager to be "genuinely helpful" at NYU. He saw many things that should be done. The worst aspect of the situation was the lack of an accessible mathematical library, which made it virtually impossible for students to do independent mathematical work. He was also soon aware of a need for financial support on the part of some students if they were going to be able to concentrate on their education. He began to talk to Hirschland and other wealthy German-Americans with whom he had become acquainted about "the reservoir of talent" at NYU and the desperate need for a *Lesezimmer*, for *Assistentstellen*—the German words came more easily. Some of these men gave him small sums in dollars. Some also offered blocked marks—the "play money"—which they had in German banks and were unable to spend outside Germany.

All during his first year he mulled over the problem of the frustratingly blocked funds. These could be used to purchase German books and journals only if they were matched with American dollars. He worked out a plan for cooperation between NYU and other universities but abandoned it as too complicated. He also wrote to the American Library Association suggesting that it might be feasible for the organized American libraries to obtain from the German authorities the right to pay a considerable part of their orders in blocked marks and so reduce the cost considerably. This suggestion did in fact later bring results, but it was considered more diplomatic by the association not to give Courant credit for it.

Unobtrusively, so he thought, he tried to establish contacts which might turn into future sources of funds. The semester had hardly begun when he hunted up Warren Weaver, the director of the Division of Natural Sciences of the Rockefeller Foundation. He also recalled himself to Henry Goldman, a wealthy German-American Jew, a friend of Born and Franck who had at one time contributed some money—"not very much"—to the physics institute in Göttingen. (Rich people, Courant always said, have no real conception of money.) With his introductions from Flexner, he met various wealthy Jews in the city.

Ultimately, he knew, it was going to be *most important* to upgrade the faculty at NYU in mathematics and in the related sciences. Breit was no longer in the physics department; and except for the selfless Flanders, whom Courant was always to refer to as "a saint," he felt very much alone in his new position. A few months after his arrival, he heard that the Rockefeller Foundation was willing to sponsor Siegel's coming to the United States. He tried to arouse interest on the part of the NYU administration in providing a place for the German mathematician; but instead Siegel received an appointment at the Institute for Advanced Study.

Courant went eagerly down to the pier to meet him when he arrived in January 1935 and transported him out to New Rochelle to stay until he had to report in Princeton.

"The most important news," Nina wrote to her family in Germany, "is the arrival of Siegel, one of Richard's first assistants, a postman's son, big, very Aryan. . . . In Richard's opinion, a mathematician of Hilbert's stature."

Less than a month later, Siegel came back to New York to deliver a lecture at Washington Square to Courant's students and invited guests. This lecture was only one of a number of lectures by well-known mathematicians which Courant arranged during his second semester at NYU. Although most of these were set up by him to give his students insights into areas of mathematics outside his own specialties, the subject of one lecture that semester was in the area which most appealed to him and was, in the course of the coming year, to bring him back at last to mathematics.

This was the lecture in March 1935 by Jesse Douglas, a mathematician at MIT who had attracted worldwide attention by solving Plateau's problem, one of the oldest and most famous problems in the calculus of variations.

Courant's invitation to Douglas to speak at NYU was to a certain extent an olive branch. Since coming to America, he had made a determined effort to remove what he called "dissonances" in his relations with American mathematicians who had earlier left Göttingen feeling somewhat offended by their treatment there. Douglas was one of these. Arriving in the year 1929–30 at the end of a European tour as a National Research Fellow, he had proposed to talk upon his not yet published work on Plateau's problem at the weekly meeting of the *mathematische Gesellschaft*. The problem had been around for a long time. Many outstanding German mathematicians, including Riemann himself, had worked on it. The members of the *Gesellschaft* simply did not believe that an American had solved it. As it happened, Douglas's solution, which was highly original, did not appear to be in completely rigorous form. When he finished his presentation, some of the members of the *Gesellschaft* took him severely to task on almost every detail of his proof. He left Göttingen deeply offended but determined to show the people there that his argument had been correct. When he finally did publish his work in 1931, he laid out the chain of reasoning in an unassailably rigorous fashion.

Remembering this past history, Courant did his best to make Douglas's lecture at NYU an event. He sent out notices to other colleges and universities in the area and saw to it that Douglas's former teachers at Columbia and CCNY received personal invitations. In view of the level of the students, however, he suggested that Douglas try to keep his remarks "as elementary as possible." It was a long way from a meeting of the *Gesellschaft*, but afterwards there was coffee and discussion as in the old days. Courant asked Douglas to send him enough reprints of the famous work so that every member of his seminar the following year could have one.

After I learned some of the details of Courant's first year in America, I became curious what idea of the future had been in his mind when he came here. I asked him if from the beginning he had hoped to build up another institute at NYU like the one he had had to leave behind in Göttingen.

"Well, that is difficult to say," he replied. "Everything is so different in the perspective of so many years. When I came here I felt some kind of— I don't like such words—but some kind of *patriotic* urge to do something for this country. I was deeply impressed by America and what was being done during the first years of the Roosevelt administration. I was very enthusiastic. I felt very loyal and thought very much about what was needed. The best I could imagine that I might be able to do was to bring my experience in Göttingen to bear upon the situation here."

EIGHTEEN

COURANT was not kept long in suspense as to who was going to have the opportunity to develop "the really rewarding project" which he saw as existing at New York University. In June 1935, at the end of his first academic year in America, he received a note from Chancellor Chase informing him that when his temporary appointment expired the following June the university was prepared to offer him a permanent place. In the note Chase stated he believed that, in conjunction with the Institute for Advanced Study, there was a real opportunity for Courant to develop at NYU a strong graduate department of mathematics in ways that would be interesting to him.

"I am saying this now," he wrote, "because I want you to feel whatever sense of security that it may bring."

It was the moment when, in Courant's opinion, "a continuous work" should begin. There was still no space at Washington Square for an office or a library, but there were some encouraging developments. Goldman had offered to give $500 for the library if Courant could raise another $2500 on his own. The ever helpful Flexner had agreed that members of his institute at Princeton could give individual lectures and even full courses at NYU during the coming year if Courant could guarantee their train fare. George Blumenthal, a banker who had sat beside Courant at an official university dinner, had donated $1500 for fellowships in the fields of Courant's mathematical interests.

The money from Blumenthal was the first Courant had had the opportunity to distribute in America, and he took up the task with enthusiasm. One fellowship went to Irving Ritter, German-born but an American citizen, who had been a graduate student at NYU during the preceding year. He was older than the other students, married, and working as a nightwatchman. He later became a professor at University Heights and is now retired. Another fellowship went to Max Shiffman, the outstanding mathematics student in the current graduating class at the City College of New York—now at the California State University at Hayward.

Having awarded one fellowship to a German-born American and the other to an American who was Jewish, Courant concluded—and others agreed with him—that the third fellowship should go to some "really good 100 per cent American." At the beginning of 1935 he sent out a number of letters to other mathematicians asking for recommendations. By August, when no suitable candidate had turned up, he suggested to Flanders that for the third fellowship they push Rudolf Lüneburg, one of his students from Göttingen, who had left Germany very early, even though he was not Jewish. At the last moment J. D. Tamarkin of Brown University, himself Jewish and a Russian immigrant of an earlier day, came up with the suggestion of a young man with the completely satisfactory name of Tom Confort. Courant managed to find some other way to support Lüneburg, who after a year at NYU went into industry and made outstanding contributions to optics. I asked several people what happened to Confort, but nobody knew. Later I found out that the young man was actually Edwin H. Comfort, now professor emeritus at Ripon College. Since his parents were English and Canadian, he was not quite so "100 per cent American" as his name suggested to Courant.

In spite of Courant's efforts in connection with the development of a graduate mathematics program at NYU and his active concern about placing friends and former students from Germany, his top priority in the summer of 1935 was still the completion of the second volume of Courant-Hilbert. Already many of his letters in English had assumed a format which was to become standard with him: "I have a very bad conscience that I have not answered your letter . . . which I must admit slipped between some other mail and was lost for a time . . . but. . . ." During the first year in America the "but" introduced the statement that he was "under very much pressure" to finish Courant-Hilbert II. In the months since his arrival he had made almost no progress; however, in August 1935—at the end of the German summer semester—he expected Friedrichs for a visit. He was again optimistically certain that with Friedrichs's help he would at last be able to finish the book.

Friedrichs had a reason for coming to the United States other than helping Courant. He was in love with a Jewish girl whom he had met just four days after Hitler had become chancellor. Although the Nuremberg laws prohibiting marriage between Aryans and non-Aryans had not yet been enacted, all official National Socialist newspapers were regularly publishing the names of non-Jews who were alleged to have had relations with Jews. In a number of towns individuals had been sent to concentration camps as a result of such charges. To protect their family and friends as well as themselves, Friedrichs and Nellie Bruell had been able to meet only clandestinely during the past year. They were considering whether they could emigrate.

Courant was delighted to have his former student and assistant in New York, to talk to him, to show him around, to introduce him to the important people he had become acquainted with, to explain America to him. He produced several new odds and ends of manuscript and the first two chapters, which had already been written when Friedrichs had first started to help with the book ten years before. "Now we will finish!" Courant insisted. But nothing much was accomplished that summer either, except that the first two chapters were finally "really written."

The day before Friedrichs was to return to Germany, he and Courant went for a long walk. Friedrichs announced that he had decided he could be happy living and working in the United States. Courant advised him to emigrate immediately—the fact that there was as yet no academic position available for him should not stop him.

"The traditional way to come to America," Courant mumbled, "is as a dishwasher."

Friedrichs answered mildly that he was not yet ready to exchange mathematics for dishwashing.

By the time Friedrichs left to return to Germany, the American semester had already begun. He remembers that Courant was very excited about the seminar he was conducting on conformal mapping, minimal surfaces, and Plateau's problem. The students were interested, intelligent, and willing to work; and one of them—young Shiffman—was in Courant's opinion exceptionally gifted. Plateau's problem was attractive to Courant; and even before Friedrichs left, he was showing signs of interest in working on it himself.

"But he didn't *want* to get involved in that," Friedrichs told me. "He *wanted* to finish the second volume of Courant-Hilbert."

In addition to trying to finish Courant-Hilbert II, to lecturing and conducting the seminar, to searching for sources of support for his enterprise at NYU, Courant was constantly concerned with the personal and professional problems arising out of National Socialist policies toward German Jews. Both his parents were now dead. His remaining brother, Fritz, and his family were in Italy, where Mussolini had not yet adopted Hitler's anti-semitic policies. Courant's interest and activity usually did not extend to his Courant cousins and aunts and uncles still in Germany, but he *was* concerned about his friends. He mounted a massive campaign to get a visa for Ilse Benfey, who had come originally as a visitor. Helping her was complicated by the fact that she was not a relative. Courant, however, managed to get Flexner, Flanders, Hirschland, Goldman, and a number of other important people to write to the effect—as Nina described it—"that Richard is a really splendid fellow and in a position to give an affidavit of support for Ilse, who also because of her own capabilities is not likely to become 'a public charge.'" Eventually, under the force of all this ammunition and Ilse's own frequent calls, the American consul in Hamburg yielded, so flustered that he signed Ilse's last name instead of his own on her visa.

Letters asking for help and advice came "by the dozens" from mathematicians in Germany. No one who had known Courant there could believe that he was not so knowledgeable and effective in New York as he had been in Göttingen. He worried the most about the future of former students. Since March he had fretted over the case of Fritz John, whose grant from the Academic Assistance Council in England was going to expire in June, leaving him and his ailing young wife virtually destitute. Then, in the fall of 1935, Fritz John received an unexpected appointment at the University of Kentucky. Courant rejoiced. It was an exceptionally good position for a man so young and with so little experience!

Back in Germany, in Göttingen, things seemed to be settling down.
When the students had placarded the Mathematics Institute with signs proclaiming that Hasse, who was the treasurer of the *Deutsche Mathematiker-Vereinigung*, "permitted Jews in the mathematical society," Hasse had gone to Theodor Vahlen in Berlin and had demanded that something be done about the situation in Göttingen.

He told me when I talked to him in San Diego:

"I tried to appeal to his mathematical soul—not his political feeling. Then he said to me—I can see him sitting there, a broken man—'Yes, I see all that, but I can't help. I am a weak man.'"

Tornier, nevertheless, was shortly transferred to Berlin. There he embarrassed the mathematics faculty by being pictured in the newspaper walking on a fashionable boulevard with a notorious prostitute on his arm and a tame tortoise on a leash.

In the fall of 1935 Davenport reported to Courant after another visit to Hasse:

"The number of students is extremely small. . . . Witt is Hasse's personal assistant, and he certainly is a good mathematician. . . . I gather that Teichmüller, who was the ring leader of the opposition to Hasse among the students, has become reconciled to him. Hasse thinks he is quite a good mathematician, but I am unable to judge. . . . Kaluza, of Kiel, has been appointed to one of the chairs at G. (I think yours) and is starting this term. . . . Hasse is trying to get Deuring to G. . . . , but all attempts meet with great delay in the ministry of education. In fact, the university business as a whole seems to be greatly neglected. . . ."

Outside every village and town there were notices posted: *Juden sind nicht erwünscht*—"Jews not wanted." There was also a box in which citizens could deposit accusations against those who had any association with Jews.

"How decent people can tolerate such things is absolutely incomprehensible to me," Davenport wrote. "If it were not for my friendship and mathematical interests in common with Hasse, I would not dream of going to Germany. . . . In fact, one has the feeling that the greater part of the population is mad—one of the characteristic features of such madness being that victims are normal and (frequently) delightful people—on all subjects but two or three."

It is hard to understand how at this time Courant could have considered a trip to Germany, even to take advantage of the blocked marks which friends had contributed for mathematics at NYU; but in his files there is a letter to Alwin Walther, a former assistant and a professor in Darmstadt, describing plans for such a trip in November or December 1935. Across the letter the word *Nicht* has been scrawled with a red pencil.

That same November for which he had planned his trip, Courant received news of the death of Julius Stenzel, the friend and mentor of his youthful days in Breslau.

"I was truly touched," Nina wrote her mother, "by the letter Richard wrote to Mrs. Stenzel, in which he said that the friendship with Julius had been a high point of his life."

Although Stenzel had not been Jewish, he had been removed in 1933 from his position as professor of philosophy at Kiel because of earlier disciplinary acts against pro-Nazi students while he was rector of the university. He had then later been sent to the university at Halle. His widow was the former Bertha Mugdan, whom Courant had once tutored in mathematics and physics. She was Jewish according to Nazi racial laws, although she had been a convert to Christianity since before her marriage to Stenzel. In his letter of condolence, Courant urged that she and her four children emigrate to the United States immediately. He offered to furnish the necessary affidavits guaranteeing their support.

Siegel had returned to Frankfurt after he had learned that Werner Weber, who had forced Landau out of the lecture hall, had been sent to that university. He had felt, according to Courant, that it was his duty to go back and fight for his colleagues, especially Max Dehn and Ernst Hellinger. But by the time Siegel arrived in Frankfurt, both Dehn and Hellinger had been removed from their positions.

In conjunction with Oswald Veblen, who spearheaded the American effort to find places for refugee mathematicians, and Harald Bohr, who was the European contact, Courant began to think also about getting friends who were not Jewish out of Germany. He was particularly worried about Artin, whose wife was half Jewish.

It was against this background of concern about his friends that Courant finally gave way to the attractions of the Problem of Plateau. The problem—which gets its name from that of a nineteenth century Belgian physicist who conducted physical experiments in connection with it—is very deep mathematically. It can, nevertheless, be stated in a way that it is intuitively clear even to people who are not mathematicians. In its simplest form it is *to find the surface of least area that spans a given closed contour in space.* In spite of this easily grasped formulation, it is far from obvious how to determine the solution analytically; and, except for very simple contours, it is not even clear, intuitively, how the surface should look.

Courant had been very much impressed by Douglas's achievement in finding the first complete solution of Plateau's problem for single and double contours. The American had ingeniously employed a peculiar minimum problem, to which he had been led by the transformation of yet another minimum problem related in turn to the minimum problem of Dirichlet. It was a highly original piece of work, and in Courant's opinion Douglas had justly earned credit for it all over the world. Much as he was impressed by Douglas's achievement, however, he found the method the American had used unappealing. It was not, in his view, nearly so simple and direct as the nature of the problem deserved. Douglas's approach seemed to Courant "roundabout" just as, more than a decade before, Weyl's approach to the problem of the Lorentz conjecture had seemed "roundabout."

"Courant would never tackle a problem just in order to solve it," Friedrichs explained to me. "The tools used to solve the problem had to appeal to him too. He could never separate them from the problem. They were always part of the deal. If the problem was beautiful, the tools had to be beautiful too. Otherwise, there was no appeal at all."

Courant was aware that mathematicians had long recognized the connection of Plateau's problem with harmonic functions and the Dirichlet problem. In his seminar in the fall of 1935, with his gift—reminiscent of Klein—for weaving connections in mathematics, he glimpsed a way in which the minimum problem related to that of Dirichlet could be employed in a straightforward manner for the solution of Plateau's problem. Again he reached for his most effective tool, the lemma implicit in his dissertation of 1910 and later explicitly stated in 1914. It was the same lemma which he had also used so effectively in the work on finite difference equations. By February 1936—a little less than a year after Douglas's lecture at NYU—he was able to write enthusiastically to Siegel that he had been able "really importantly" to simplify and extend the solution of Plateau's problem so that the whole affair—in his words—"is now no more difficult than the fundamental theorem of conformal mapping."

At home he played a great many Bach fugues.

"There is nothing but good news to report from here," Nina wrote. "Richard has done more scientific work than in a long time—including in the nice Göttingen time—and he is very happy because he has found something."

Although Nina was a mathematician's daughter and had more than an ordinary talent for the subject, she had long ago given up trying to understand her husband's mathematics.

"To be able to work with one's husband in his own subject must be really wonderful!" she wrote to her mother. "Had I been more energetic (and poorer!) I would have learned to do Richard's secretarial work, but I am afraid that I would have done it wretchedly. Still, this working together of husband and wife is very modern. I think it would be splendid!"

"My relationship with mathematics is rather like that of a small child making its first discoveries and *playing* with things," she told me.

"It enjoys with fresh wonder and with huge joy that a marble will roll or that the pieces of a puzzle fit together. That is how I felt when I first learned to have letters represent numbers or geometrical units in formulas and equations. And then, having seen some mathematical facts, I was not quite satisfied with that, but I wanted to know: why is this so, what is behind it, what follows from it? However, I immediately forgot my results and remembered only the exhilaration they had given me. Being a lazy person, I was never tempted to work my way into higher mathematics.

"What I really loved was to be present when Richard was having mathematical conversations with Friedrichs or Fritz John. Not understanding a word they said (nor trying to), I could feel their eagerness to learn from each other—there was a kind of peaceful agreement and mutual respect between them which I never saw between people in other fields."

In his professional and personal activities up to the time of his work on Plateau's problem, Courant had made a considerable effort to efface himself. He was well aware of the native American's distaste for anything that could be construed as "meddling" by foreigners, especially when the foreigners were also Jewish. He had also observed how recent immigrants seemed especially threatening to immigrants of an earlier day. During his first years at NYU he almost always put forth his proposals through Flanders, "a genuine 100 per cent American," whose ancestors had actually been in America before the *Mayflower* arrived.

In spite of these efforts "not to offend," he received a disturbing letter from Franck in the winter of 1935–36. From people "irreproachably loyal," Franck had heard that in Princeton and New Haven "and perhaps other places as well" there was a feeling that Courant was being too forward for a foreigner.

"I know that your goodness and helpfulness cause you to be concerned about people," Franck assured him, "and I myself have often enough called upon you for help in the case of my own children, but in spite of that I beg you most sincerely to take this report to heart and hold yourself 100 per cent back."

It seemed to Franck that hostility to foreigners was growing—he and others often felt there was more anti-semitism in the United States than there had been in pre-Hitler Germany—and that Courant should not in the future write to places other than New York University unless he had been asked expressly to do so.

"If you don't follow this advice, you may be able to help a couple of people that otherwise you couldn't help, but the damage you will do to yourself and your friends will be greater in total."

Perhaps, Franck conceded, a similar feeling against himself existed.

"But by accident I have heard these rumors about you, and they must serve for both of us as a warning to be enormously careful. We dare not forget that we already once before deluded ourselves about the firmness of the ground on which we stood."

Carefully examining his actions since coming to America, Courant came to the conclusion that there was really no *objective* basis for the criticism Franck had reported.

"I can't believe that they are 'the' mathematicians at Princeton or at Yale. In Princeton I am with Weyl, Veblen, and Neumann on *absolutely* friendly and open terms. Likewise with Lefschetz. With others I have never talked about questions concerning German emigrés. But there is naturally in Princeton another group with whom I have much less contact. These people, who are in opposition to Veblen's group, were seemingly against Siegel's appointment and may have connected me with that. I have never been at Yale and know only Hille closely. He has always been a candid friend and has often spoken with me in general about the hostility toward foreigners."

It seemed to him that his conscience was clear.

"But I understand completely that I must shun the slightest appearance [of meddling] and that my propensity for harmless utterances that signify nothing is dangerous. I will, therefore, seek to be still more cautious. . . ."

The resolution was scarcely made when Courant found himself in an unpleasant position with an American mathematician as a result of his work on Plateau's problem.

Courant's approach, in addition to giving solutions for problems of single and double contours, as Douglas's had, could more readily be carried over to higher cases in which the boundary consists of any number of distinct contours. When Courant's results establishing this were published, Douglas insisted that the same thing could be done by his method, that it was in fact implicit in his work. He pointed out a paper, which he had already submitted, in which he had applied his method to the case of several contours.

The situation resulted in some bad feeling. The American was in a fragile mental condition at the time and was shortly to be institutionalized for a period. Between 1936 and 1938 he had no academic position.

In 1937 Courant rather sweepingly gave Douglas full credit for his method for the general case and stated that his publications did not contest Douglas's claim of priority.

During the first half of 1936, Courant published three papers on Plateau's problem. It was the beginning of one of the great productive periods of mathematical creation in his career—when he was almost fifty—and the first real mathematical achievement since the work on finite differences.

"You see, in the last years in Göttingen I had lost my contact with mathematics a little bit," he explained to me. "But then here in this country I had to get on my own feet, and I was very happy that I started doing mathematics again. That was very satisfactory to find that I had not lost my competence and was able to do something."

A few months before his death, I asked him what he recalled as the most mathematically exciting event of his life. He placed the return to mathematics in the 1930's in the same class with his youthful contact with Hilbert.

"So in a way I was really grateful for that. I thought I was the beneficiary of Hitler."

He continued to make little progress on the second volume of Courant-Hilbert, but he had reason to be pleased with developments at NYU. The number of graduate students had grown as his presence in the city had become known. Many who attended his classes were men already established in fields that involved the applications of mathematics. Immigrating scientists and foreign visitors stopped to attend his lectures as they passed through New York. The arrangement with the Institute for Advanced Study had resulted in a number of exciting individual lectures and a broadening of the graduate offerings. Also, and most important, Chancellor Chase had promised that in September 1936 space for a library and an office for Courant would be made available at Washington Square.

It was at this time, during his second spring in America, that Courant first began to talk to people with access to large amounts of money about his plans for a graduate center of mathematics at New York University. One of these was Dr. Frederick Keppel, to whom he had been introduced by Flexner and through whom he hoped to interest the Carnegie Foundation. Keppel was friendly, but he told Courant that *just* building up a mathematics department at New York University was not going to appeal to the Carnegie people.

"Therefore, he suggested that one should prepare the attack a little more on tactical lines. . . ," Courant explained to Flexner, "how we are attempting, so to speak, a new experiment in graduate education . . . and reasons why just New York University and why mathematics. Fortunately what we have in mind can, without any artificial effort, be presented in such a way."

"I wanted to do something like what Flexner had presented in his book on universities, which I had read and which had very much impressed me," Courant explained to me. "This was not done at his own Institute for Advanced Study, because there was no teaching there and no contact with the applications."

Interwoven with Flexner's ideas was Courant's own personal conception of Göttingen as a place where no distinction was made between mathematics and its applications and where advanced students and faculty were like a family. The "twist" would be to transplant this essentially elitist institution with its noble scientific tradition to a mediocre, business-oriented American university with no tradition at all and a student body composed largely of the sons and daughters of working-class Jewish immigrants.

In June 1936 Courant received the permanent appointment that Chase had promised and an increase in salary of $1000 a year. Chase also announced in his annual address to the Board of Trustees the intention of the university to establish "a graduate center for mathematics" with Professor Courant as its director.

Space for the new center was obtained in a building recently purchased on the north side of Washington Square. It was only three small rooms upstairs in a house that had been on the Square since the days of Henry James, but Courant was happy. Mathematics at NYU now had an address. Remembering from his Göttingen days the power of a name on a letterhead, before he left for the International Mathematical Congress being held that summer in Oslo, he ordered stationery printed:

NEW YORK UNIVERSITY
The Graduate Center for Mathematics

NINETEEN

THE NEW Graduate Center for Mathematics had a faculty of one—Courant himself—but already, as he set off for Europe in the summer of 1936, the combination of circumstances which would provide him with two young colleagues perfectly cast for his purposes had been put into motion.

During the preceding year he had heard from his Göttingen friend Heinz Hopf, by that time a professor in Zurich, about a young American engineer, an assistant professor of mechanics at the Carnegie Institute of Technology, who had taken his Ph.D. under Hopf with a thesis on a topic in differential geometry. The combination was rare: an American, an engineer, an interest in pure mathematics. Courant had immediately written to J. J. Stoker: "In case you should come to New York sometime at the end of this or at the beginning of the next academic year, I should be glad if you could get in touch."

That same spring, Courant had been trying to arrange some sort of position that would enable Friedrichs to come again to New York. Although he had been successful in obtaining only enough money to defray the younger man's living expenses, he had dispatched a formal invitation in English for a few lectures at NYU. To his surprise Friedrichs had wired back in German, "Unfortunately I am not in a position to accept your friendly invitation." Mystified by the refusal, Courant was looking forward to seeing Friedrichs at the International Congress in Oslo and receiving an explanation.

When Courant arrived in Copenhagen, he found Friedrichs already there. The young man explained that the fact that Courant's invitation had been in English had brought a rebuke from the official in the ministry to whom he had had to apply for a leave.

"I can find no explanation," this man had written angrily to Braunschweig, "why Professor Friedrichs has not already declined the invitation from Professor Courant, one of the Jewish emigrés who, although he knows German, writes in English. I want a copy of Professor Friedrichs's refusal of the New York University invitation sent to me immediately."

Under the circumstances Friedrichs had thought it would be more discreet to meet Courant in Copenhagen rather than at the congress in Oslo, where their meeting would be observed. He told Courant that he was now prepared to emigrate if Courant could arrange a professional position of any sort for him. Courant was not optimistic. There were so many Jewish mathematicians, who *had* to emigrate, and so very few positions. He mentioned again that the traditional way to come to America was as a dishwasher, but Friedrichs again declined that alternative.

The congress at Oslo was marred for Courant, Weyl, and other German refugees to the United States by the absence of a number of colleagues who had remained in Germany. On the American side there was also a notable absence. Jesse Douglas and L.V. Ahlfors had been chosen as the first recipients of Fields Medals, but Douglas had not been able to afford the trip to Oslo. His award had to be accepted for him by Norbert Wiener.

Although Courant had given Nina "a thousand reasons" why it would not be wise for her to visit Germany at this time, he himself planned to pass through that country on his way to Carlsbad, a favorite spa in Czechoslovakia. He found it strange and a little bit frightening to enter Germany again; but it was the summer that the Olympic Games were being held in Berlin and Hitler was showing off to the world the accomplishments of three years of National Socialism. And so, as he said, "There was nothing difficult."

In Hamburg he saw Artin, who had not been given leave to attend the congress in Oslo. Artin now wanted to visit the United States with his family and see how he liked it before deciding whether to emigrate.

From Hamburg, Courant went to Berlin, where he made arrangements for blocked marks in his account to be placed at the disposal of his friend Stenzel's daughter Anna so that she could emigrate. He also saw Springer, who was apparently still doing business very much as usual. Although the Nazis had forced a party member upon him as a "partner" in place of his cousin Julius, who had been found to be more than half Jewish, there had been no further interference in the activities of the firm. Courant was still editor-in-chief of the Yellow Series. It was still planned that Springer would publish the second volume of Courant-Hilbert. There were still Jewish names on the title pages of Springer journals. An obituary of Emmy Noether, who had died in the United States in the spring of 1935, had appeared in the *Mathematische Annalen*.

I asked Courant if he had gone to Göttingen on this trip.

No, he said—he had called Hilbert from Oslo, but he had not gone to Göttingen.

When he arrived at his hotel in Carlsbad, he was informed that a man had telephoned, asked when Courant was arriving, and then hung up without leaving his name. Courant was certain that the caller had been Friedrichs, and the next morning Friedrichs appeared and announced—Courant told me—"I will even go as a dishwasher."

"I don't remember saying that," Friedrichs said with a smile, "but, well, Courant likes to tell it that way."

Courant returned to America in September 1936 full of plans for the future. In Carlsbad he had also met David Sarnoff, the president of RCA— "one of the most interesting men I have ever become acquainted with," he reported to Franck—and the acquaintance must be pursued, both for himself and for Franck. People must be contacted about an invitation for Artin. Preparations must be made for Friedrichs's arrival, some sort of position found. He had decided that it would be good for Flanders, who seemed to have lost his scientific and personal impetus, to spend part of his upcoming sabbatical year with Harald Bohr in Copenhagen—and that too must be arranged. Anna Stenzel would need support. The new rooms of the Graduate Center must be furnished.

Back at NYU, while he was still installed in his old office at the Heights, a little Jewish girl presented herself with a letter of introduction from Flexner. Courant looked doubtfully at her and asked if she could type and take shorthand.

"I *could* type, but I never learned shorthand very well," Bella Manel, now Kotkin, confessed to me, "but since Courant spoke very haltingly in English and since I was not afraid to ask him to repeat if I missed something, I was able to manage. After I gave my first report in the seminar, his attitude toward me changed completely. I was accepted then and asked to lunch with him and other students."

She still remembers very well Courant's first days in the little rooms upstairs at 20 Washington Square North. She acted as his secretary and receptionist, answering the telephone, giving instructions to students, supervising the small library—which consisted mainly of Courant's books and journals and his personal collection of reprints. People from all walks of life came to the office, she told me—famous scientists "and even royalty." It was all very exciting to a young student, but it was a relief when in the spring Courant got her a Blumenthal Fellowship, and she could begin to concentrate on her dissertation.

Sometimes on weekends she and other students, including Max Shiffman, to whom she was later married, were invited out to New Rochelle. There they were put to work doing chores in the yard.

"It is true that Courant utilized people to do things for him," she conceded, "but never without giving something in return. We were welcomed into the warmth of his family by him and Nina. That was wonderful, and it was worth a lot to us."

With his appointment permanent, Courant began to try even more intensively to bring his experience in Göttingen to bear upon the situation at NYU. During the fall semester 1936–37 he offered a general course for mathematics teachers as well as for graduate mathematics students, which he called "Elementary Mathematics From a Higher Viewpoint." In his opinion, Felix Klein's lectures under this title had contributed greatly to the improvement of secondary-school mathematics teaching in Germany and had also given mathematics students there a broader understanding of the connections between higher mathematics and elementary problems and between the different branches of their science. Something similar might have a similar effect in America.

That same fall G. H. Hardy, who was then a visitor at Princeton, came to NYU and gave some talks. He was much impressed by the responsiveness of his audience and later wrote Courant to that effect. Tremendously pleased and encouraged, Courant sent copies of Hardy's letter to a number of people he was trying to interest in the Graduate Center.

He continued to be fascinated by Plateau's problem and began to perform soap-film experiments which were more sophisticated than those which had been performed by Plateau.

For such experiments a thin wire frame, shaped to a given closed contour, is dipped into a viscous liquid similar to soapsuds and carefully withdrawn. If one ignores gravity and other forces which interfere with the tendency of the film to assume a stable equilibrium by attaining the smallest possible area, the film that then spans the frame is the physical representation of the minimal surface for that particular contour.

Although the physical existence of a solution to a physical problem does not establish the existence of a solution to the corresponding mathematical problem—as Weierstrass had pointed out in connection with Dirichlet's principle—the recent work of Douglas and Courant had established mathematically that such a surface does exist for every closed contour in space. Many other interesting mathematical questions about minimal surfaces were, however, still unanswered—among them, questions concerning the uniqueness and continuity of solutions.

Courant began to explore some of these questions by means of the soap-film experiments. Most of these are difficult to visualize without demonstration, but the description of a few can give some idea of the way in which the physical solution can suggest and sometimes answer mathematical questions. One wonders, for instance, if for any given closed contour there is only one possible minimal surface and discovers that a frame shaped like a headset (circular earphones joined by a curved band to go over the head) permits not one but three minimal surfaces. Another question concerns the effect of deformation on a contour. For such experiments little handles have to be attached to the frames. Then, after one has dipped a circular frame into the viscous liquid and drawn out a simple two-sided disk, he can gently give the frame a slight twist. Suddenly the surface of the film becomes a one-sided Moebius strip. Further deformation returns it equally suddenly to a two-sided strip. To explore other questions concerning the effect of free boundaries, Courant utilized frames which were flexible in part.

He found "the scope and informative value of soap-film experiments with minimal surfaces [much] wider than the original demonstrations by Plateau." For the next four or five years he was to play with them—in the seminar, in lecture halls, in his office, at home—with the absorption of a child.

The experimentation with soap films was only one example of a quality of playfulness in Courant—what Friedrichs described to me as "the ability to be fascinated." Although he could never understand his friend Hardy's passionate interest in cricket, he himself bought one toy after another—cameras, cars, phonographs, radios. Later he had the most sophisticated "hi fi" and tape recording systems.

Automobiles, beginning with the "thoroughbred" Röhr in Göttingen, were a particular delight.

"Courant was always having amusing, unique accidents," Friedrichs recalled. "Once, I remember, on a road in the mountains where the traffic was permitted to go up in the morning and down in the afternoon, he stalled his car for most of the day, half way up."

Hans Lewy conjectured that Courant played so intensely with his toys because he had never had any in his childhood.

During 1936–37, while Courant was giving lecture-demonstrations on soap films at various eastern colleges, news of Göttingen continued to come from time to time. Otto Toeplitz—who with other Jewish professors remaining in Germany had recently been arbitrarily retired—made what he described to Courant as "a sentimental journey" to visit Hilbert at the New Year. Hilbert's memory, which had been greatly affected by his illness, seemed to Toeplitz to have improved.

"He definitely became the most animated when I told him about your activity. He wanted to know everything about it, and I could not tell him enough.... The liveliness of his interest in this part of my account which concerned you was the most lovely proof of how very much he was attached to you. When I just in passing mentioned that you had now done some beautiful things in the calculus of variations, he parried immediately, 'With that I have not concerned myself at all.'"

In closing, Toeplitz reminded Courant not to forget Hilbert's seventy-fifth birthday on January 23, and added: "About what has become of his enterprise in Göttingen, he is perfectly clear."

But for the most part, by 1937, Göttingen was in New Rochelle. "It seems to us that it is you who are in exile," Nina had long ago written to her mother, "and we who are at home." Time after time the house was filled with old friends, visiting colleagues, former students—such occasions always described by Nina as *sehr göttingisch*.

Anna Stenzel had been added to the household now in place of Hilde Pick, who had found "a real job." Courant put her to work typing the third chapter of Courant-Hilbert II. Suddenly, after thirteen years, the book was beginning to move. By December half of the manuscript had gone to Springer. In the spring Stenzel's daughter was sufficiently confident of her "American English" to take over from Bella Manel at the Graduate Center.

With both his writing and his mathematical work going well, Courant could still not escape the feeling that he should try to do something more for the situation of mathematics at NYU. Unlike many private universities, it had no significant endowment. Almost all expenses were met by student fees. Clearly, any new development could come about only with financial support from the outside.

There was already in existence a model for Courant of what could be done. The interest of public-spirited New Yorkers in art had been utilized by Walter S. Cook to provide NYU's Institute of Fine Arts with the funds to hire a stellar group of art historians who had been forced to leave Germany. Courant was optimistically certain that if New Yorkers would support art, they would also support mathematics which, in addition to being an art, was of uncontested practical value. Whenever an opportunity offered, he tried to establish contacts with wealthy and influential people, obtaining introductions, parlaying one acquaintance into another, presenting himself to virtual strangers by some tenuous connection. A sample of his approach in these first years is contained in a series of letters to a Mr. Henry Gaisman, to whom he had been "encouraged" to write by Percy Straus, the president of Macy's.

Mr. Gaisman, according to Courant's description of him, was over sixty years old, "a little bit bashful, very wealthy and sometimes generous." Starting out as a newsboy, he had become a successful inventor and businessman.

"I need not say, indeed, how delighted I should be if you would permit me to discuss matters with you," Courant wrote to Gaisman. NYU had a number of gifted students—"many Jewish, earning their living by hard work, some of them original personalities." One of his students was an ingenious inventor. The whole situation was such that "with help from the outside on a comparatively modest scale an enormous progress of high public usefulness could be achieved."

When after two weeks there was still no reply from Gaisman, Courant wrote again:

"I suppose that a man like you must be subject to every kind of pressure to contribute to various things of different merits. Therefore I wish to assure you that although I hope to enlist your interest, I certainly shall not ask you for any active help which you should not deem appropriate on your spontaneous judgment. Asking you for a chance to discuss the matter personally might already appear as some kind of intrusion; but the cause seems to me sufficiently worthy from many points of view, not the least of which is the Jewish angle. It is this consideration which makes me overcome the natural hesitation I feel about asking you again for the favor of a personal discussion."

For more than half a year Courant persisted in trying to interest Gaisman. During this time the only response he elicited was a note from Gaisman's secretary to the effect that his employer gone away for the winter.

For a while Courant also had high hopes of obtaining a grant from the Carnegie Foundation after his talk the previous year with the friendly Keppel, but this too failed to materialize. Somewhat hesitantly, he turned finally to Warren Weaver.

"I really do not feel impelled by personal ambition to embark again on a work of organization as I did in Göttingen . . . ," he assured Weaver in December 1936. "But seeing the need and the chances for the proposed development and feeling my deep indebtedness . . . I simply have to try. . . ."

The ultimate goal at NYU should be "a strong and many-sided institution . . . which carries on research and educational work, not only in pure and abstract mathematics, but also emphasizes the connection between mathematics and other fields as physics, engineering, possibly biology and economics, and which cooperates in helping to develop better standards in high school instruction."

A "highly efficient" use could be made of any money, even a small amount. For the library $4000 would be sufficient to make it satisfactory, $6000 would make it really good. "But also less than $4000 would be a great help."

Weaver was, by virtue of his own background, very sympathetic to Courant's ideas. He had taken a degree in civil engineering and had then done his advanced work in mathematics. His most recent teaching position had been as head of the mathematics department at the University of Wisconsin. In 1932, when Max Mason, whom he admired more than any other scientist, had asked him to come to the Rockefeller Foundation, he had taken a cool look at his abilities and concluded that he lacked "that strange and wonderful creative spark that makes a good researcher. . . . There was a definite ceiling on my possibilities as a mathematics professor." He had accepted Mason's offer.

Weaver was convinced that Courant had a "sound and important" plan:

"There are few places in the country where applied mathematics is being emphasized in any adequate and competent way; and a fine development of this sort in the New York area would seem to be clearly indicated."

Unfortunately a development of the kind Courant had in mind was, as Courant himself knew, "quite outside of the program and possibilities of The Rockefeller Foundation." In spite of this fact, a small grant to NYU was subsequently made and regularly renewed for a number of years.

"We make such exceptions only in cases which are judged, on the basis of the best evidence we can gather, to be of very unusual merit and importance," Weaver was always to remind Courant. "You are therefore justified in viewing this assistance, even though of very modest amount, as a real evidence of our interest."

One of Courant's main concerns when he talked to Weaver was that without additional funds existing chances in personnel might slip away. Among these he included J. J. Stoker, the assistant professor of mechanics at Carnegie Tech.

Courant had finally met Stoker in December 1936 when that young man had come to New York and given a talk before Courant's seminar. The interest, aroused by Hopf, had been immediately confirmed. Stoker was the son of an immigrant Englishman who had worked his way up from ordinary miner to superintendent of all the coal mines of Bethlehem Steel and had later become a mining inspector for the state of Pennsylvania. Young Stoker had originally studied engineering as it related to coal mining and after graduation had worked for a year as a mining engineer before returning to Carnegie to teach mechanics. There he had shortly come to recognize a serious lack in his education. In four years of college he had learned nothing of mathematics except a little elementary calculus. He decided that he should go back to school and study mathematics as it related to mechanics and physics.

At that time there was no university in the United States which offered the kind of work he wanted, Stoker told me when I talked to him in his office at the Courant Institute, which—if it had existed then—would have suited his purposes exactly. A few individuals at different places combined mathematics with applications. One was T. L. Smith, with whom he had taken some courses at Pittsburgh. Smith had studied in Göttingen after getting his degree at Harvard, and Göttingen was where he advised Stoker to go for what he wanted.

"And I would have gone except that by then—it was summer 1932—it was obvious that things were not good in Germany."

As an alternative to Göttingen, Smith suggested the Eidgenössische Technische Hochschule; and in what was probably the worst year of the Depression, Stoker took his pregnant wife, Nancy, and his two-year-old daughter and set off for Zurich to study with Ernst Meissner, an applied mathematician who had been the first new friend Courant had made during his student days in Göttingen.

Stoker planned to work with Meissner in elasticity, a subject with which he was already somewhat familiar; but after attending lectures by the various professors in Zurich, he decided to work with Heinz Hopf instead. "I found Heinz Hopf's lectures and his whole way of doing mathematics—well, very very attractive." One of Hopf's courses that year was on differential geometry, which involves notions that also play a role in the theory of elasticity. "I had never studied it, but I found it so really beautiful—he did it so well—that I asked him to give me a subject in it for my dissertation."

Somehow—Stoker doesn't really know exactly how—after their meeting in December 1936 Courant managed to arrange that Stoker was hired as an assistant professor in the mathematics department of NYU's College of Engineering at University Heights. It was agreed that he would also give a regular course of lectures at the Graduate Center at Washington Square.

"And did you ever get to Göttingen?" I asked.

"No, I have never visited Germany at all. I have been in Europe a lot, but after the war I never had any desire to go to Germany. The Germans never harmed any of my relatives, I'm not Jewish, but I felt so angry with them, I felt why bother visiting places like that. On the other hand, I have a bad conscience about it, because the individual Germans are like anybody else and I have liked some of them very much. Anyway, I had no desire to go to Germany. And so I have never been to Göttingen."

At the same time Courant was arranging Stoker's employment at NYU, he was also trying to find a position for Friedrichs, who—in Germany—was setting up what he now thought of as his "escape."

Since Friedrichs was of an age for military service, it was necessary for him to get permission from the military to leave the country even for a short time. Fortunately he had a sister living in Paris, and he was able to apply legitimately to visit her during his spring vacation. To go to the United States, he also had to obtain a visitor's visa from the American consul. This required affidavits that he had residence and employment in Germany. He was sure that his landlady and the clerk at the Technische Hochschule where he taught (who wore a swastika prominently displayed in his lapel) guessed his plans, but they signed the required papers without a question. Everything costing money had to be done before leaving Germany, for he would be permitted to take no more than ten marks out of the country. As surreptitiously as possible, he purchased passage on a steamship that went from France to the United States.

He and Nellie Bruell, who was French by birth and thus had a French passport, had arranged that she would stay in Germany until he was safely out of the country. Then she would go to France and wait at her father's in Lyon until Friedrichs had obtained a position and could send for her.

In Paris, Friedrichs telegraphed Courant that he was on his way. He arrived in New York on March 4, 1937, quite penniless. Friends lent him money, and a few days after his arrival he sent a letter to Germany formally resigning his position as a professor. He took it with him to Princeton to mail so that it would not bear a New Rochelle postmark. Courant found him a room and paid him to help with the second volume of Courant-Hilbert, which was now finished except for the final chapter.

"So I was his assistant again. That was fine with me. Most immigrants to this country start at the bottom. I felt perfectly natural about it."

Courant wrote letters about Friedrichs's presence in the United States to everyone he knew who was interested in the development of applied mathematics. He emphasized the two years that Friedrichs had spent at the aerodynamics institute in Aachen and presented him as "a mathematician in the style of C. Runge." He was in fact so active on Friedrichs's behalf that even Hans Lewy began to be afraid that his efforts to place Friedrichs might jeopardize his own position at NYU.

"But you don't need to worry," Courant reassured Lewy two months after Friedrichs's arrival. "It is actually so that the Dean of the Engineering School, the Dean of the Graduate School, the Head of the Aerodynamics Institute, etc., are all of one mind that we should not let pass this splendid opportunity to cultivate applied mathematics."

In June 1937 Friedrichs received a temporary appointment at NYU as "professor of applied mathematics" in the graduate department. Half of his modest stipend of $3,000 came from the Rockefeller Foundation and the other half from private individuals whom Courant referred to as "friends" of the Graduate Center. As soon as Friedrichs had a position, he wired Nellie Bruell—now Mrs. Friedrichs—*to come!*

Thus it happened that by the end of his third academic year in the United States, Courant had at this side the two colleagues who would work with him until his retirement—and both of them had come to him by way of Göttingen.

TWENTY

THE YEAR 1937 saw the publication in Germany of Courant's last book to be written in German and the publication in the United States of his first book to be translated into English. Both books, as always with Courant, had stories behind their publication.

To Friedrichs's amazement, when he had arrived in March 1937, he had found the second volume of Courant-Hilbert already in proof—except for the last chapter.

"It was like a miracle!" he still marvels. "Of course it had been in his mind all the time; but the fact that I would now come and we would go over it together and fix it up and so on, and of course write the last chapter —that was what did it."

I had heard references to the seventh and final chapter in the Courant family—how little Lori had once stamped her foot and cried, "Never mention that seventh chapter to me again. It has ruined a year of my life!"— and so it was with interest that I asked Friedrichs what had happened.

It had been planned from the beginning, he explained to me, that the concluding chapter would present a general existence theory for the solutions of elliptic partial differential equations. When the chapter had first been discussed, Friedrichs had been Courant's assistant. Their thinking on the subject had been very close. In the years since then, Friedrichs had moved toward an increasingly abstract approach to the ideas which were to be presented in the chapter. In 1930, while reworking a paper which he had written earlier in the general area of Hilbert's spectral theory, he had come across von Neumann's basic paper on Hilbert space, in which Hilbert's spectral theory becomes a special case in a general theory. That paper had been a revelation to him, and he had immediately translated his paper into von Neumann's Hilbert-space language. When he had returned to Germany after his visit to Courant in 1935, he had seen still another approach to the whole complex of problems, also involving Hilbert-space methods, but going much farther than he had gone in his earlier work.

"So when I came back in 1937 and started to work with Courant on the last chapter of Courant-Hilbert II, I had a conflict. I didn't think it should be written the way Courant saw it—in the style of 1925 or 1926, say. On the other hand, I knew that to write it the way I saw it would be unnatural for Courant and it wouldn't then be Courant's. So we discussed it. Courant insisted he had already done it my way. He hadn't really. He had some basic notions of Hilbert-space theory, and he recognized the importance of von Neumann's work. But it was not natural for him."

The two men struggled into the summer of 1937 with the chapter, working on the side porch of the house in New Rochelle. Courant had a great deal of trouble settling down; but, once settled, he could go on long after Friedrichs was exhausted.

"Richard sighs and groans and is very irritable," Nina reported. "From week to week he says, this week it must now be finished. But then it is not."

In July the landlord proposed to sell the house. Horrified at the thought of having to gather up manuscript and galley proofs and notes, Courant promptly bought it.

Finally, in the middle of the summer, Courant announced to the family that the seventh chapter was at last finished. To celebrate, he took them all to Radio City Music Hall. They had a wonderful time, but then in the middle of the night they heard him prowling around again, muttering to himself. It was all wrong, it was hopeless, it wouldn't do.

In the end the published chapter was a curious compromise.

"People always shook their heads about that chapter," Friedrichs said. "It was clear it wasn't just Courant's, but it wasn't just mine either."

Oddly enough, in the published book there is no mention of Friedrichs in spite of his dozen years of association with the project.

"That was on purpose," he explained. "The book was being published in Germany, and we thought it would be better not to associate me with Courant. You see, I still had family in Germany. It probably wouldn't have made any difference, but at that time we all had a tendency to overreact."

In spite of his delicacy in not mentioning Friedrichs's help, Courant audaciously proposed to dedicate the second volume of Courant-Hilbert to Harald Bohr, who had been the anathema of pro-Nazi mathematicians since his response to Bieberbach's speech on "Personality Structure and Mathematical Creation." At the last moment he gave up the dedication to Bohr because it might make difficulty for Springer.

The other book by Courant which appeared in 1937 was the American edition of the English translation of his calculus—the first of his books to be published in direct contact with a publisher other than Ferdinand Springer.

Springer had given the English-language world-rights to Courant when he left Göttingen for Cambridge in 1933. During that unhappy year Courant had given them to Blackie, an English publishing firm in which a Cambridge professor who had been friendly to him had had an interest. Blackie had no American co-publisher, and Courant had taken on the job of approaching various American publishers about issuing the book in the United States under an agreement with the English firm. He had considered making McShane, by then a quite well-known young American mathematician, a co-author. He had then abandoned the idea.

I asked McShane if he had ever received any royalties from the translated version of the calculus. He said no. The agreement between him and Courant had been that he would be paid for the translation and would get credit for it. "I never felt that Courant owed me anything," he said. "But he did support me considerably in my early career as a mathematician."

In 1936, while Courant was still engaged in dickering with American publishers about the calculus, he met Erich Proskauer, the editorial adviser of a German publishing house. Proskauer was in America to investigate the possibility of founding a European-style scientific publishing company in that country. The result was the organization in 1937 of the Nordeman Company (which in 1940 became Interscience Publishers). Although Courant had just negotiated successfully with a well-known American publisher to issue his calculus, he decided to give the book to Nordeman instead.

Proskauer, with whom I talked one morning in 1975 in his apartment facing the southern end of Central Park, is still amazed at "the mixture of naiveté and shrewdness" with which Courant gave his book to the new company. One attraction, he now thinks, may have been the attitude toward publishing which Nordeman represented. "In Europe the author was king while in the United States the reader was king."

The European viewpoint may have been especially attractive to Courant at that time; for he was having an unpleasant experience with the publishing of his first long paper in an American journal. Proofs had been returned to him by Solomon Lefschetz, one of the editors of the *Annals of Mathematics*, who had let up a howl at the "extraordinary" number of corrections. In addition Courant had asked for page proofs. Page proofs! "Owing to the delay that all this is causing us, we may be obliged to postpone your paper. . . ."

Proskauer smiled wryly when I recounted Lefschetz's reaction, remembering a time much later when the first five chapters of the English translation of Courant-Hilbert, already set up in type, had to be killed.

The agreement with Nordeman represented an important change in Courant's relations with his publisher. After the experience with Blackie, he was "very deliberate," according to Proskauer, and "very loathe to give up any of his copyrights." In addition to their agreement regarding Blackie's rights, he and Proskauer signed another agreement according to which author and publisher shared equally in profits. It was under this agreement that the first volume of Courant's calculus appeared in the United States in 1937.

There were by this time quite a few outstanding students at the Graduate Center. The little library had been augmented by the addition of a number of volumes belonging to Emmy Noether. Flanders, in Copenhagen for the year, was missed; but Friedrichs and Stoker furnished the companionship and support which Courant felt he required.

The two younger men—both then in their thirties—were as different from Courant as it is possible to imagine, and as different from each other. Stoker was straightforward, outspoken, "a moralist" in the words of later students and colleagues. He shared Courant's educational and mathematical philosophy and his sense of mission, but from the beginning he struggled against Courant's oblique way of operating. Friedrichs, who came from a lawyer's family, was judicious and, as Courant often complained, "legalistic." He had grown away from Courant mathematically, but he had a sympathetic understanding of "Courant's way" and was able to accept it. Still, even Friedrichs felt a need to protect himself if he was to survive. After he was married and living near Courant in New Rochelle, he established definite rules about times when he was available.

"My way of handling the problem was so alien to Courant that he was constantly annoyed by it," Friedrichs told me, "but he somehow or other knew it was necessary for me."

It was typical of the flexible organic development which Courant envisioned for the Graduate Center that in those early years Stoker, who was an applied mathematician, was "professor of mathematics" while

Friedrichs, who in spite of his two years with von Kármán was basically a pure mathematician, was "professor of applied mathematics." Both men subscribe to Courant's thesis that there is no boundary between "pure" and "applied," but they frequently use one word or the other quite specifically in conversation. I asked Friedrichs once how they made the distinction.

He replied that the distinction is made in different ways at different places by different people. He cited as an example two courses which he and Courant taught during those first years.

In Courant's lectures on mathematical physics, he discussed mathematical problems which had arisen originally from problems in physics. But he handled these problems in a purely mathematical way. He was concerned with general theories, with existence and uniqueness of solutions rather than with special methods of determining solutions concretely. His treatment was "pure," although he also referred the mathematics to the applications throughout the course.

"Now a very pure mathematician will call mathematical physics applied mathematics regardless of how it is taught, so to him Courant's course would be applied mathematics. We who work in the field call it pure mathematics."

In Friedrichs's lectures on fluid dynamics, on the other hand, although he formulated the problems mathematically, he discussed a number of special solutions—"technical methods for getting the answer"—and rarely tried to prove anything rigorously but simply referred the applications to the mathematical theory from time to time.

"I considered that what I was doing on fluid dynamics was applied mathematics. Of course somebody from the engineering department would probably have considered my treatment pure mathematics. So you see what I mean—it depends on who is making the distinction."

That same fall of 1937 when Friedrichs and Stoker joined Courant at NYU, Artin and his family finally arrived in the United States.

The mystery of Artin's long silence in regard to several American invitations had been cleared up when Flanders, in Copenhagen, had received news of him from a Danish mathematician who had been in Hamburg. The German ministry of education had turned down Artin's application for a leave of absence during the summer of 1937 on the grounds that he was indispensable. Other German mathematicians had urged him to remain in Germany. Among these was Hasse, who had recently applied for membership in the National Socialist Party.

"My endeavor at that time was to keep up Göttingen's mathematical glory," Hasse explained to me. "For this I needed the consensus of party functionaries at the university whenever I wanted to get some distinguished mathematician to fill a vacancy in Göttingen. Among these functionaries I had one close friend and one who was leaning towards helping me. They asked me to join the party so that they could help me better. It is true that I gave in and applied for membership. But on my application I put that there was a Jewish branch in my father's family. I was almost sure that this would lead to my application being declined. And so it was. The answer, which I received only after the outbreak of the war, was that the application was not going to be acted upon until the war was over. In the meantime, however, I had been able to help several mathematicians who were having political difficulties at other universities, by offering them positions in Göttingen."

On a visit to Hamburg in 1937, Hasse suggested to Artin that it might be possible for his children, who were only a quarter Jewish, to be declared Aryan.

The Artins lived at the end of a cul-de-sac; and whenever Natascha Artin, who was half Jewish, heard the sound of an automobile at night, she was sure it carried the Gestapo. But Artin, who still held his professorship in Hamburg, could not bring himself to accept the only permanent American position which had been offered to him—a professorship at Notre Dame. He had been born a Catholic, and he feared that the Catholic fathers in South Bend would find this out and try to bring his children into the church. In this situation—only a few months after his request for a temporary leave had been denied on the grounds that he was indispensable—he was removed from his professorship.

As soon as Courant had heard these details from Flanders, he had immediately suggested that Flanders write to Chancellor Chase, explaining the situation and the desirability of a "non-sectarian" position for Artin. Chase, who had already provided a place for Friedrichs that fall, had not felt he could take on another refugee. Artin had had to accept the Notre Dame invitation.

When Artin and his family arrived in New York on October 1, 1937, Courant and Hermann Weyl were at the pier to meet them.

"I will never forget our welcome by Courant and Weyl," Natascha Artin Brunswick told me with a smile at the memory. "You cannot imagine a greater contrast—Courant very little and not the sort of man you would look at twice—I think his face became much more interesting with age, and suffering too, I suppose—and Weyl, who was most impressive—a very, very interesting-looking man."

The Artins accompanied Courant out to New Rochelle, and a few days later Courant decided to give a party to introduce his friends to some of the big people in the administration at NYU. On the day of the party he and the Artins returned from the city about four o'clock. They found no one at home. Absolutely nothing had been done to prepare for a party. The house had not been cleaned or even straightened, and there was practically no food in the place. Courant was beside himself. Then finally Nina arrived, completely unperturbed, and said when he began to sputter, "But, Richard, you know this is the day I have my rehearsal."

Everybody, including the Artins, pitched in. By the time the guests arrived, the house had been vacuumed and food purchased and prepared. The party went off very well.

A few days later the Artins left the Courants' and went to Princeton, where there was to be another party in their honor at Weyl's home. His was "a very polished household"; and his party, an elegant affair with literary luminaries and famous philosophers as well as scientists among the guests. The country lad of Courant's student days had become a sophisticated man of the cultural world.

It was a great satisfaction to Courant to have Artin in the United States, and whenever that mathematician was in the east he regularly lectured at the NYU Graduate Center.

"The charming Artin family has been here," Nina wrote at Easter 1938. "He is teaching at a university in the Middle West . . . but Richard intends to get him to Princeton or someplace like that as soon as possible."

Courant spent the summer of 1938 writing what he described to Max Born as "a little book for teachers, not really popular, on higher mathematics from an elementary viewpoint. Very different from Klein's book and, as I think, quite necessary."

By September 1938, the beginning of Courant's fifth year at NYU, the activities of the Graduate Center of Mathematics had burgeoned into such an operation that more space was required than the tiny rooms on Washington Square North. Then, just as the semester began, Courant received a notice that even these rooms were to be taken from him—mathematics did not really require physical space beyond the lecture hall! Courant was very upset, but he still felt helpless in relation to the giant university. Thanks, however, to the intervention of a friendly young law professor, who went before the administration to plead his case, he was given some other rooms—a "suite" in a girls' dormitory next to the Judson Church on the south side of Washington Square—on the condition that the mathematics faculty and students would pass with appropriate decorum through the lobby of the dormitory on the way to their quarters.

Bedrooms were promptly transformed into offices, the living room into a library and lounge for the meetings of the mathematics club, a bathroom became a kitchen as well. In the long hall in front of the elevator, a ping pong table was installed.

A student who first saw Courant in these quarters was Charles De Prima, now a professor at the California Institute of Technology. In the fall of 1938 De Prima was twenty years old, the son of Italian immigrants, a boy who had grown up across the Hudson in New Jersey. After some college and work experience, he had come to NYU vaguely planning to go into journalism. He took a few courses in mathematics and physics, which had always fascinated him. At the suggestion of one of his teachers he went to a seminar talk at the Graduate Center, where Courant introduced the speaker. De Prima had heard Courant's name and had studied calculus from his book, but he had never seen him before. Yet somehow Courant seemed very familiar.

"*How* did I know him? Well, in the middle of the lecture it occurred to me that he was typical of the drygoods salesman who used to go from house to house at that time in the mixed Italian-Jewish neighborhood in which I had grown up. I had seen that man over and over again all my life. *Courant was that man.*"

After the lecture, when De Prima was introduced by his teacher, Courant mumbled something to the effect that the young man should come and see him sometime. Taking the remark as an instruction, De Prima turned up a few days later at Courant's office in the Judson Dormitory.

"He didn't remember who I was, but we started to talk. After a while he said something like, 'Well, fine. Do you know New Rochelle?' So vague arrangements were made that I should come out to his house there in a couple of weeks, and I did that too."

In New Rochelle, while Courant picked up a few sticks and talked, De Prima chopped wood, mowed the lawn, cleaned up the front and back yards, and found himself "simply entranced" by the way Courant talked about mathematics and Göttingen.

"It was so darn human. I had always thought of mathematics as something where you sit down at a desk and apply yourself. But now I found out you could do it while cutting grass."

De Prima, who left NYU relatively early in his career but has always remained in touch, has a particular perspective on Courant. In the course of our interview in the faculty club at Caltech, he talked to me about the influence of the second volume of Courant-Hilbert, which had just appeared when he was a student. Now, more than thirty-five years later, he feels that something happened with the second volume which was similar to that which happened with the first volume: "Only this time it happened for mathematicians rather than for physicists."

For some time, there had been a steadily growing and worldwide development in mathematics which emphasized abstraction, generality, and axiomatization. All his professional life Courant was to oppose this development, not because he opposed the aspects of mathematics which it represented, but because he felt that one side of the subject was being increasingly overemphasized at the expense of the other, equally essential side.

"Living mathematics rests on the fluctuation between the antithetical powers of intuition and logic, the individuality of 'grounded' problems and the generality of far-reaching abstractions," he said on the 100th anniversary of Hilbert's birth. "We ourselves must prevent the development being forced to only one pole of the life-giving antithesis."

In the subject of Courant's special interest, the trend to which he objected is seen in functional analysis, a development which received one of its early impulses from some of Hilbert's work during Courant's student days in Göttingen. Courant never liked even the name functional analysis. "What does it mean?" he would demand. "Analysis is analysis. You always deal with functions in analysis. What do you mean *functional analysis*?" And he would announce: "I do not specialize in generalities."

"I think he thought that freedom was being lost by the axiomatization of mathematics," De Prima said. "I think he was wrong in certain respects. I think that where initially you lost a freedom you gained others that you hadn't had before. But—anyhow—some of this had repercussions in the second volume of Courant-Hilbert." He looked cautiously at my tape recorder. "Maybe we ought to turn that thing off, because I am going to try to put into words something that I have never voiced before."

But he went on talking.

"You see, Courant-Hilbert II was one of the very few 'almost coherent' discussions on partial differential equations that existed in the late thirties and the forties. There were some others, but I think Courant was really the one who began to see the beginnings of a theory in this subject. It is so broad and so very loosely held together, yet certain things were standing out. Also there in Courant-Hilbert II, Friedrichs had written a chapter where the beginnings of the functional analytic approach had begun to show up."

"The notorious seventh chapter?"

"Yes, that's it. But it was done in a way that was, from the point of view of a functional analyst of ten years later, very clumsy. But what happened was that in the late forties people who were not at all trained in this Courantian way of thinking about partial differential equations suddenly saw the light through all of this and saw, my gosh, we have all the tools we need to attack these problems and give this thing some coherence. There were several important things that occurred during this period, and also some things later on at the Courant Institute. But by then it had taken a jump beyond Courant and maybe beyond Friedrichs too."

De Prima's comments have the advantage of more than thirty-five years of hindsight. At the time of publication, when Hermann Weyl was reviewing the second volume of Courant-Hilbert, he found it "comforting":

"When one has lost himself in the flower gardens of abstract algebra or topology, as many of us do nowadays, one becomes aware here once more, perhaps with some surprise, of how mighty and fruitbearing an orchard is classical analysis."

Shortly after the second volume of Courant-Hilbert was published in Germany, Courant began to talk to American publishers about the little book for teachers based on his lectures on elementary mathematics. By now he was considering *What Is Mathematics?* in place of Felix Klein's title, *Elementary Mathematics From a Higher Viewpoint*, and had broadened his idea of the audience for whom he intended the book to "the educated layman."

While Courant was thus occupied with publishing plans in the United States, scientific publishing in Germany was deteriorating very rapidly. In October 1938 Ferdinand Springer wrote to Neugebauer in Copenhagen, where Neugebauer was still editing several Springer journals, and instructed him to remove the name of Tullio Levi-Civita from the masthead of the *Zentralblatt* on the grounds that, as a result of Italian racial legislation earlier that year, Levi-Civita was no longer a university professor. Neugebauer promptly resigned his editorship of all Springer journals.

There were immediately many questions in America. Should the American editors—Courant, Tamarkin, and Veblen—also resign? Or should they, merely by threat of resignation, try to support Springer against orders they knew came from the German government? What of Neugebauer's professional future and what of the future of the *Zentralblatt*, which was the most important international review journal? Could the plan—in mind for sometime—of publishing a similar journal in America now be put into effect?

"I have the idea that pretty quickly you should come over here for a consultation and a few lectures," Courant hastened to advise Neugebauer. "That certainly is easily arranged—I have already some available cash on hand, which together with a few other invitations ought to finance the trip.... At the end of December there will be a big meeting of the American Mathematical Society, and it could very easily turn out that your presence there would result in a speedy solution of the problem."

After much correspondence the three American editors sent a joint letter of resignation to Springer. It was mailed rather than cabled because, in Veblen's opinion, a cable "would be a slight overemphasis."

The time had come, Courant now decided, for him to resign as editor-in-chief of the Yellow Series. During the past year, in agreement with Neugebauer, Bohr and Hardy, he had done everything possible to make easier what he recognized as Springer's "delicate position"; but recently contracts had been cancelled and plans dropped for books by a number of American authors. The end of such accommodation seemed to have been reached. Courant communicated his intention to resign to his fellow American editors, George Birkhoff and Marston Morse; but he did not ask them to join him.

Courant's letter of resignation to Springer was dated almost exactly twenty years after the two men had sat down in Berlin, in the midst of the post-war revolution, and had signed the contract for the Yellow Series. The books could now be found in every corner of the mathematical world.

(Two years later Max Dehn, fleeing the Nazis across Russia, went to inspect the library in Vladivostock during a train stop. In the entire library there was but one shelf of mathematical books—the Yellow Series.)

"I need not say to you how deeply this development, which lies outside your and my sphere of influence, disturbs me," Courant wrote to Ferdinand Springer on November 20, 1938. "But I would like to emphasize that my personal loyalty and friendship for you and my readiness to be at your disposal with my advice remain unchanged.... No matter how things may shape up, I will always look back with pleasure on our long collaboration."

TWENTY-ONE

THE ACADEMIC YEAR 1938-39 began with the Munich agreement. It was Courant's fifth year at NYU. He carried on his teaching duties, lecturing and conducting his seminar. He continued to concentrate on Plateau's problem and other minimal-surface problems. He worked with various students on *What Is Mathematics?*, paying them out of a $1500 grant he had obtained from the Rockefeller Foundation for that purpose. He also approached the Philosophical Society about a grant to write a book on Dirichlet's principle. He gave demonstrations of soap-film experiments and a talk on the radio on "Infinity." He participated in the early stages of the negotiations which resulted in Neugebauer's coming to America as editor of the new abstracting journal sponsored by the American Mathematical Society—the *Mathematical Reviews*. He continued to seek financial support for the library and for fellowships at the Graduate Center, to arrange to bring important mathematicians there to lecture, to try to broaden the base of interest in New York City as a whole.

It was at this time that Morris Kline, who had spent three years at Princeton as the assistant of the topologist J. W. Alexander, returned to NYU as an instructor in the undergraduate department at Washington Square.

"When I came back from the Institute for Advanced Study in 1938, Courant made it pretty clear that he couldn't afford to expand in many fields," Kline recalled, "and so I began to shift gradually to applied mathematics through my own studies, even during the period from 1938 to 1942."

I asked Kline why he thought Americans had always been so inclined toward pure mathematics.

"I think that when the Americans first became research-conscious they took up the newer fields which didn't require so much background; for example, they went into abstract algebra and topology. In the early 1900's those were new fields. One could understand what had been done without too much background. One could see any number of problems that were open. If they had chosen to go into analysis, they would have had to acquire much more extensive backgrounds before they could do original work of their own. I think that people in our country got started the easiest way at the time."

(Such a narrow purism was not characteristic of many of the outstanding native-born mathematicians. Birkhoff, speaking in 1938 of Benjamin Peirce—"a kind of father of pure mathematicians in our country," also spoke for himself: "In his deep appreciation of the elegant and the abstract we may recognize a continuing characteristic of American mathematics. In his concern with its many applications there resides a virtue which we are finding it more difficult to realize, because of the trend toward professional specialization. Without doubt, however, there is a spiritual necessity upon us today to regain a similar breadth of outlook.")

All of Courant's activity during the academic year 1938–39 took place against a background of involvement in what was happening to his friends still in Germany and to those who had already emigrated.

Except for his brother Fritz, he continued to brush off problems of relatives. But for Marianne Landau, who had emigrated after her husband's death early in 1938, he hastened to help, firing off a number of letters on her behalf. One of these went to Paul Muni, who had played her father, Paul Ehrlich, the discoverer of the cure for syphilis, in a motion picture. And for Otto Toeplitz, who was trying to obtain a visitor's visa to the United States in order to establish contact between Jewish charitable organizations in Germany and similar organizations in the United States, he offered in a letter to the American consul "to assume personal responsibility by posting bond if necessary, [and] to extend to Professor Toeplitz my personal hospitality."

Up until the beginning of November 1938, many Jewish professors, like Toeplitz, thought that although "emeritized" they could continue to live and work in Germany, since they still received their pensions. Then the government announced that on November 9-10 male Jews under sixty years of age would be arrested and imprisoned for a period. On that night, which became known as the *Kristallnacht*, the sound of the shattering windows of Jewish businesses and homes was heard all over the country. A few days later Courant received a telegram from Siegel announcing that Hellinger had been sent to a concentration camp. There was no news of what had happened to Toeplitz.

Two weeks after the *Kristallnacht*, the American Mathematical Society, which had been founded on November 24, 1888, as the New York Mathematical Society, concluded its first half century of existence. The anniversary had earlier been celebrated at a meeting of the society in

September. On that occasion George Birkhoff, the leading representative of native American mathematics, had summarized the history of the subject in the United States and Canada.

The extraordinary contrast between 1888 and 1938, he pointed out, was manifested by the increase in competent mathematicians in the country. Among these he directed the attention of his listeners to two special groups. One was those mathematicians who had shown "the rare quality of leadership" by their participation in the activities of the mathematical society; the other was made up "of mathematicians who have come here from Europe in the last twenty years, largely on account of various adverse conditions."

This influx had recently been especially large.

"And we have gained very much by it. Nearly all of the newcomers have been men of high ability, and some of them have been justly reckoned as among the greatest mathematicians of Europe."

Their coming had certainly worked to the advantage of the general mathematical situation in America. As a teacher of young American mathematics students, however, Birkhoff saw the continuing arrival of outstanding foreign mathematicians as a danger against which he felt he must warn his countrymen.

"With this eminent group among us, there inevitably arises a sense of increased duty toward our own promising younger American mathematicians. In fact most of the newcomers hold research positions, sometimes with modest stipend, but nevertheless with ample opportunity for their own investigations, and not burdened with the usual heavy round of teaching duties. In this way the number of similar positions available for young American mathematicians is certain to be lessened with the attendant probability that some of them will be forced to become 'hewers of wood and drawers of water.' I believe we have reached a point of saturation, where we must definitely avoid this danger."

The words were underlined in the minds of the refugee mathematicians who heard Birkhoff. Many of them were sure that "foreigners" meant "Jews." A number of Americans, including many who had no Jewish blood, resented the distinction between "American" mathematicians and those equally "American" but born elsewhere. Birkhoff himself was only two generations removed from Europe. But many quietly applauded.

In Birkhoff's defense it should be said that the situation of gifted young American Ph.D.'s in mathematics was desperate in 1938. Many were married and already had children. The lucky ones were National

Research Fellows, who received $2400 a year. Those who had been unfortunate enough to choose fellowships over jobs when jobs were still available now tried to live on stipends of $1000 a year at the Institute for Advanced Study. In some colleges there were a few "teaching assistantships" which paid $750 a year. The less gifted, of course, did not have jobs of any sort.

I asked Courant if he felt that Birkhoff's remarks about foreign mathematicians were a result of the fact that he was generally considered, as someone tactfully put it in a memoir after his death, "not pro-semitic."

"I don't think he was any more anti-semitic than good society in Cambridge, Massachusetts, used to be," Courant objected. "His attitude was very common in America at that time. I think Birkhoff was narrow and certainly he was wrong—but he was a very good mathematician."

A glimpse of the situation of Jewish mathematicians in Germany after November 10 is given in the communications of Hermann Weyl, chairman of the German Mathematicians Relief Fund, which he had organized in 1934 with Emmy Noether.

"According to an alarming telegram, Alfred Brauer (Berlin) seems to be in serious danger in connection with the recent pogrom," Weyl told the other members. "In an attempt to bring him over to this country at once, I pledged $300 from the Relief Fund and beg you to endorse my commitment."

This left a balance of $43.75 in the treasury.

In less than a month an assistantship for Brauer had been arranged at the Institute for Advanced Study.

"[And] we have invited Hellinger, Hamburger, and Arthur Rosenthal to come over without a stipend," Weyl now reported. "Affidavits have been prepared for these men and also for Dehn. Dehn obviously will try first to go to England or the Scandinavian countries. It is sure that Rosenthal and Hellinger are in concentration camps; probably also Hamburger. We are in touch with [Hellinger's sister], Siegel, and Mrs. Dehn. Nothing has been heard of, or done for, Toeplitz. Remak is in a concentration camp but we could see no way of helping him; however, I am told that the English will try to do something for him. . . ."

A month later Courant received a communication from Toeplitz, who did not make any mention of his experiences on November 10. He wrote merely that the trip to the United States would have to be given up. The American Consul kept making new stipulations "and does not have the courage to say no."

"We thought we were unlucky when Richard was one of the first professors to be dismissed," Nina said to me, "but we were the lucky ones."

After the *Kristallnacht* older mathematicians like Dehn had to be got out of Germany, too. The need to help the younger, unestablished men continued. Pleas for help were also now coming from places other than Germany, especially Italy. Even people who had already successfully emigrated to the United States were often without resources after their original temporary appointments had expired.

There were some particularly desperate cases. One of these was that of Eduard Helly, who had so disastrously interrupted Courant's report in the Hilbert-Minkowski seminar of 1907. Helly had not obtained a professorship in his native Austria. Instead, although he continued to do mathematics as an avocation and to make important contributions, he had become an actuary. As such he had more difficulty finding a job in America than even a young scholar. Someone sent him to Friedrichs, who had a small and temporary tutoring job available.

"I don't actually remember that it was Courant who sent Helly to me," Friedrichs says now, "but it seems most likely that he was the one. It would have been like Courant."

Then there were, as Courant wrote unhappily to Harald Bohr, "problems not connected with mathematics"—like the case of his brother and his family, who were still in Italy. He expressed the hope that a more liberal handling of immigration might permit scholars without promised positions to enter the United States, but he very much feared that for the masses of Jews in Europe there was no solution: "The future development of the present situation can hardly be imagined."

Christmas 1938 was not happy in New Rochelle. The children complained that Papa was dreadful about money.

"Richard is uneasy about money now for two reasons," Nina explained to her mother. "It seems advisable that his brother Fritz and his family leave Rome so—for a couple of years—they may be dependent on his support.... And then the responsibility for the people for whom Richard has given affidavits, and where he is in fact implored for still further guarantees to help them, burdens him also. He does not permit himself any concert or theatre—and would indeed hardly be in the frame of mind for it. He is often in a terrible mood."

The year 1939 began with a visit from Niels Bohr, and that brightened the gloom for Courant; for to him both Bohr brothers were kings.

In the course of Bohr's visit to the United States, he also took Courant with him to call upon his friend Lewis L. Strauss, a banker who was interested in scientific matters. Later, writing to Strauss, Bohr commented that for the most part the refugee problem was being treated by the Americans in a way helpful for the organization and development of scientific research in America: "In that connection it is of course very important that all sound undertakings of such kind find proper support, and I thought I might take the opportunity to express my great sympathy with the endeavors of Professor Courant to utilize his unique experience and the traditions of the former great German schools of mathematics in advancing the closest possible cooperation between pure and applied mathematics in America."

About the same time that Courant thus became acquainted with Strauss, Flexner asked him to summarize the changed situation of the mathematical-physical faculty of Göttingen since the Rockefeller Foundation's grant.

Siegel was currently in Göttingen with Hasse and Kaluza, Courant told Flexner. The places in physics had been filled by two "rather good" men, Becker and Joos, neither of whom, however, could be considered comparable to Franck or Born. Prandtl's aeronautics institute had been completely dissociated from the university and put under military control —the expansion was enormous. The Rockefeller Foundation's gifts to mineralogy and physical chemistry had been negated by subsequent Nazi dismissals.

"In judging the effect of change," Courant reminded Flexner, "one must not be deceived by the fact that still there are quite a few very good scholars in Göttingen. They are now isolated, mostly unhappy and dissatisfied individuals, and nothing seems to be left of the old spirit of closest cooperation and comradeship also with the younger members of the staff and with the older students."

By late 1938 a position at Brown, in addition to the editorship of the *Mathematical Reviews*, had been arranged for Otto Neugebauer. Courant could not have been happier as he went down to the pier in February 1939 to meet Neugebauer and transport him out to New Rochelle. Neugebauer was not only one of Courant's heroes, but also an intimate friend and longtime associate who would now be only four hours away by train.

"It is wonderful how more and more of Richard's friends are here in the vicinity," Nina wrote. "That pleases Puss in Boots and makes him purr."

At this same time, to make him even happier, Courant received the opportunity which he had so long desired—an invitation to spend a summer term as a visiting lecturer at the University of California in Berkeley. That spring of 1939, as Hitler occupied Czechoslovakia, Courant began to plan *un grand voyage d'instruction* for his family, crossing the United States by the southern route and returning by the northern.

At the end of April, Hitler denounced the German-Polish non-aggression pact and the Anglo-German naval treaty. In May he announced the understanding with Italy was now a "pact of steel."

Courant had tried everything he could think of to arrange a position for his brother Fritz in the United States. This had included approaching Donald Flanders's oldest brother, Ralph, who was at that time president of the Federal Reserve Bank of Boston. Fritz was, according to Courant's description, "a businessman of wide engineering experience and knowledge [who has] been extraordinarily successful." (Until recently he had been establishing and equipping railroads in Ethiopia and Libya.) But even with such recommendations and such contacts, Courant was not able to place his brother in the United States. In the spring of 1939 Fritz Courant emigrated with his family to Brazil.

In spite of the events of the spring, at the beginning of June when the Courants and their three older children set off for Berkeley, it did not seem to Courant that there would be war in Europe.

But the problems of the refugees pursued him even to what Nina described as "the magic land" of California. One of the most difficult cases was that of his gymnasium classmate Wolfgang Sternberg, who had emigrated but had been unable to obtain a position on the east coast. Courant felt that Sternberg was an extremely gifted man who had been prevented from realizing his potential by deep-rooted psychological problems. Since 1933, however, he had been constantly plagued by Sternberg's pleas for advice and assistance. He found repugnant any reminder of his youth in Breslau. Yet he remembered gratefully how the Sternberg family had invited him regularly to a *Freitisch*, the weekly meal an orthodox Jewish family shared with the poor. At the same time the memory of having been the recipient of a *Freitisch* made him very uncomfortable.

Because of Sternberg's personal problems, he felt that he could not recommend him for a teaching position; but before leaving Berkeley, he wrote kindly and at length suggesting that Sternberg come to California, "where the weather is better," and study applied statistics under Jerzy Neyman "and hope that something in connection with statistics and applications, not with teaching, might develop."

At other times he avoided Sternberg and did not answer his calls and letters.

There were some refugees who were difficult to help because of their critical attitude toward everything American and their feeling that a distinguished position was only their due.

"You do not seem to have any feeling what it is all about," Courant scolded one man. "The only safe method for you is: keep your mouth shut, never talk about yourself and your sad fate, never complain, never criticize anything, weather, climate, subways, women, administration in a country that has not asked you to come. . . . Forget your European-German superstition that scientific standing establishes a claim in society or from society. . . . This is a friendly warning which better comes now than later when it might be too late . . . you may accept or reject my statements. Discussion would be idle talk, for which I have neither inclination or time. . . . You are fairly well at the end of the rope, and it depends solely on your tact and wisdom whether the chance given to you now will result in making your talents available for useful service in society or whether you are heading for ultimate failure."

On August 5, 1939, writing a letter to her mother while sitting under a giant redwood tree in Muir Woods, Nina noted below the date: "25 years ago war broke out." A few weeks later, the Nazi-Soviet non-aggression pact was announced. While the Courant family was driving back to New Rochelle, they heard the news that the German army had marched into Poland. Two days later Great Britain and France declared war on Germany.

Ernest Courant remembers his father calling his tribe together and telling them that from that moment forward they were not to speak German in public.

It was shortly after this instruction that they were eating at a roadside restaurant. As Courant got up to go for a second cup of coffee, Nina called after him—in German—to bring her a cup too.

"I do not understand you," Courant replied softly but distinctly in English.

Nina repeated her request, still in German.

"I do not understand you," Courant hissed fiercely in English.

Nina asked a third time, still in German, but now loudly and disstinctly.

"I do not understand you!" Courant shouted. *"Du weisst ganz genau, was ich meine."*

At the time that war began in Europe, Courant had still not met the residence requirement for becoming an American citizen. But he felt like one. He was sure that the United States would be in the war eventually, and he wanted to do everything he personally could to help prepare the country. He was convinced that the kind of mathematics and the kind of mathematicians he was trying to develop at NYU would be needed.

Very shortly after they had come to NYU, Friedrichs and Stoker had begun a collaboration which represented for Courant exactly the kind of thing which he had hoped would occur in the kind of mathematical situation he was trying to develop. This work had originated when Stoker had discussed with Friedrichs a problem in Stoker's favorite subject—elasticity.

"Stoker knew the engineering significance of the problem, which involved the buckling of plates, much better than I," Friedrichs explained to me, "but we worked on it together. Worked very hard and got some results and developed a rather rounded theory for a special case. This was not exactly the realistic case. The realistic case was the buckling of a rectangular plate, we took a circular plate—it was mathematically simpler —but what we did clarified the whole situation. I had more mathematical background, Stoker had more engineering, and we worked very well together."

By the time classes began at NYU in late September 1939, the Germans had virtually subjugated Poland. One German triumph followed another. Courant worried about Siegel, still in Göttingen. Then in March a cable arrived from Harald Bohr saying that Siegel was in Copenhagen and eager to return to Princeton. The Institute for Advanced Study offered an emergency appointment for one year, and Siegel left Scandinavia three days before the Germans invaded.

Now it was the fate of the Bohrs that was of concern to Courant and their other friends. With a special grant, the Rockefeller Foundation made it possible to invite both brothers to America; but, although expressing their appreciation, they both refused to leave their institutions and their friends in Denmark.

Before the war began, Toeplitz had left Germany for Palestine. At the beginning of 1940, he died there. Hellinger had also managed to leave Germany, and he was in the United States by 1940.

On May 14 of that year, Courant wrote a long newsy letter to Born, who had remained in England. Up until the declaration of war, the two men had continued to carry on their correspondence in German; but now they wrote, still a little awkwardly, in English.

The eastern part of the United States was literally flooded with scholars from Europe, Courant told Born. Hellinger had at last "a very tiny and insecure position" at Northwestern.

"I sometimes think that Toeplitz's lot was perhaps not so tragic. He might easily have faced terrible disappointments all around, and none of us can really hope to live to see a somehow not completely crazy world."

Between the time that Courant wrote his letter and Born received it, Holland and Belgium surrendered to Germany. Nevertheless, on the day the letter arrived, Born had been prepared to answer with characteristic cheerfulness "in spite of the grave situation." Then, the next day, he heard that France had capitulated.

"Now there is no way of being cheerful, for even if America would come in immediately there would be not much hope. . . . I stand to the British as long as there is any possibility. But I must reckon with the other possibility that a particular fate is in store for us, Lublin or something equivalent. . . . But as nobody likes to go without any resistance I wonder whether you and our friends, Einstein, Franck, etc., could find a way out for [us]. . . . If you write that we shall live in a crazy world I am afraid that is an understatement—if we shall live at all."

During that first disastrous year of war in Europe, there was beginning to be more activity at the Judson Dormitory. The mathematicians were able to take over another suite from the girls. Since Courant's arrival more Ph.D.'s in mathematics had been awarded by him at New York University than in all its first century of existence. Although Friedrichs and Stoker and some other men from the Heights gave courses at the Graduate Center, it was Courant who seemed to be always around, talking to the students, asking them about what they were doing, telling them they must come out to New Rochelle sometime, becoming involved in their lives, realizing always that at times personal problems could be more important than mathematical problems.

He loved to tell stories of Göttingen. He told so many stories about Hilbert (who was still alive in Germany) that when someone facetiously referred in print to Richard Courant's *Complete and Unexpurgated Book of Hilbert Stories*, many people took the existence of such a book as a fact.

"I suppose every math department has its stories," De Prima reflected, "but Courant's were part of the attraction of NYU for me. I graduated in 1940 and with the war, the impending war, I didn't know what I should do. I applied to various places and got some acceptances, including even the offer of a fellowship to Princeton. Then I had this long session with Courant, and—that is how well he can sell things, you see."

"How did he 'sell' you?"

"Well, he is both obvious and subtle in these things. He kept pointing out what could be done at NYU. You see, the reason he was in New York—and he would hint at this on many occasions—why should he of all people stay in New York? He had no school, no staff besides Friedrichs and Stoker—and they were up at the Heights mainly. The thing was actually not a school at all. There were so many things that made it unreal, and yet Courant believed in it. He believed very strongly in it, because he felt that in the city—in the whole area—there was a tremendous amount of natural talent."

"A reservoir of talent?"

"Yes. That's the word he used. *A reservoir*. This tremendous reservoir of talent. He felt that there were many gifted people in the area—they might even be working, doing other things, but the reservoir was there, we had to tap it. Also Courant was keenly aware that war was coming, we were going to be involved. Even if we were not directly involved, we were going to have to do a great deal in this country scientifically. And then there were the stories about Göttingen. All these things just impressed me so. I felt so close to Courant. I had the feeling that I was part of the family. And so I stayed at NYU," he concluded, "and had the gall to turn Princeton down."

In the spring of 1940 Courant became an American citizen. He felt that he was now no longer a member of that class of "foreigners" who must always hold back their opinions. He began to speak out quite forthrightly about what he thought should be done in science to prepare the United States for the inevitable war with Germany.

TWENTY-TWO

DURING 1940 and 1941 Courant was deeply concerned with two projects, both of which he conceived as contributions to higher education in America by one who was "profoundly grateful" for the opportunity offered him by that country. The first was a plan for a *national institute of basic and applied sciences*; the other was the book tentatively entitled *What Is Mathematics?*

Since his arrival in 1934 he had been pondering the scientific needs of his new country. He saw these as extending far beyond the capabilities of any single university. In the United States there should be a national scientific center—like the École Polytechnique in France—which would produce a responsible, well-trained elite of scientific manpower for the coming war and for the difficult years which would follow.

The first two drafts of a memorandum entitled "A National Institute for Advanced Instruction in Basic and Applied Sciences," are without dates, but the third is penciled "Winter 1940–41."

From the first paragraphs in all versions, the voice of Felix Klein rings out. Klein had intensely admired the educational philosophy and program of the École Polytechnique and had always regretted that the ideals it represented—pure and applied science in close connection, teaching and research combined, personal contact between faculty and students—"had never taken proper root in German soil." From the time he had arrived in Göttingen, he had endeavored to establish these admired ideals there.

In his memorandum Courant stressed the national situation out of which the École Polytechnique had developed. After the Revolution, France had been "economically disrupted, intellectually and morally uncertain, at war with the strongest powers of Europe, her educational facilities disorganized." Scientists "of vision and initiative" had conceived a plan for an institution of higher learning "at a level unheard of before," the students democratically but carefully selected, the best scientists in the country as teachers. The new facility had soon fulfilled "the highest expectations" of its founders; and within less than two years Army, Navy, Industry, and Government had begun to receive a supply of men whose education was superior to anything then available in the world—"men equipped for productive work and not dulled by a dead routine of training."

Courant's view of the national institute he proposed was a broad one—"something very universal," he told me, "not only mathematics but also physics, other sciences, and also history and philosophy." He did not see it as competing with the Institute for Advanced Study or with any other existing facilities. Rather it was "to supplement them, in close cooperation."

After working his ideas over several times in 1940—the changes were almost always in language rather than in content—he decided at the beginning of 1941, now that he was an American citizen, to circulate the proposal. It seemed to him that such an institute could be put into operation quite soon if it began simply with a program in mathematics and physics. At the end of the memorandum dated winter 1940–41, he optimistically set the opening for September 1941.

During the academic year 1940–41 Courant was also devoting time to the other project which he conceived as a patriotic service, the book *What Is Mathematics?* He had been working on it off and on for almost five years and had drawn a number of his students into the project. In addition to David Gilbarg, who wrote up the notes of the original lectures, seven young men, including Courant's son Ernest, are mentioned in the preface as helping "in the endless task of writing and rewriting the manuscript."

By spring 1939 Courant had apparently decided that the subject matter of the book was a little narrow, too restricted to his own interests. On a visit to Princeton, he had asked various people for recommendations for an instructorship which was open at the Heights. Marston Morse had suggested his assistant, Herbert Robbins, a young topologist from Harvard; and Courant had stopped by Robbins's office to meet him.

When I talked to Robbins in 1975 in his Riverside Drive apartment just outside the gates of Columbia University, he did not remember whether it was at this first meeting or one shortly afterwards that Courant brought up the subject of also working on *What Is Mathematics?* He told me that he came to NYU in the fall of 1939, taught elementary subjects at the Heights during the day and gave advanced lectures at the Square at night. Courant handed over to him what earlier assistants had already done with *What Is Mathematics?*, talked about the concept of the book, and asked him to go over the manuscript and improve and amplify it.

I asked Robbins how he and Courant had worked together.

"It's hard to say," Robbins answered. "He had a set of mimeographed notes which were a course he had given sometime previously, which were written up by someone who had taken the course, a student, and they formed about a quarter or a third of the material that finally ended up in the book. Some chapters were there in their final form, some weren't there at all.

So I would suggest, say for example, that we have a chapter on topology, and we would discuss what should be in it. On some things he had very definite ideas, and on some things I had very definite ideas. Over a period of two years I would just work away and show him the results; and he would comment and criticize and I would re-do them. . . . Sometimes he would think of clever ways of doing something, and sometimes I would. . . . I don't know that there was anything particularly striking about the method of collaboration. It was pretty close collaboration, although we never sat down to write together."

Robbins stopped to explain that he never took the course as Courant gave it, or any other course taught by Courant.

At first—Courant told me—he found young Robbins not much help—"in fact, he impeded me by not doing very much work and pulling a little bit in a different direction"; but then, after a frank discussion between them, Robbins settled down and was "quite useful."

Finally, as Robbins recalls, Courant told him that he was very pleased with the work to date. The $1500 from the Rockefeller Foundation, out of which he had been paying Robbins, who made $2500 a year as an instructor at NYU, had by now been used up. Courant suggested, according to Robbins, that the two of them make *What Is Mathematics?* "a joint authorship production," larger and more comprehensive than originally planned.

"Are you sure that that was what he said?" I asked, for I had heard many tales of people leaving Courant's office not at all sure from his mumblings what had been decided.

"I was never in any doubt," Robbins said emphatically. "My initial cooperation was largely because I wanted to make some money. The reason I wasn't so happy [at first] was that [the book] was taking a lot of my time, and for someone who's just got his Ph.D., as you know, his future reputation depends on research and not on popular exposition. So I was sort of hesitant to put in another year and a half or so of what I would regard as a distraction from what I was really interested in doing. . . . The original thing was merely . . . helping someone else with his book. I didn't expect to be a joint author. . . . But when he suggested making it a joint book, then I agreed. I too had gotten sort of engrossed in it by that time."

With the new plan, the book suddenly began to move. Proskauer advised Courant to seek a more general publisher than Interscience for the audience he wanted to reach; and at the beginning of 1941 Courant opened negotiations with McGraw-Hill, who had earlier expressed interest in *What Is Mathematics?* It occurred to him that the book might also serve as a wedge for interesting a big established American publishing house in some other ideas he had. Barely acknowledging a proffered contract, he began to sketch a plan for a series of advanced mathematical textbooks patterned on his Yellow Series.

While Courant was working with Robbins on the book, he was also concerned about the young man's future as a mathematician. He felt that although Robbins was very talented, he was not making the progress in his chosen field of topology which he should be making. (Robbins explained to me that he was not able to get his thesis into print until March 1941, because he was being thoroughly overworked at NYU and in New Rochelle.) The previous year an arrangement to have Will Feller, by then at Brown, give a course on probability and statistics at NYU had fallen through; and Courant had assigned the course to Robbins.

"I had just a few weeks' notice until it started," Robbins recalled, "and up until that time I had not the faintest acquaintance with or interest in either probability or statistics."

Courant was very impressed by the job Robbins had done with the course. He thought it would be desirable for the young man to study statistics and probability "at the source." The source, as far as Courant was concerned, was Jerzy Neyman, the famous Polish authority in the field, then recently come to Berkeley. In early spring 1941 Courant wrote Griffith Evans, the head of the mathematics department at the University of California, "with perhaps a strange kind of suggestion which I hope you will deem justified by the extraordinary general circumstances."

It was Courant's suggestion that Robbins, who was quite poor and the sole support of his mother and younger sister, be given some "modest" financial support by Evans so that he could work with Neyman during the summer.

"I am sure that this will be of great benefit for Robbins, and maybe for statistics."

Knowing that Robbins is now one of the outstanding men in statistics and probability, I asked him if he had in fact been given the opportunity to go to Berkeley in the summer of 1941 to work with Neyman.

"No," he said—he had not even known that Courant had made such a suggestion to Evans. "Had I done so, my life would have been quite different, I'm sure, because I didn't meet Neyman until much, much later."

During that spring of 1941, Courant was extremely busy. He was trying to finish the book, trying to interest people in his idea of a national scientific institute, trying also to set up at NYU a summer series of defense-oriented mathematics courses. Thus harried, when the people at McGraw-Hill expressed some slight skepticism about the commercial possibilities of *What Is Mathematics?*, although still insisting upon their willingness to to publish it, he was very upset. He was unwilling to make any concessions about the book's handling, and he was determined that it must be in print by the fall of 1941.

He had already got a taste of publishing in his profit-sharing arrangement with Interscience Publishers. Now he decided to become his own publisher. In the summer of 1941 he borrowed money from a wealthy friend and contracted with the Waverly Press to set the manuscript in type. He then signed a contract with the Oxford Press to handle the distribution.

During this same period Courant continued to press his idea of a national scientific institute. As the year 1941 progressed—the Balkans occupied, the British pushed back in North Africa, the Soviet Union invaded—it became in his view an *emergency* institute.

At first, rather than soliciting directly for funds, he tried to arouse interest in the project among well-known public figures who could provide him with access to additional non-academic support. One of those to whom he wrote was Justice Felix Frankfurter, whom he had recently met through Flexner. Another was Dorothy Thompson, the newspaper columnist, to whom he had been introduced by Donald Flanders's brother Ralph "about two years ago . . . in the Savoy-Plaza dining room." He asked her also for an early opportunity for a quiet personal discussion "to seek your advice in an educational matter that might become of non-negligible importance for the pressing problems of the day and for the future problems of reconstruction."

Justice Frankfurter responded regretfully. Dorothy Thompson's secretary gave him an appointment and then cancelled it.

What disappointed Courant most, however, was the lack of enthusiasm and cooperation on the part of the fellow scientists whom he had particularly counted on to support him, such as von Kármán, who—Courant told me—thought the institute idea was just "a cheap imitation of Felix Klein."

"Kármán knew Courant very well and he realized that Courant was not really an applied mathematician," Friedrichs explained. "Yes, mathematical physics—*Courant-Hilbert,* fine—but that is not applied mathematics. Kármán had come to the United States before Courant. He had the engineering background. He also had a very good understanding of *applying* mathematics—he wouldn't have said 'pure' mathematics but rather 'sophisticated' mathematics—to engineering. For all these reasons Kármán felt that *he* should be the one to develop applied mathematics in this country, not Courant; and he sometimes expressed himself in a not too friendly way about Courant. But, as you can see from Kármán's autobiography, he could not have done what Courant did. In fact, in later years he reluctantly agreed with me that this was so."

The most effective opposition to Courant's plan came in his opinion from R. G. D. Richardson, the dean of the graduate school at Brown University and one of the relatively few Americans active in placing refugee mathematicians. Since 1933 he had arranged positions at Brown for such former Göttingers as Lewy, Neugebauer, and Feller as well as for Willy Prager, who had been a *Privatdozent* for applied mechanics in Göttingen from 1929 to 1933.

Richardson was also a key member of a group of mathematicians who had shown the quality of leadership by their participation in the activities of the American Mathematical Society. With extraordinary energy and ingenuity these men and their predecessors had organized the mathematical society, solicited members, obtained support from business and industry, collected libraries, published journals, prepared the necessary equipment for mathematics in America so that it was already there when the refugee mathematicians from Europe began to arrive in 1933.

After graduating from college in Nova Scotia and serving as a high school principal for several years, Richardson had taken a second A.B., an M.A. and a Ph.D. in mathematics at Yale. He had then gone abroad for additional study. He had arrived in Göttingen in the year 1908 just after Courant, ten years younger than he and still a student, had become Hilbert's assistant. During that year Richardson had given Courant a mathematical paper to pass on to Hilbert for publication in the *Mathematische Annalen*. No one I spoke to really knew what happened. Courant said only, "I was very young and careless." Richardson's paper never appeared in the *Annalen*. In 1931, reviewing the second edition of the first volume of Courant-Hilbert, Tamarkin, a colleague of Richardson's at Brown, commented pointedly: "The bibliographical references are a little more complete in the present edition [1930] than in the first one. In this connection the reference to an unpublished paper by R. G. D. Richardson should be welcomed. . . . It contained numerous points of contact with the results of Chapter VI. . . ."

When Courant came to America, he tried immediately to remove the "dissonance" in his relationship with Richardson. On the surface he was more or less successful. The two men exchanged polite letters about the placement of refugees and the need for applied mathematics. But in 1941 the feeling of each that he was the champion of this subject so long neglected in America brought them into collision.

Since the outbreak of war in Europe, there had been increased interest in the applications of mathematics on the American side of the Atlantic. In September 1940, at a joint meeting of the American Mathematical Society and the Mathematical Association of America, Marston Morse, as chairman of the War Preparedness Committee, had recommended "that graduate schools extend their courses in applied mathematics . . . and that advanced students be urged to become highly qualified in one or more fields of applied mathematics." By 1941 plans were being made by the Office of Education for a defense-oriented summer program. Courant thought the proper place for such a program was New York City, told Richardson so, and began to make arrangements at NYU. Richardson went ahead with his own plans for setting up a program at Brown. In a short time he was able to make a public announcement of the creation at that university of "the nation's first center where engineers, mathematicians, technicians, and other specialists in defense production can devote their full time intensively to problems of higher mathematics as applied to industry."

The program at Brown in the summer of 1941 was "a very high level affair," according to Friedrichs, who taught a portion of one of the courses. Most of the faculty had come originally from Europe. Richardson was especially proud that he had obtained the advisory and instructional services of von Mises, the former director of the Institute for Applied Mathematics at the University of Berlin, who had recently emigrated to the United States from Turkey.

That same summer Courant circulated his proposal for a national science institute, and he sent Dean Richardson a copy. Richardson responded politely, saying that although he thought that such an institute was an important idea, he was thoroughly convinced that it should be attached to some existing institution. Still he was "vitally interested . . . ready to give time and thought to the advancement of the cause, whatever is finally chosen as a plan."

Shortly after receiving this letter from Richardson, Courant, chatting in the office of a friend, perceived lying on the desk among some other papers a letter which he recognized, although it was upside down, as coming from the dean at Brown.

"We had a long talk, this man and I, in the course of which we shuffled some papers," Courant recalled. "Afterwards I went on a little trip to the Adirondacks and when I unpacked my briefcase there were some papers in it which I didn't know. So I had by mistake swiped some of this man's papers, and there was a document by Richardson which discussed this proposal of mine and was really very hostile. He was very much against it —not against it, but against me doing it. He didn't think that foreigners should come in and make such proposals."

"Would that letter from Richardson be in the files at the institute?" I asked Courant.

"Oh no," he said. "You see, I stole it."

By summer 1941 most of *What Is Mathematics?* was in type, and Courant was preparing to write the preface. He had produced what he felt was a different kind of popular book on mathematics. The 1930's had seen a number of such books in English; but, in Courant's opinion, all of them had a serious flaw. The basic premise on which they were written was wrong. The understanding of mathematics could not be transmitted by painless entertainment any more than the appreciation of music could be conveyed by even the most brilliant journalism to those who had never listened intently to music. To understand mathematics, one must *do* mathematics. In his book he proposed to give his reader "actual contact with the *content* of living mathematics."

What Is Mathematics? presupposed only knowledge that a good high school mathematics course could impart. Technicalities had been avoided, and also emphasis on routine and "forbidding dogmatism which refuses to disclose motive or goal and which is an unfair obstacle to honest effort." On the other hand, the book was not "a concession to the dangerous tendency toward dodging all exertion." The promise which Courant held out to his readers was that in his book they would be able to proceed "on a straight road from the very elements to vantage points from which the substance and driving force of modern mathematics can be surveyed."

The title of the book still caused Courant some concern. It seemed "a little bit dishonest." At a party one night at Weyl's, he consulted Thomas Mann. Should the title be *What Is Mathematics?* or should it be something like *Mathematical Discussions of Basic Elementary Problems for the General Public*, which was more accurate but "a little bit boring"?

"Then Mann told me that he couldn't advise me, but he could tell me about his own experience," Courant said. "Among his German books published in English there was one—*Lotte in Weimar*—Goethe's Lotte—and shortly before the book was to be published Mr. Knopf came to him and said, 'Oh, now we should select a title; and my wife, who has a very good sense about these things, thinks we should call it *The Beloved Returns*.' Then Mann said he felt a little uneasy about this title—after all, the title *Lotte in Weimar* was just as good in English as it was in German. And Knopf said, 'All right—but I just want to make one remark. If it is published as *Lotte in Weimar*, then we will sell ten thousand or maybe twenty thousand copies; but if it's *The Beloved Returns*, we will sell one hundred thousand copies—and of course the royalties will correspond.' Mann said, 'I have decided—it will be *The Beloved Returns*.'"

Courant thanked Mann and went immediately to the telephone to call the printer.

Although young Robbins had been reading galley proofs and had made several trips to the printer in Baltimore, he had not yet seen a proof of the title page. Then, toward the beginning of August 1941, he saw for the first time "*What Is Mathematics?* by Richard Courant."

"When I saw the title page, I suddenly said, 'My god, the man's a crook. It was just a sort of a cold bath. What I regretted was not only that I was not going to get my name on the book—because I immediately knew that I was—[but that] it was the end of my feeling that Courant was a decent, honorable person who wished to foster young people's work without worrying about his own prestige, etc. . . . Later, of course, I heard over and over again 'Dirty Dick.' People found it very easy to believe this had happened because they had heard other stories. I had not heard any other stories. I knew nothing but good about Courant. . . . I had rather loved him."

Robbins consulted some of the other people at Washington Square about what he should do.

"And they said, 'Well, you have to understand that in Germany this has not been uncommon. Many books are purportedly written by some well-known professor but are in fact written by some young student of his as part of the educational process.' I said, 'Well, first of all, I am not a student of his; second of all, this isn't Germany; and third of all, I don't like it.' "

It is impossible at this date to determine the chronology of events not recorded in the correspondence between Courant and Robbins; but at some point, Robbins told me, he spoke to Hassler Whitney, under whom he had taken his Ph.D. at Harvard.

"When I told Whitney of this, he was highly indignant and he said, 'Well, you tell Courant that if he goes through with this, I will bring the matter up at the next meeting of the American Mathematical Society, and we will expel him from membership.'"

Robbins had "promised or perhaps threatened" as he reminded Courant in a letter dated August 17, 1941, to state his general sentiments concerning the wording of the title page in writing. He had postponed doing this for some time because his feelings on the subject were so intense that it seemed unlikely he and Courant could have a quiet discussion and "a heated argument might have had a bad effect on the book, or at least on its publication date." In the letter of August 17 he stated that although he recognized that the book was in all essential ways Courant's book, he himself had worked so hard on it and had become so emotionally involved with it that having his name on the title page equally with Courant's had become very important to him. Besides, he felt that while the practice might be different in Europe, this was the standard American custom in such cases. Everyone recognized that the first name given on the title page was that of the real author of the book and the second name that of the assisting younger colleague. As to the financial arrangements, he was perfectly willing to leave those entirely in Courant's hands. He simply asked that the title page read "by R. Courant and H. Robbins."

Neither Whitney's threat nor Robbins's determination to take legal action is mentioned or implied in Robbins's letter. It is Robbins's feeling that Courant learned of these from others.

At any rate, after receiving Robbins's letter, Courant agreed to change the title page.

Robbins explained to me that he wrote his letter of August 17 in the way he did to make the idea of coauthorship more palatable to Courant. It had that effect. When Courant showed me the letter, he told me that he felt that it expressed exactly the situation that had existed between himself and Robbins.

In the next few weeks of the autumn of 1941, however, Courant apparently became aware of the feeling on Robbins's part that he was in fact a coauthor. On September 28 he wrote a long, stern letter to the young man.

"I am under the impression that you have permitted yourself to drift or to be pushed into an awkward psychological position. I think it is imperative that you be fully aware of the actual situation. It is not a question how much time, energy, devotion you have spent on your collaboration. By conception, by planning, by content, by original mathematical ideas the book is my child and more than any other of my publications expresses my very personal views and aims. You are enough of an individual to have your own outspoken views that need not fully coincide with mine, and it would be only natural [for] some marked deviation . . . to develop in the future. For this reason I was and I am anxious that the matter of authorship should be clearly understood by everybody, in the first place by yourself. This is not a question of ambition, it is a matter of scientific responsibility. Naturally I depended on assistance. . . . Your cooperation far exceeded what I could expect from a competent mathematician with a high standing of his own. By no means do I wish to withhold praise and public recognition. When in your letter of August 17th, you insisted on having this recognition expressed on the title page, I consented immediately. Your letter reassured me that there was not and never could be an essential misunderstanding between us concerning the basic issue of authorship and that misleading impressions on the public were not intended. . . ."

What Is Mathematics?—by Richard Courant and Herbert Robbins—turned out to be much more successful than anyone, except perhaps Courant, ever expected it to be. As of the date of this writing, it has been translated into a number of languages and has sold well over 100,000 copies. It is often described admiringly as "a mathematical best seller." Robbins told me that he never received a regular accounting of the book's sales, but simply from time to time a personal check and a note from Courant saying, "Dear Robbins, Enclosed is a check representing your share of the sales of *What Is Mathematics?* for the year 19—."

He is, as he says, "heartily sick" of the whole matter. When people ask him who wrote *What Is Mathematics?*, he sometimes tells them, "Courant wrote the book, but he put my name on it so it would sell."

"The important thing," he told me, "is that it's a good book, and it's had a good effect."

Courant himself was always "a little bit disappointed" in the book; for in spite of its success it never reached to any appreciable degree that general public of "educated laymen," to whom he had hoped to be able to convey something of the beauty of mathematics.

By the time the academic year 1941–42 began, Courant had achieved one of his goals—*What Is Mathematics?* was in print—but he had not achieved his other goal of a national institute of basic science. He always believed that the opposition by Richardson was decisive. In retrospect, though, he said that he also found it "of course childish and unrealistic" that he had even thought he could carry off such a grand project at a time when he had not been in the country very long and had not understood the degree of resistance to government support of science which existed in the United States at that time.

"So did you give up the idea?" I asked.

"I didn't give it up," he objected. "I thought I would try other means."

TWENTY-THREE

IN SPITE OF seven years of effort, Courant had developed only a very modest operation at NYU by September 1941. A folder of that date announcing the establishment of an Institute for Applied Mathematics in connection with the Graduate Center lists an impressive number of regular and emergency courses; but when one examines them, the new "institute" turns out to be just another name for Courant, Friedrichs, and Stoker.

By the end of November, the German army had effectively occupied most of the continent of Europe. The Committee in Aid of Displaced German Scholars had long since substituted the word "Foreign" for "German" in its name. Since 1933 approximately 130 mathematicians, mathematical physicists, and statisticians had come from Europe to the United States. Not included in the total were the many European children who also came with their families during those years and grew up to be contributing members of the American mathematical community. One of this group is Peter Lax, the present director of the Courant Institute. As a 15-year-old boy, Lax was on the high seas with his parents on their way from Hungary to the United States on December 7, 1941.

In the United States on December 7, which was a Sunday, the broadcast of a concert by the New York Philharmonic was interrupted by the announcement that the Japanese had attacked Pearl Harbor. On Monday the United States declared war on Japan. By Thursday Germany and her European allies had declared war on the United States.

In spite of the fact that he had long expected and even hoped for the United States to enter the war, the actuality was traumatic for Courant, as it was for most refugees. In the midst of the abrupt adjustment, he wrote to Heinz Hopf in neutral Switzerland and asked him to arrange for "a beautiful bouquet" to be delivered in Göttingen on January 23—the card to carry the message that the flowers came to Hilbert on his eightieth birthday "from his American students and friends with all their love and respect."

After Pearl Harbor, Courant continually paced up and down the long corridor between the offices at the Judson Dormitory. When De Prima brought lecture notes to be stenciled in the office of the secretary, he would stop the young man and get him to pace back and forth with him.

"I am sure he would do that with other people too. He was always on edge. It was very noisy there on Washington Square South so you could just barely hear a little jingle when the phone rang. But he would always hear it and immediately jump over to his phone, in spite of the fact that there was a secretary there to answer it, and say, 'Courant speaking! Courant speaking!' And I came to the conclusion—although he never told me so—that he was just waiting to hear from people in Washington to whom he had offered his services."

As soon as the United States entered the war, Courant began to grab for the talent he was sure he would soon be needing. Eleazer Bromberg remembers that in the spring of 1942 he came to see Courant about completing an unfinished course at NYU. He had already visited Brown, where the long-established and influential Richardson had been able to develop a program in applied mathematics which was much bigger than the one at NYU. He had pretty well decided to go to Brown. "Why not here?" Courant demanded and immediately arranged a job for Bromberg as an instructor in the physics department.

As the United States began to give indications in early 1942 that it was preparing to strike back against the Japanese in the Pacific, Courant was quite delighted to receive a grant of $3500 from the Carnegie Foundation, which would enable him to support a few more people. He conveyed the news proudly to Flexner and others, including Warren Weaver.

Weaver had been one of the early appointees of Vannevar Bush's National Defense Research Council (NDRC), set up in June 1940. His first assignment—as the bombing of Britain began—had been to take charge of the "fire control" section of the NDRC, which concerned itself with devices and procedures to make certain that a projectile would hit the intended target. He had later moved with Bush to the Office of Scientific Research and Development (OSRD), which was created by executive order of President Roosevelt in June 1941 to coordinate defense research—as Bush said—"wherever it might be."

The emphasis on the design and production of hardware had tapered off even by 1942, according to Weaver, because in general new devices could not be conceived, designed, built, tested, improved, standardized, and put into service in time to affect the outcome of the war. On the other hand, the need for analytical studies had increased. By the summer of 1942, Weaver was recruiting mathematicians.

One of those whose services he obtained was Stoker, who with his student Bromberg was sent up to Columbia to join a group composed mainly of statisticians working directly with Weaver on anti-aircraft artillery problems. Stoker was impressed by the talent Weaver had assembled, but he was enormously unhappy at Columbia.

"How did you happen to leave NYU anyway?" I asked.

"Oh, it was one of the bargains Courant drove with Warren Weaver," the forthright Stoker replied. In return for Stoker, Weaver agreed to support Courant, who with Friedrichs was already doing work for the Navy on the transmission of sound under water. "There were just not very many applied mathematicians in the country at that time." This was no wonder, since a report issued in 1941 had concluded that American industry could not absorb more than ten applied mathematicians a year.

In the fall of 1942 Courant ran a seminar at NYU on the theory of stationary shocks, important in aerodynamics. He was particularly interested in the theory because of Riemann's pioneering work in it. That December he happened to run into John von Neumann in Washington. Von Neumann had been doing some work on moving shocks, important in explosion theory. Since he was about to leave for some new and secret project, he offered to have Raymond Seeger, a co-worker in the Navy's ordnance group for fundamental explosive research, brief the NYU people on his results. This was the beginning of extensive work at NYU on shock waves, mainly by Friedrichs. The early contact with Seeger, both while he was with the Navy and later when he was with the National Science Foundation, was to be of inestimable value to Courant and his enterprise.

By this time the OSRD had carried out a reorganization which had shifted the fire control problems to a new Division 7 and created another OSRD agency called the Applied Mathematics Panel. The panel was to be of general assistance in connection with the fast developing analytical and mathematical problems, not only for Division 7 but for all the other divisions of OSRD as well and, even more broadly, for all the military services and the war effort. Weaver was made its chief scientific officer. The call from Washington for which Courant had been waiting came very shortly. It was Weaver inviting him to become one of the half-dozen members of the new panel.

The Applied Mathematics Panel cut across all divisions of OSRD. It was planned as an organization of civilian mathematicians who would provide mathematical help to other scientists involved in military work or to parts of the military needing such help. It was hoped that by this means all the leading mathematicians in the country not already employed by the military would become fully involved in the war effort.

For his panel, which was officially known as the "Committee Advisory to the Scientific Officer," Weaver selected, in addition to Courant, T. C. Fry, L. M. Graves, Marston Morse, Oswald Veblen and Samuel S. Wilks. Years later, Weaver recalled the serious discussions which had preceded the extending of the invitation to Courant—"in view of the classified nature of many of [the] projects, whether it was prudent to include in the top governing committee, a man—however distinguished and able—who had been a member of the Imperial German Army during World War One."

At the first meeting of the panel, before Courant arrived, Fry of the Bell Laboratories said, "We must, from the very outset, make Courant realize that we view him, with no conceivable reservation, to be *one of us*."

While Fry was saying this, there was a soft little knock on the door.

Fry, who like the others had always addressed him in the past as "Professor Courant," sought a way to express his friendly acceptance of another American.

"Come in, Dick!" he said.

Courant took naive delight in his new importance and influence as a member of the Applied Mathematics Panel. He bustled back and forth between New York and Washington, dropping well-known names not quite inaudibly, mumbling vaguely of important secrets, glancing constantly at his watch, complaining apologetically about "the almost inhuman burden of work under which we all suffer at this time."

He was constantly generating ideas. A stream of papers labeled "Memorandum to WW" began to flow from his desk.

Many of Courant's suggestions to Weaver had to do with personnel. One mathematician he immediately proposed for the panel was Griffith Evans of the University of California. He was also eager to bring in von Kármán "as a foremost representative of applied mathematics and applied science in this country," and a lengthy memorandum on the "psychological background" to be considered in approaching Kármán went to Weaver.

Evans did become a member of the Applied Mathematics Panel. Kármán, who was the scientific advisor to the Army Air Corps, indicated interest in the group; but he was prevented from joining by illness and the approaching end of the war.

Somewhat later Courant recommended to Weaver a young woman named Mina Rees, a junior mathematics professor at Hunter College, for the position of technical aide. This position was to turn out to be more important than originally expected, since for a time during the war Weaver was incapacitated by an ear infection which affected his sense of balance. Mina Rees then became in practice his representative on a day-to-day basis in relation to the panel and increasingly represented the panel in connection with its activities. The job was for her the beginning of an outstanding career in government and university administration.

"I really don't know how Courant happened to think of me," she said when I talked to her in her office in the handsome new building of the Graduate School of the City University of New York. "He hardly knew me except for mathematical meetings and things like that. I have always thought it may have been because a very good student of mine later went to NYU. Her name was Bella Manel—and it just happened through an accident of scheduling that almost all of her undergraduate work in mathematics was taken with me."

I told her Friedrichs had suggested to me that Courant had recommended her to Weaver simply "on the hunch" that she had great potential as an administrator.

"I think Friedrichs is probably right," she agreed after a moment's thought. "I think Courant had a real genius for identifying people's potential. He knew what you wanted, what you were really interested in, sometimes even before you did."

The Applied Mathematics Panel operated through "contracts" with various universities. The idea of supporting scientific work by means of such contracts was new in America—"and one of the great inventions which came out of the war," according to Mina Rees. A total of 194 studies were conducted. These were summarized in four published volumes after the war. In just one instance—the development of powerful new statistical techniques which improved the efficiency and lowered the cost of testing war matériel—more money was saved by the military in a few months than the panel had expended during it entire existence. The group working at NYU was only one of a large number of groups set up at institutions from coast to coast; but for Courant and his co-workers the period of connection with the Applied Mathematics Panel was more important than for most. It was the turning point in the long struggle to establish a scientific center at NYU.

With the prospect of an Applied Mathematics Panel contract which would provide a generous amount of money, Courant began to hire talent he had long coveted for his group. The first person he drew into the new work was Max Shiffman, who since obtaining his Ph.D. had been teaching mathematics at CCNY. Courant considered Shiffman the best student he had had in America—one of the best he had ever had—and he had long sought a way to bring him back to the Graduate Center. The second person he hired was J. K. L. MacDonald, whom he arranged to take on loan from the nearby Cooper Union. He saw MacDonald as having a mind "both analytical and inventive" and "embodying a unique combination of mathematics and theoretical and experimental physics and a keen sense of applications."

In addition to young people already connected with the Graduate Center, Courant brought in young mathematicians from other places. One of these was Bernard Friedman—"a first-rate research worker," in Courant's opinion, "of enormous versatility and power." Everybody at the Judson Dormitory recruited friends and acquaintances. By the end of the war the applied mathematics group there numbered more than thirty.

The first work of the AMP group at NYU was a continuation of special Navy assignments at the beginning of the war in the two different fields of underwater acoustics and explosion theory.

"Automatically," as Courant saw it, "[the work] developed on a rather broad front. For example, the whole field of interferences of nonlinear waves was studied, the phenomena of the problem of gas motion after underwater explosion explored, its connection to surface phenomena studied with the consequence of progress in the air-water entry problem. The explosion theory led to general studies in supersonic gas dynamics which [then] drew the group into work on jets, rockets, and jet propulsion in general."

What pleased Courant most was that in this way "a state of affairs was reached where knowledge and experience of the group as a team [were] utilized for very different agencies." This couldn't have happened, he maintained, if the group had been scattered or divided into sub-groups, each dealing with a circumscribed assignment. The *group* was important.

Courant

During the war, at NYU, as at other American colleges and universities, the number of full-time graduate students—there were only a very few even in normal times—declined drastically. There were, however, some part-time students who worked during the day in New York in government or defense-related industry.

In spite of the load of increasing war research and the few students, Courant insisted on continuing to teach. As the war continued, his lectures became less and less prepared. Finally they were not prepared at all. Notations would be changed ten or twelve times in a single lecture. At one point he was teaching a class in the calculus of variations on Tuesday nights and a class on partial differential equations on Thursday nights. The two subjects being related, when he found that he could not prove a certain theorem in the Tuesday class, he would mumble, "Ja, ja, well, I will prove that theorem in my Thursday class." Two nights later, when the necessity for proving the theorem again came up, he would tell the students, "Ja, ja, well, I will prove that theorem in my Tuesday night class."

"His were the most incredibly diffuse and complicated lectures I have ever heard," said Harold Grad, now director of the Magneto-Fluid Dynamics Division of the Courant Institute, then one of the few full-time students. "But beautiful. It is hard to describe why. But you learned. You always learned something that was more interesting maybe than what he had originally planned to teach. I always enjoyed listening to him."

The war period was an exciting time, remembered with nostalgia by everyone I interviewed who had been at the Judson Dormitory then. Suddenly, after more than seven years, the tiny enterprise on the second floor began to expand. Suite by suite the girls were pushed out. Finally the mathematicians had three floors of the four-story building.

Even during the war Courant continued to feel that scientific writing should not be abandoned. One of the most important contributions the group made to the war effort was the writing of a manual by Courant and Friedrichs on shock waves. The editorial assistant for this project was a bearded young man with a degree in English with whom Courant had scraped up on acquaintance while sharing a bench in Washington Square.

Courant also continued to think about the little book he had long wanted to write on Dirichlet's principle, but there was so little time—that would have to wait.

There was suddenly a fresh demand for Courant-Hilbert from scientists working with the military. Since all German properties in the United States, including copyrights, were subject to seizure by the Alien Property Custodian, the two-volume book could be photographically reproduced by an American publishing company under a license obtained from the custodian. In 1943 Interscience Publishers issued a photo-offset edition. Even though it was in German, it sold 7,000 copies.

In Germany—Courant later learned—a request by the research director of the Luftwaffe for a reprinting of Courant-Hilbert had at first been rejected on the grounds that one of the authors was Jewish. It had finally been granted—two months before the end of the war—with the stipulation that the run be limited to 500 copies for official use only.

The same year that Courant-Hilbert was published in the United States, Hilbert died in Göttingen, a few weeks after his eighty-first birthday. It fell to Hermann Weyl to evaluate Hilbert's life and work for that part of the world which was at war with Germany.

"It was appropriate that Weyl should do it," Friedrichs said. "Hermann Weyl was the mathematical son of Hilbert—he understood Hilbert as a mathematician much better than Courant did. It's true that Courant always considered himself the son of Hilbert—and he always played down what he owed to Felix Klein—but in fact he was the son of Klein. He learned from Hilbert, and he was greatly influenced by Hilbert. But he always took what he got from Hilbert and transformed it in the spirit of Klein. In a way it was that combination—Hilbert in the spirit of Klein—which was *Courant*."

As pressure on technological centers to provide numerical solutions of partial differential equations increased, there was also a call for another old piece of work from Göttingen.

At Los Alamos, where Donald Flanders was heading the computing department, an ingenious system of punch cards was being utilized in order to solve—by means of finite-difference schemes—the hyperbolic partial differential equations for fluid flows that build up in the course of time. At first, inexplicable errors appeared in the results. When the mathematicians turned to the old Courant-Friedrichs-Lewy paper on finite difference equations, they found there the explanation of their difficulties in an observation made by Lewy. Seizing upon the fact, which he had learned in

Courant's course, that disturbances of the solutions of hyperbolic equations, produced by initial disturbances, always travel with a definite finite speed, Lewy had recognized that the solution of such an equation cannot be approximated by that of the finite difference equation if the propagation speed for the latter equation is less than that for the differential equation. He accordingly had drawn the conclusion that severe restrictions must be placed on the choice of the mesh-width of the lattice in designing a scheme for obtaining the approximate solution of the hyperbolic partial differential equation by means of the finite difference equation.

"Once it is said, it is clear to anybody," Friedrichs said of the point made by Lewy, "but Lewy was the first to say it!"

At NYU the computation required for AMP contracts was organized into something of an operation by Eugene Isaacson, a former student who had been working for the Bureau of Standards since he took his master's degree. There, as part of a group which computed tables for military research, he had begun for the first time to see mathematics as a possible career. He had thought vaguely of going back to school after the war. A friend from the group at NYU had encouraged him to go and see Courant about a job.

"I remember almost too clearly," Isaacson told me with a rueful shake of his head, "I had to come back and see Courant on about five or ten different occasions, because he would see me and mumble something and it was never quite clear to me whether I had a job or not. He would always be fumbling with some papers while he talked to me. He had the property of never saying 'yes' or even 'maybe yes' unless he was absolutely sure that he could fulfill the commitment. So it was only after he had seen me about ten times that he finally told me that I would have a job here."

Isaacson's job was to set up a system for numerical calculation similar to the one under which he had worked at the Bureau of Standards. Today, in the age of electronic computing machines, it sounds trivial; however, it permitted a few experienced people with hand-operated calculators to supervise the work of a large number of less experienced people, who performed the bulk of the calculations with paper and pencil according to step-by-step mimeographed instructions which contained built-in checks of their work. To Isaacson's surprise Nina Courant volunteered her services to his group. He found the daughter of Carl Runge "extremely intelligent in how she looked at the subject" and very interested in learning some of the theory behind what she was doing.

While Isaacson was running the computing department, "learning a bit from the scientists involved about where the problems originated," he began to attend "classes." These were seminars or lectures given by some members of the group for other members. Everybody was a student, and often a teacher as well.

For Courant this maintenance of training during the war was an important principle.

"I *fought* for that," he told me proudly. "I opposed Mr. Bush, who was the head of the OSRD. He said I should not waste time and energy teaching when I could be doing research work for these war purposes, but I opposed him. I said it was self-defeating. So anyhow I organized it so that the people we hired for research work here should also be members of seminars, give courses and seminars themselves. I think that was really one of the most constructive things I did, that I resisted the idea that the Applied Mathematics Panel should only solve problems. I thought the purpose of the university, the teaching, the seminars should not be reduced. On the contrary, maintaining them would be very important for the future."

Toward the beginning of 1944, Courant became acquainted with E. S. Roberts, the chief engineer of the Chemical Construction Company, a subsidiary of American Cyanamid. Roberts, who was interested in rockets and intrigued by newspaper reports of "secret weapons" in Germany, played an active role in stimulating American research on jet propulsion during the war. He suggested to the NYU group that they investigate the flow through the nozzle of a particular kind of rocket. The result was a series of studies of flow problems in rockets and related structures and the incidental development of a computational procedure which turned out later to be very useful.

Courant was tremendously enthusiastic about Roberts, "a very dynamic, superbly intelligent and well-informed and inventive man, . . . one of those engineers that are not interested in money but in achievement and social progress." Friedrichs remembers how he and Courant and Roberts sat late in Courant's living room in New Rochelle plotting the future of the NYU group and of science in America. They were especially concerned about the continuation of government support of science after the end of the war.

"What we were really talking about," Friedrichs says now, "was something like a national science foundation."

The war work at "Courant's shop" (as Mina Rees called it) was carried on in a style somewhat different from that at other AMP contract institutions; Courant—even during the war—insisted on maintaining the basic character of the research program there. This was in opposition to the general attitude in Washington, Mina Rees told me.

"We—the Americans—were pretty much committed, in contrast to the English, to the notion that we would work on problems of immediate use rather than on basic research; and almost all the applied mathematics contracts were let out on that basis. Courant was impatient with the idea and insisted that he had to let his people and himself work on basic materials. So that Friedrichs's work, for example, was heavily on the development of the whole shock wave and fluid mechanics theory.

"But," she pointed out, "Friedrichs was indispensable in using the insights he got out of this basic stuff when people had practical questions. I remember one time when he and I went out to Caltech because they were having trouble—they had a rocket contract out there and there was something the matter—the rockets didn't take off properly. Because of his understanding of the flow mechanism, Friedrichs was invaluable in making suggestions. . . . Yes," she said in answer to my question, "they did finally get the rockets to work as a result of his help."

"Well, it wasn't quite like that," Friedrichs smiled. "I didn't ever actually solve their problems for them. Maybe I acted as a kind of catalyzer. Since I did not know too much about the engineering aspects, I had to ask them quite a number of questions. Naturally my way of formulating questions as a mathematician was quite different from what they were used to. This forced them to look at their problems in a different, more fundamental way. That probably helped them. In the end, of course, they solved their problems themselves."

Friedrichs's remarks illuminate the way in which for the most part the applied mathematics group at NYU provided assistance for the military and for "real applied mathematicians"—i.e., engineers. Although the group dealt with such practical-sounding subjects as explosions, detection of underwater sound, flow of air through jet nozzles, design of ramjets, supersonic flow and shock waves, they usually treated these subjects from a rather theoretical point of view.

There were cases in which this approach was directly and practically very effective. But for the most part the people with the problems read the NYU reports on the mathematical theory involved—often in the so-called "ideal" or "generalized" case, which was not at all the same as the specific case they were dealing with—and as a result they were able to recognize the mathematical situation which concerned them and—as Friedrichs said —to solve their own problems.

By the beginning of 1944, it was clear that an Allied victory was inevitable, but on both sides of the world the war went on.
"Unfortunately, I cannot be so optimistic as to count on a foreseeable end...," Courant wrote in January 1944 to Flanders at Los Alamos.

Six months later, on June 6, 1944, the Allied Forces landed on the beaches of Normandy. By the middle of August the west flank of the German army had collapsed.
Courant wrote again to Flanders:
"The time has come for thinking of post-war plans."

TWENTY-FOUR

ON MARCH 23, 1945, the United States Ninth Army and the British Second Army crossed the Rhine, one half of a pincer operation planned to trap the industrial heart of Germany. On April 8, as they swept eastward, Göttingen fell in the path of the Americans. "There was one unpleasant half hour during which we were attacked by artillery," Franz Rellich, who was living with Herglotz at the time, wrote to Courant. In Herglotz's garden a shell exploded, broke a window, and deposited an apple tree on the balcony of the second floor. "Ten minutes later American soldiers were in our house and asking for water. The Thousand-Year-Reich was over!"

Four days later, in the United States, Roosevelt died. He had been president in 1933 when the first refugees from Hitler had begun to arrive, and they had never known another president.

Fighting still went on in the Pacific, but everybody at top policy levels in Washington agreed that the Applied Mathematics Panel should be phased out as soon as the war was over. It was intended, however, that the practice of government encouragement of scientific research by contract would continue under the newly organized Research Board for National Security (RBNS). Courant continued to push the ideas he had advocated while a member of the Applied Mathematics Panel. The new group should "continue and expand work of *a basic character* . . . the wisest policy," he maintained, "would seem to be the most generous." It should also tackle the problem of training of personnel—otherwise "no financial support, no panel, and no organization [will be able to] achieve the desirable results." Utmost flexibility should be one of the guiding principles. "No one has sufficient wisdom to determine . . . what fields ought to be attacked by scientists."

In the year following the European victory, the terrible story of what had happened to relatives, friends, and former students in Germany gradually began to be pieced together from letters and messages.

Of Courant's aunts and uncles, six had been alive when the war began. Three of these had emigrated, one had managed to live in Berlin throughout the war, and two had died in Theresienstadt, the "model concentration camp" for the elderly and the well known. Nineteen of Courant's cousins had left Germany to settle on five different continents. Four had survived with their parents in Berlin. Two had committed suicide. Five had died in the gas chambers.

Among the established mathematicians in Germany, there had been several suicides but almost no deaths directly at the hands of the Nazis. Nearly everyone who had been in danger had been got out, in some way, before the war began.

Courant was relieved to hear that Rellich, who had been a professor in Dresden at the time of the fire-bombing of that city, was safe.

"Dear Courant," Rellich had written ten days after the American occupation of Göttingen. "Since in the later part of this letter I will talk about myself all the time, I will at the beginning at least express the hope that you and your family are well and that the same is true of the larger family which consists of Friedrichs and Lewy and Neugebauer and Busemann and Feller, etc., etc.

"The following happened to me during the war:

"Since the summer of 1939 I have been in Dresden . . . and the great number of students there made it possible that I was not inducted into the army and also not completely submerged in 'war mathematics' as were most of the colleagues of my age. Since '41 I have had a proper apartment with four rooms, an amazing achievement for a bachelor; but this Dresden idyll came to an end with the air raid on the 13th and 14th of February 1945. Since I was staying with a colleague that night, everything of mine was destroyed. . . . I have not a single one of my papers, and actually I no longer know who I am.

"Anyway I myself did not burn, and this alone is (for me) quite pleasant. More pleasant things are to come. I received an order to continue my 'war-important' mathematical researches in Göttingen, and so here I sailed into the Fourth Reich. . . .

"There have been many moves to Göttingen, the trend from east to west is quite apparent . . . and these have resulted in an abundance of mathematicians here—which almost reminds me of times past, at least as far as quantity is concerned. The Mathematics Institute, however, sleeps like Sleeping Beauty and obviously awaits the kiss of the Prince. It would be wonderful if you felt like playing the role of the Prince."

There were many details in the letter of colleagues, deaths, illnesses, "existence."

"Of myself," Rellich concluded, "I can say very little at the moment except *sum*. In many ways this condition of being unburdened is comparable to that of *anno* 1926 when I first arrived in Göttingen.

"Only this time I miss you very much."

Courant would have liked to go to Göttingen immediately to see what he could do to help Rellich, other old friends, and members of Nina's family; but the transition from war to peace in the United States, the closing down of the Applied Mathematics Panel, the reorganization of the government's relation to science, the new possibilities in obtaining personnel—all these required his presence. Yet the problem of the rehabilitation of Germany occupied a great deal of his thought.

He got a letter on this subject from Harald Bohr in July 1945. It was his first direct communication from Bohr since 1943 when the two Bohrs— Jewish on their mother's side—had found it necessary to flee Nazi-occupied Denmark for neutral Sweden. Since Niels Bohr, under the name of Mr. Baker, had been almost immediately spirited away to the United States to assist in what was then referred to as the Manhattan Project, Harald Bohr had felt it "a kind of duty" to remain in Sweden even though arrangements had been made by Courant and others through the Rockefeller Foundation for him also to leave.

"During the time in Sweden," Harald Bohr wrote to Courant in English, "I often thought of writing a real letter to you, and in a way I cannot quite understand why I did not do it, but everything was so abnormal and absurd that in fact I did not do many things which in themselves would seem most natural and obvious."

Bohr too was concerned with the problem of Germany.

"To put it briefly, I should say that in my opinion it is both necessary and just to try to distinguish strongly between the notions of 'German' and 'Nazi.' Or to express it another way: The belief very common, for instance, in this country, that Nazism is something deeply and fundamentally fitted for human beings of German origin, seems to me very dangerous in two respects. On the one side it gives a quite absurd credit to the cruel and inhuman Nazi-movement to consider it as something 'natural' for a rather great proportion of mankind, and on the other side it gives a base for considering a possible reduction [sic] of the German people as of beforehand hopeless."

Although Courant agreed with Bohr in principle, he found himself repelled by the tone of almost all the letters which began to come to him from Germany. They were querulous and critical of the Allies. They conveyed no sense of responsibility for what had happened to the Jews or for the holocaust that had been brought upon the world by Hitler. Letters like Rellich's—uncomplaining and full of hopes and plans for the future—were rare. Courant wrote to his former student, "I cannot tell you how much I admire your attitude!"

As soon as the war ended in Europe, young scientists who had not yet finished their education began to think of returning to school. In summer 1945 Louis Nirenberg, who was working in a physics laboratory in Montreal with Courant's daughter-in-law Sara, asked her to get him some advice from her famous father-in-law, "just a suggestion as to where to go to study physics in the United States." She returned with an offer of an assistantship in mathematics at NYU. It was Courant's idea that the young Canadian spend a year or so learning mathematics, get his master's degree in that subject, and then go on and get his doctor's degree in theoretical physics.

Although Nirenberg had never heard of NYU, he accepted Courant's offer. Today he is a very pure mathematician.

"I haven't done any work at all in connection with physics," he told me. "I think people who turn from physics to mathematics are drawn to it because of its more abstract nature, and so they tend to go all the way."

Like most of his German-born colleagues, Courant differed from American mathematicians of his age in that he had personally experienced combat in the First World War. This fact enabled him to empathize with the men still fighting in the Pacific but also made him determined to save gifted young scientists from their fate. In 1945, when the Hungarian-born Peter Lax was drafted, Courant hastened to pull a few wires; and Lax very shortly received orders to report at Los Alamos.

For Lax, the Manhattan Project was a fascinating and formative experience. Later, after he got his Ph.D. at NYU in 1949, he went back to Los Alamos for a year and then, after that, for several summers as a consultant.

"That's really how I got into applied mathematics," Lax told me. "It's true I picked up a lot at NYU, but it made a crucial difference to be associated with a project that had some very definite technical goals. On the basis of that experience, I would say that's the way to learn applied mathematics. It's not just an armchair business."

Courant and many people in his group suspected, although none of them knew for sure, what was going on at Los Alamos. Stoker still remembers how, in summer 1945, running into Robert Oppenheimer on a train going west and being somewhat annoyed by Oppenheimer's condescending attitude toward the novel he was reading, he snapped: "You know, what you're trying to do out there, you are never going to be able to do it!"

Shortly afterwards, the first atomic bomb was dropped on Hiroshima. On August 14, 1945, the Japanese accepted the terms of the Potsdam Proclamation—the total surrender of all military forces.

The war was over.

On September 17, 1945, Göttingen reopened its doors, the first German university to do so. A second letter to Courant from Rellich, dated September 26, described the event.

Allied lawyers, busy interrogating the population, were using most of the rooms in the Mathematics Institute. A few days before, they had been instrumental in having Hasse dismissed from the faculty.

"This did not take place, as it seems, on the strength of the questionnaire of the military government but rather following some conversations Hasse had with the Americans. Herglotz and Kaluza sent a letter to the rector protesting the dismissal and pointing out the great mathematical achievements of Hasse, but without success."

When I talked to Hasse some thirty years later, he told me that he himself did not know whether he was dismissed on the basis of remarks he had made to two Americans who had come to his home to obtain a manuscript concerning his wartime investigations of ballistic problems—or on the basis of some things he had said in the first faculty meeting:

"My political feelings have never been National-Socialistic but rather 'national' in the sense of the Deutschnationale Partei, which succeeded the Conservative Party of the Second Empire (under Wilhelm II). I had strong feelings for Germany as it was created by Bismarck in 1871. When this was heavily damaged by the Treaty of Versailles in 1919, I resented that very much. I approved with all my heart and soul Hitler's endeavors to remove the injustices done to Germany in that treaty. It was from this truly national standpoint that I reacted when the Faculty more or less suggested that such a view was not permissible in one of its members. It was also the background for my remarks to the Americans. They were talking about reeducating Germany, and I said some strong things against this. It irked me that everything against Hitler was desirable, and everything that he had done was wrong. I continued to be a national German, and I resented Germany being trampled under the feet of foreign nations."

That first post-war September Göttingen was very lively. Rellich, lecturing in the place of Siegel, had more than three hundred students attending his calculus lectures. But the return of Siegel would be decisive for Göttingen, Rellich told Courant. "More important, it would give some hope again to all mathematicians in Germany, a hope that not everything is lost here forever."

The happenings he was reporting in his letter were, he wrote, "just the first flowers on a grave."

"Unfortunately it is not only thoughts of the past that are depressing. It is terrible to think of the fate of my friends in Dresden and Leipzig, many of whom will have starved to death in another six months."

At NYU, that same September, Courant faced a new problem in relation to the group he had built up during the war. There was a general movement to continue government support of science in the universities and a feeling on the side of the universities, which had in the past been suspicious of such aid, that some sort of support from the federal government was going to be necessary to keep American institutions on an equal footing with the state-supported institutions of Europe and Russia. It seemed certain that a bill establishing a national science foundation would be enacted and would make sizable sums available for scientific research. At the same time, closely connected with this development, efforts were being made to secure the continuation of some specifically military research with the gap filled by the Navy, which needed science more than the Army. At the moment, and in the foreseeable future, it was going to be easy to obtain money under contracts from industry and from the military; but such a situation was not going to last forever—and in the long run, in Courant's opinion, it was not desirable.

The numerous memorandums which he had composed in 1940 and 1941 for "a national institute"—"a basic science institute"—"an emergency institute"—lay in his files. He was still convinced that America had need for a high level scientific facility combining research and teaching in some central location—or even perhaps, as an alternative, a group of smaller institutes in different parts of the country.

Now, at the end of war, he thought he saw a different way to approach this goal, one which would solve the problem on a personal as well as on a national level. The type of regional center he had in mind could be put into operation immediately at New York University—"a natural development," as he put it, "of consistent though unobtrusive efforts" which had been being made for a decade.

Asking Warren Weaver for names of people to approach for the necessary outside financial support, he wrote a little wearily: "After many disappointments it may seem foolish for me not to give up and rather concentrate on my own research work. However, experience during the war and observation of the present trends have further strengthened my feeling of responsibility to go on with my efforts even if they have only a very limited objective."

In spite of Courant's fears, he did not lose any of his key personnel during the year 1945–46. Support for the group was beginning to come increasingly from Navy contracts.

"Our work now is very interesting and we need not compromise our scientific conscience if we go along the same lines," Courant wrote to Flanders. "Nevertheless, we are all gradually trying to take up mathematics for its own sake."

Ultimately, the Navy's interest in supporting research led to the establishment of the Office of Naval Research (ONR). Courant and his people were drawn into the new group by their friend Raymond Seeger.

Stoker told me that he has always felt that Courant deserved a great deal of credit for the post-war intervention of the government in research, particularly that of the defense departments.

"That started with the ONR; and we had, I believe, the first such contract in mathematics. The way that was formulated set the pattern for all the others afterwards."

Since the war Courant had heard very little about what had happened to his old friend and publisher, Ferdinand Springer. Then, on a train in November 1945, he picked up a copy of the *New Yorker* and found in the middle of the "Report from Berlin" the following sentences:

"Not long ago, I had the luck to talk with a German who, just after the fall of Berlin, was permitted by the Russians to look not only at a corpse that some Germans think may possibly be Hitler's but also at one that was undeniably [that of Goebbels]." The German to whom the *New Yorker* correspondent had spoken turned out to have been Ferdinand Springer.

The report went on to say that in late February 1945, having been evicted from his firm by the Nazis, Springer and his family were living with friends in Pomerania. When the Red Army arrived on its way to Berlin, he was taken into custody by the Russian secret service, the NKVD, and asked to list all journals and books he had published. He was hardly into the onerous task when the major questioning him suddenly demanded to know if he had published a genetics journal. Springer replied, "Ja, *Der Züchter*." The major, a professor of animal husbandry in a Siberian university, beamed: "*I* was one of your contributors!" It was, as the *New Yorker* correspondent commented, an example of the best in publisher-writer relations. From that moment on, Springer was treated as a friend. He remained with the major and the NKVD, which followed directly behind the front, and entered Berlin under the fire of German machine guns.

The first letter which Springer sent abroad went to Courant on April 11, 1946. By then the 65-year-old publisher was able to report a certain amount of progress in "rebuilding." He planned to decentralize his firm, not only on account of the four zones into which Germany had been divided by the Allies, but also because of his expectation that in post-war Germany culture itself would be decentralized. His first publishing ventures would be made in Heidelberg, in the American zone, and Göttingen, by now under the British. Mathematics had first place in his publishing plans.

Although Courant could not say how much help Springer could expect from other former authors now in the United States, he wrote that he himself would be happy to give him the publication rights to *What Is Mathematics?*

In his own operation at NYU during this period, Courant was making a very great effort "to stabilize" by obtaining academic appointments for the people who were working under the contract with the Navy. He was also eager that Flanders, still at Los Alamos, should return to the faculty. The secret wartime development of the electronic computer had been made public at the beginning of 1946, and Courant hoped that Flanders would take over a more advanced computing program at NYU.

"All the questions related to the big new fashion of machines, etc., seem now to be much more interesting and fascinating from a mathematical angle than I originally thought," he wrote to Flanders.

But Flanders was hesitant about returning.

"I hate to refer to my inferiority complex again, but I think that is the real basis of my interest in staying here. At the university I am constantly oppressed by the feeling that I have gotten myself into a situation with which I am unable to cope, and I count the years until I can retire, hoping that at least I can get my children educated before I or the situation crack."

Courant hastened to respond.

"Of course, we are all looking forward to the day when you will come back . . . , [but] any ideas of leaving us in the lurch . . . are out of place. . . . Our working philosophy has been, and is becoming more strongly so, that there should not be job assignments to be attended to by whomever one can find to fill the demand but rather that the existing human beings and teams are the primary element and projects and assignments should be adjusted to them."

After a visit from Courant in Albuquerque and a long weekend by himself in Berkeley, Flanders finally wrote that he had made up his mind to return to New York.

Chancellor Chase, under Courant's prodding, had earlier appointed a committee to evaluate Courant's idea of establishing an institute for advanced training in mathematics and mechanics at New York University. The chairman—at Courant's suggestion—had been E. S. Roberts, whose wide interests in science and education continued to impress Courant and seemed to him to more than make up for the fact that Roberts lacked the contacts of someone like Rear Admiral Lewis L. Strauss, whom he also considered as a possibility for the chairmanship.

By the end of the academic year 1945–46, the evaluating committee had submitted its report. Although recognizing much in the NYU situation that would not be favorable for the kind of advanced scientific institute Courant proposed, the committee recommended that the plan be put immediately into operation:

"The program envisages research integrated to a much higher degree with teaching than is ordinarily done in American graduate schools. In this respect, as well as in emphasis on the connection between mathematics and applications, and in emphasis on teamwork, [it] differs distinctly from the average graduate school program. It is hoped that the development of the New York University Group will stimulate similar developments in other graduate schools."

In spite of this encouraging conclusion, the university took no immediate steps toward establishing the proposed institute.

It is impossible to convey in a linear narrative the multi-dimensional nature of Courant's activity in the years following the Second World War, and it is not actually relevant to this narrative. Still, no one who lived through that period with the group at NYU can separate even the larger events of the time from Courant. He seemed to them to be involved in everything that related in any way to the scientific community. In July 1946, when the first peace-time test of an atomic bomb was carried out near Bikini, Courant was among the scientific observers.

Later that same summer he went to England on a scientific mission connected with the Office of Naval Research. There was a possibility that he might get to Germany, but the press of duty and the lack of time prevented his going. When he returned to New York, he found that what he had feared had begun to happen. His group was breaking up. While he had been in England, De Prima had accepted a position at Caltech.

At the same time Courant lost De Prima, however, he gained Fritz John as a regular member of the NYU faculty. It was the first real addition he had been able to make to his enterprise since the hiring of Friedrichs and Stoker almost ten years earlier.

As soon as their wartime manual on shock waves was declassified, Courant and Friedrichs began to prepare an enlarged version for publication. In the fall of 1946, Courant hired Cathleen Morawetz, the daughter of the applied mathematician J. L. Synge, and put her in charge of collating sections that had been farmed out to various young people and of checking Courant's and Friedrichs's English.

She told me that at the time she—like Courant's daughter Gertrude—had recently married. Her father and Courant, running into each other at a mathematical meeting, had bemoaned the fact that now their daughters would not go on in their chosen fields of mathematics and biology, respectively.

"Ja ja, well, you can't do anything about my daughter," Courant had sighed, "but maybe I can do something about yours. You should send her to see me sometime."

Cathleen Morawetz is now a professor of applied mathematics at the

Courant Institute and a trustee of Princeton University as well as the mother of four grown children. When she came to work at NYU, she had no intention of continuing her studies—"After all, I was married!"—but Friedrichs was lecturing on topology that year and "everybody" was going to hear him and so she went too.

Courant's work on the shock-wave manual came to her in his scrawling, almost illegible handwriting or, sometimes, in typescript. She told me that he was the only person that she ever knew whose typing was characteristic. Letters were struck one on top of another, words were x'd out—typed in— and x'd out again—perhaps with a question mark. The question mark would be very light, because the key had been hit indecisively. Several words—all synonyms—would be typed in order, blanks would be left, or sometimes a word typed in but then enclosed in parentheses to indicate the tentative nature of its choice. In contrast, although Friedrichs wrote and rewrote constantly, he always gave her a very neatly copied manuscript. If anything was crossed out, it was thoroughly inked over; otherwise, Friedrichs knew, Courant would read what had been crossed out and not what had been left.

I asked her how such different people had been able to collaborate at all.

"The way it was," she recalled, "one or the other of them would take a section from the manual and rewrite it. If Courant did it, then it went to Friedrichs. And Friedrichs would look at it and grumble that it wasn't sufficiently exact. He would rewrite it, and it would become all *if's* and *but's*. Then Courant would take it, and he would mumble and groan that it was much too complicated. Then he would rewrite it. Then Friedrichs would take it back and say it wasn't precise enough. The process went on many, many times for each section. When it came back to Friedrichs, he would put in again some of what he had had before, but not so much. Then the next time Courant wouldn't take out so much. They were both pretty determined about what the end product should be, and they were both quite willing to do an awful lot of work. So they were really great at cooperating, but that is the way the cooperation took place. I never remember a single session where they both sat down together over the manuscript."

By June 1947 a year had passed since the evaluating committee's recommendation that an institute of mathematics and mechanics be established at NYU. With the exception of Fritz John, none of the younger people had been given academic appointments of any significance. All over the country the need for personnel was urgent. What Courant referred to as "attacks from the outside"—i.e., offers from other universities—were becoming more frequent and more difficult to ward off. Money was no problem. Government support of research seemed assured for years to come; but young men like Max Shiffman and Bernard Friedman, approaching or in their thirties, wanted academic appointments. The university administration hesitated to commit itself by hiring them. What if government support for science were to be suddenly withdrawn?

Stoker, especially, urged Courant to push the matter of proper support with the university, which was profiting greatly from the group's government work.

"Courant would go out and get money from the outside," Stoker told me, "but he would never ask the university for anything, unless he was pretty sure he was going to get it. Up until the time that I succeeded him as director, almost all of us—including Courant and Friedrichs and me—were paid, say, not more than one quarter of our salaries by the university. The rest of it came from many, many contracts. The university itself was absolutely no help to us in getting these or in carrying them out. It was a bitter fight all the time to get space or any of the things we needed. I found that ridiculous and used to urge Courant—'That's no way to do—it's out of line with what's done everywhere else—you don't improve your bargaining position by giving way to people all the time.' But Courant always preferred to take the most roundabout way instead of attacking things frontally. I used to tell him, 'If there are ten ways to do a thing, you will choose the one which is the least direct.'"

I wondered aloud why that was.

"Oh, that's just Courant, and just the opposite from me. In a way he liked that kind of intrigue, that was fun for him, he enjoyed it. And he always disliked making decisions. He would postpone decisions as long as he could and would even then always, if possible, fix things so that he would have a way to retreat. Even when he was quite sure he knew what he wanted. But his effects were very positive. That you have to say. Oh yes, he knew how to do things!"

In June 1947 the university did finally provide larger quarters for Courant's group, renting the entire second floor of the Bible House near the Astor Place subway stop. The first floor housed the Bible Society; and the third floor, the staff and equipment of the *New Masses*. Again, a new address was accompanied by a new name. Courant liked the concept of *an institute*—according to the dictionary definition, "a unit within a university organized for advanced instruction and research in a relatively narrow field"—and he always chose to use the word *institute* instead of *department* because, as he told me, "it permitted us to keep the options open for the future."

Before he left for Europe that summer—for his first visit to Göttingen in thirteen years—he again ordered a new letterhead:

NEW YORK UNIVERSITY
Institute for Mathematics and Mechanics

TWENTY-FIVE

COURANT'S trip to Germany in the summer of 1947 was financed by the Office of Naval Research. His assignment was to visit a number of universities and technical schools and to determine the extent of German progress during the war in the development of computing machines. In his new importance he was able to convince the ONR that he needed someone to assist him. It was arranged that Natascha Artin would accompany him on the trip.

The year before, when Artin had received an appointment at Princeton, Natascha had sought Courant's advice about finding a job in the east. He had responded by creating a position for her at Washington Square. She would sift through the mathematical reports that were constantly being received from the government and the military, and see that they were directed to those members of the staff to whom they would be of interest. Although she had no degree in mathematics, she had studied with a galaxy of famous mathematicians and physicists at Hamburg. These included—in addition to Artin—Blaschke, Hecke, Jordan, Pauli and Schreier. But she had married before getting her degree and had felt she should not go back to school to her husband's colleagues.

On June 14, 1947, she and Courant left New York for London, their journey sped along by Donald Flanders's good friend in the State Department, Alger Hiss. After a few days in England, where they found Hardy very ill but optimistic about the future of science in Germany, they flew to Frankfurt.

On June 20 they stepped out of their American plane onto German soil. In the journals which they kept of the journey, they both described how their eyes searched for the landmarks they had known. There were almost none. But the physical devastation of the city was much less affecting than the demoralization of the population. They were shaken by the sight of great crowds of ragged, hungry Germans, many of them begging.

During the next six weeks Courant and Natascha traveled extensively over the American and British zones of Germany, usually with a jeep and a driver, going as far north as Hamburg and as far south as Munich and, in Natascha's case, Vienna. They talked to innumerable people—former colleagues, former students, former friends in the non-academic world, relatives, military personnel, industrialists, young men and women, children, drivers, clerks, "a sad and complaining usherwoman" at a performance of "Theseus" in Göttingen.

Natascha was always to be amazed at Courant's ability to converse with people.

"He has a very good insight into the person he talks to," she told me. "He knows immediately what to talk about, and people become interested. It's certainly not his looks nor his superficial behavior. But there *is* something about him. I would still put a big question mark about what it is. I really don't know. There is no doubt that it exists. Because I've seen it!"

Warren Weaver had asked them to look for German scientists between the ages of twenty and twenty-five who might benefit from spending time in the United States. At the university in Darmstadt, the buildings of which had been almost completely destroyed, they found more than 2,000 students devoting half a day every two weeks to the construction of new buildings—a condition of their enrollment. However, the rector of the university—the physicist Vieweg, who in 1943 had publicly refused to join the Nazi Party—told them of great difficulties with the young people because of poor preparation and almost complete lack of ethical standards. For some, it seemed that the only thing the Nazis had done wrong was to lose the war. They began to feel that Warren Weaver would have to raise his age limits.

In Heidelberg they visited Ferdinand Springer. Courant noted "slight lack of resonance" at first. But he found Springer as optimistic and energetic as ever—forty-two journals were "appearing," Courant's calculus was being reissued. In Heidelberg they also visited the philosopher Karl Jaspers, who told them of the great need for contemporary books.

In Marburg they climbed the streets of the old town to see Kurt Reidemeister. Of all the people they talked to, he seemed physically the weakest; but he was full of plans for reconstruction.

"Reidemeister and his group are exponents of what we consider the proper attitude," Natascha wrote in her journal. "He is afraid of people leaving Germany permanently, but very much in favor of half-year or year fellowships. . . . [But he] thinks that there are no young people from 20-25 who would fall into the class of people [Warren Weaver asked us to look for] . . . . [These] are only now beginning to study because of the war. He thinks that people from 30-35 should be considered. Here the difficulty is that most of them are *belastet* [a German word meaning "subjected to a burden"—thus, carrying the stigma of having been National Socialist

Party members]—but he is sure that many of the *belastet* have not mentally been Nazis. [They] had to join the Party in order to keep their positions.... [Still] he said that 90 per cent of the population, including academic people, are dangerously but not hopelessly nationalistic. The natural science people in general much less, however."

From Marburg the visitors returned to Frankfurt and then flew to Munich. There and in the rest of the south they found much less enthusiasm for reconstruction than in the middle of Germany. Courant summarized many different conversations in his journal:

"Fear of Russians. Bitterness against French. Rumors also of American mismanagement. General lack of understanding for what America actually does to help the Germans. Little contact between scientists in different towns. None with abroad, almost none with Austria.... [Criticism] of German administration. Small-time politicians, no understanding for cultural issues. University has no support from them. Complaints also about zone competition. French do not permit some scientists to travel to other zones. Americans and British likewise compete for scientists and, allegedly, impose restrictions.... [Many scientists] do not dare travel through Russian zone for fear of kidnapping, which sounds unbelievable but is universally accepted as real danger."

After Munich the two visitors separated. Natascha went on to Vienna; and Courant, with some misgivings, proceeded by plane and car to Göttingen. Arriving there on the evening of July 3, 1947, he noted only, "Arrived, slightly nostalgic, in Göttingen about 7:30."

Physically the town was almost undamaged—it had been specifically exempted from Allied air raids, according to von Kármán. But people pressed upon him from all sides. There seemed to be thousands enrolled at the university. Lecture halls were often so crowded that students had to push their way in.

Courant found association with some of his old friends and colleagues very depressing: "Absolutely bitter, negative, accusing, discouraged, aggressive. Main point: Allies have substituted Stalin for Hitler, worse for bad. Russia looms as the inevitable danger."

In contrast was a pleasant evening spent with Heisenberg. He reported, "Heisenberg very superior, quiet, not complaining, and basically positive. Quite active scientifically, which of course is basis for psychological equilibrium." But another day, discussing politics, he found that Heisenberg "came out finally with the same stories and aggressiveness against Allied 'policy of starvation' [this referred to the dismantling of

German factories] as the less cool and more emotional people." Courant concluded that the physicist was "still in need of education, which I hope will be provided [when he visits] Niels Bohr in August."

The aerodynamics institute had become "a veritable fortress." Although ill and depressed, Prandtl was mentally active. He had given much thought to analog computing machines with a view to meteorological computations. The dimensions of the machine he was constructing had been determined by the size of ball bearings found by chance among war surplus.

By the time of Courant's visit, Rellich had become director of the Mathematics Institute. I got a picture of how Rellich operated in this position from Jürgen Moser, now a professor at the Courant Institute but at that time a 19 year old student recently arrived in Göttingen from the Russian zone.

After having crossed the east-west border under gunfire, Moser had presented a letter of recommendation to Rellich from his gymnasium teacher. Rellich had been immediately interested and friendly. He helped Moser find a place to live in the overcrowded town—"One could not register as a student without showing that he had a place to live, and he couldn't get a place to live unless he was a student." Afterwards, whenever he saw the youth, he greeted him in the Austrian fashion, "Grüss Gott, Herr Moser!"

"I was always amazed," Moser told me. "One is a student, one is rather shy, and to be addressed personally was somehow so—" He could not find the appropriate English word. "And then there was the matter of food. We students were literally starving, and Rellich managed somehow to get CARE packages and distribute food to us."

"Do you feel that Rellich was in a way following the example set by Courant?" I asked.

"I think so, but that's of course hindsight, because I knew Rellich first. I was always just amazed by him," Moser repeated. "I can illustrate with another, more scientifically interesting example:

"Most of the students who came to Göttingen then were of course much older than I. One of these was a man named Erhard Heinz—he is now a professor in Göttingen. He had been a student of Rellich's in Dresden, had been drafted, then captured and sent to England as a prisoner of war. Of course prisoners of war remained prisoners beyond the war, but all the time Rellich kept up a correspondence with Heinz and provided him with mathematical problems—otherwise one gets dull right away, you know. Later Heinz came to Göttingen and caught up very quickly. I remember him running around in these horrible clothes from the prison camp and living somewhere by himself, very lonely. Then, at Easter, Rellich sent him three colored and decorated Easter Eggs, each of them with a mathematical problem, an equation, on it. It was a charming thing to do, very typical of Rellich. I cannot imagine Heinz having developed the way he did without Rellich."

At Rellich's invitation Courant gave a lecture and soap-film demonstration on one occasion and a lecture on partial differential equations on another. Afterwards there was tea at the Mathematics Institute and a walk up the Hainberg for dinner, as in the old days.

Since the end of the war, Courant had tried persistently to find out from trusted friends like Rellich how various German mathematicians had behaved in relation to the Nazis. In Göttingen also, everywhere he went, even with Nina's family, the unspoken question lay below the surface of the conversation.

"I found very few people in Germany with whom an immediate natural contact was possible," he wrote later to his friend Winthrop Bell, with whom he had debated the rebuilding of Germany after the First World War. "They all hide something before themselves and even more so from others." But he did not have the heart, he noted, to inquire in many cases about activities and motives during the war.

To his surprise he continued to hear favorable reports of many who were *belastet*. Once he even heard a strong defense of a man who had been the president of a People's Court but was sympathetically looked upon by people whom Courant himself approved. He asked in his journal: *"Where then are the bad Nazis?"*

He did not have occasion to see Bieberbach; but in Berlin he met Hasse, who was living in the American zone and lecturing at the University of Berlin in the Russian zone. Courant noted only: "Met Hasse. Mixed feelings."

In Berlin Courant also saw his brother-in-law, Wilhelm Runge, who in 1916 had helped him install devices for earth telegraphy on the Western Front.

In 1945, in the primitive conditions of the devastated capital after the war, there had been a desperate need for matches, which were no longer being produced. As soon as power lines were in operation, Runge and a few other men from Telefunken had begun to manufacture a little transformer which could be plugged into the line with a switch on the secondary and a coil made out of a paper clip. When switched on, the glowing paper clip was sufficient to start a fire, which could then be used to boil potatoes. With this beginning Runge and his colleagues had set out to rebuild Telefunken.

Courant thought his brother-in-law's optimism was "more temperamental than rationally founded." Still, he could not deny an urge to help people like Runge, Rellich, Springer, Reidemeister, and others with such "constructive energy."

Before leaving Germany for Copenhagen and the Bohrs, Courant found time to see such old Göttingen friends as Klein's daughter, Putti Staiger, who because of her loyalty to her many Jewish friends had been removed by the Nazis from her position as the director of a gymnasium for girls in Hildesheim.

He also learned the tragic fate of his cousin Edith Stein. She had become a Catholic in 1922 and had later taken the vows of the barefoot Carmelite order. In 1942, with her older sister Rose, a lay member of the order, she had been snatched by the Gestapo from a convent in Holland—part of a general retaliation against the Dutch church for its criticism of Nazi anti-semitism in Holland. Both sisters had died in the gas chambers of Auschwitz.

(Since the death of Edith Stein [Sister Teresa Benedicta of the Cross], there has been a continuing movement to have her declared a saint. Although Courant was deeply affected by her death, the incongruity of a Catholic saint among the descendants of Salomon Courant tickled him; and in later years he was always to refer proudly whenever possible to "my cousin, the saint.")

Back home in August 1947, Courant and Natascha sorted out their impressions of their journey. As they had expected, they had not discovered in Germany anything in the actual development of computing machines comparable with the results attained or likely to be attained before long in the United States. It nevertheless seemed to them that German skill and manpower might be utilized to help in the American development. On the human side, they wrote to Weaver at the Rockefeller Foundation:

"In spite of many objections and misgivings, we feel strongly that saving science in Germany from complete disintegration is a necessity first because of human obligations to the minority of unimpeachable German scientists who have kept faith with scientific and moral values. . . . It is equally necessary because the world cannot afford the scientific potential in German territory to be wasted."

"Today it seems difficult to imagine how important it was that [Courant] came back to the University of Göttingen as soon as the circumstances permitted," Claus Müller of Aachen said years later after Courant's death. "I still clearly remember the tremendous effects of his first visit."

In America, after his return, Courant continued to face the same problems he had faced during the first post-war years. His group still had no stability in relation to the university, but as a result of support by the Office of Naval Research there was a large degree of financial flexibility. In addition, to encourage the development of centers of applied mathematics in the country, the Rockefeller Foundation had made both NYU and Brown a generous yearly grant for a period of five years. Combining this money with contractual and university sources, Courant was able to find places for a number of gifted young people when they were brought to his attention.

One of these was Harold N. Shapiro, a student of Artin's at Princeton. Receiving his degree in 1947—a year when jobs for Ph.D.'s in mathematics were especially numerous—Shapiro had elected to remain at Princeton another year. Then, when he went to look for a job in 1948, jobs were suddenly very scarce. It is Shapiro's guess that Artin, who regularly gave courses at NYU, told Courant about him.

"Anyway, I got a phone call from Courant, whom incidentally I had never met. A little voice at the other end of the line said, 'This is Courant,' and then there was complete silence. I recognized the name, but I didn't say anything. Then after what seemed an endless silence, he said, 'I understand you are looking for employment. Do you think you would like to come to NYU?'"

Courant

For Shapiro—who later spent a number of years in business—the interesting thing about the way in which Courant hired him was that it represented the kind of spirit in which he did things.

"Courant gambled. He understood human beings and life. He felt that if he took somebody who didn't work out, he'd find some way to get rid of him—which usually meant getting him a job somewhere else. If he couldn't get him a job—it was sort of a joke but it was also true—he would simply move him to the office farthest back in the building—and the Bible House which we occupied then was a huge building—and hope that he would go away. Courant was 70 per cent from the outside world, 30 per cent from the academic world. He brought to the academic world this sense of timing, gambling, change and so on. And that's what made the institute."

Shapiro feels that he had an advantage in getting along with Courant during his first years at NYU because he had friends who had had unhappy experiences with him and he was thus able to avoid their mistakes.

It often happened, he told me, that Courant would become very close to a young mathematician and would shower the youth with attention, praise, support.

"He was a kind of public relations manager, too. He had this wide circle of friends in mathematics. It was thrilling, you know. I remember once I had published some paper, and he came to me and said that Siegel had been his guest over the weekend and he had explained to Siegel what I had done. Siegel knew? He cared? He had read my paper? It was fantastic!"

A sort of father-son relationship would develop between Courant and the favored young man, the "son" apparently unable to do any wrong. Then there would come a point when, instead of waiting for Courant's altruism—"which, by the way, you could always depend on"—the youth would presume on the relationship by asking for something, or by asking for too much. Such a request would always provoke a negative reaction on Courant's part. "He wanted to be benevolent, but if *you* asked him for something, he would freeze up."

The result, according to Shapiro, was not merely the rejection of the request but also, often, total rejection of the young man who had made it. In Courant's files I have several times seen letters which reveal such incidents—a young man writing bitterly, "Since [such and such a specific date] you have changed completely toward me"—and the helpless question, "*What have I done wrong?*"

1947-1949

There were also often young women students over whose studies Courant would fuss constantly. One of these was Anneli Leopold, who is now the wife of Peter Lax.

"When I was very young, I almost felt about the Courants as if they were another set of parents," she told me. "They would invite me to stay at their home, and always Courant was interested in what I was doing, my studies and so on, a very fatherly kind of interest, also very stimulating to me."

After the war, when he had more financial flexibility, Courant sometimes added to the group attractive young women who had no special job qualifications. Somehow, though, he always managed to find useful tasks they could perform. They were referred to as Courant's "flames"—the term which had been applied in Göttingen to Hilbert's many romantic attachments. The true "flames" were, like Natascha Artin, intelligent and attractive women who were to continue to consider Courant their very good friend long after his intense interest in them had passed. They were always drawn into his mathematics enterprise and always became an integral part of the group. Jerome Berkowitz, who came to the Bible House in 1949 and observed the "flames" over a period of more than twenty years, told me that it seemed to him that Courant utilized his women friends to extend his own grasp of life. Friedrichs, who admits he never really paid too much attention to such matters, sees in the "flames" yet another example of Courant's "ability to be fascinated"—"Courant was fascinated by very rich men, he was fascinated by gifted young mathematicians, and he was fascinated by attractive and intelligent young women!"

"Has anyone spoken about Courant and women?" Louis Nirenberg asked me at the end of our conversation. "If you're not there and you don't see it, you just think, here is this old man and he's enamoured with this young girl, isn't that silly; but when you see what it does to him, it's not silly. It transforms him. You know how nice it is to be in love. Well, it is just as nice for him at an old age as it was thirty or forty years before. It is a constant thing. There is always somebody filling this role for him to a different degree—the degree varies from person to person. I don't find it silly. And I don't laugh at it. He just seems more alive at that time and more giving, not just to that person. It's touching and beautiful. Well, you have to see his face."

Discussing the many women friends of her husband, Nina said to me quite simply, "He needed them." For the most part—perhaps with the example of Mrs. Hilbert before her—she brushed them away as she did the moroseness and rudeness which he often showed at home. After his death she made it possible for the two young women who had been important to him in his last years to fly to the memorial service held in his honor at NYU.

A young woman who came to the Bible House at approximately the same time as Shapiro came was Lucile Gardner, now Wolff. She is the younger sister of Clifford Gardner, who was then a member of the group. As soon as Courant learned that Gardner had a sister who was a gifted violinist, he insisted that the young man bring her out to New Rochelle.

"So we went out to New Rochelle, and there we had the family chamber music, which was pretty astonishing for anyone," she told me. "Mama always knew what note it was and where we should be and so on. Papa, on the other hand, played really beautifully but was all wrong half the time. Imagine the young student arriving and there is the family. One person understands everything—not the mathematician. The mathematician doesn't know what beat he's on, he is just—in the most romantic way—playing away, whereas the musician in the family knows all about the beats but wrecks everything by coming in and saying, 'Papa! B!' Then he would come out of his fog and go, 'Ah!'—sort of a gasp, very humble. He was always wrong. He did not play at all intellectually. He simply had *no* idea!"

After meeting Lucile Gardner, Courant immediately set out to find some sort of employment for her, beginning with mechanical drawing and ending—when it was evident that she couldn't draw a straight line—with editorial work on the English translation of Courant-Hilbert. She just happened to be proficient in German. She told me she had admired the beauty of the Courant calculus, and she was stunned by the way in which the translation and revision of the classic of mathematical physics was being prepared.

"Nobody but Courant could have done that, putting together a book by farming it out to twenty different people. He let everybody put in their two cents' worth, and then somehow he pulled it all together. Only he didn't really pull—and yet somehow in the end it was all really together."

It was very much like other things she observed at the Bible House: "Papa—Courant—collected people and he collected ideas and he collected things and he managed to put them together. He ran everything in a way that you are brought up to think is the wrong way. The amazing thing was that the operation worked, and it worked because he was in the center of it, not letting go of any of the threads really, not really letting go. I wouldn't have believed it ahead of time, and it's hard to believe in retrospect!"

Lucile Gardner stayed at the Bible House until her marriage to a man much older than she, upon whom she had set her heart even before her meeting with Courant. While she was with the group, she did a lot of rewriting and editing, struggled to improve Courant's innumerable memorandums on the *reservoir of talent* which was *out there* waiting to be *tapped* by NYU; helped Natascha with the editing of a newly established house journal; burned secret documents.

"So it turned out I wasn't really a handicap; but if I had been, I wouldn't have been fired anyway."

After the second war, as after the first, Courant was much concerned with scientific publishing. He continued his collaboration with Interscience; and in some ways, Proskauer told me, Courant and he established the relationship in this country that Courant had had with Ferdinand Springer in Germany. But although they were "close and dear" friends, the relationship was not really the same; for they were of a different generation.

"Courant always drove a very hard bargain in his sweet way, but he was always very protective of Springer's rights, very alert to the slightest disadvantage to Springer," Proskauer went on. "After the war he worried a lot about the future of the Springer firm and even tried to bring Springer to this country to start a new publishing firm here. He thought if Springer did not survive, it would be a great loss to science. Also—of course," Proskauer smiled indulgently, "Courant always wanted to be a big wheeler-dealer in the business world."

After the war Courant had the idea that the reports which his people were regularly making on their work for the ONR could be printed in a journal instead of being mimeographed for limited circulation. The work could then be shared with the entire scientific community; there would be a better job for Natascha Artin, as the managing editor; New York University would have the prestige of publishing its own mathematical journal. All of this could be legitimately subsidized by the ONR. With 200 firm government subscriptions the first issue of NYU's *Communications on Applied Mathematics* appeared in January 1948. (Originally intended as a house journal, the *Communications* very shortly began to invite "outside" contributions and by its fifth issue had changed its name to *Communications on Pure and Applied Mathematics*.)

That same year Interscience also began to publish a series entitled *Monographs in Pure and Applied Mathematics*—an American counterpart of the famous Yellow Series, but a "non-yellow" series, the bindings always red and gray. (The first volume was the Courant-Friedrichs *Supersonic flow and shock waves*.) Ultimately, the series absorbed another series, the Interscience *Tracts in Pure and Applied Mathematics*, which Lipman Bers, then at NYU, was active in developing.

In the summer of 1948 Courant visited Germany for the second time since the war.

By then, the intense economic, political, military, ideological conflict between East and West which was known as the Cold War had resulted in a considerable change in the attitude of the western victors toward West Germany, which was now looked upon more as an ally than as a defeated enemy.

After the trips of 1947 and 1948, Courant visited Germany almost every summer until his death. He did everything he could to assist Rellich in rebuilding the Mathematics Institute, accepted membership again in the Göttingen Academy of Science, and became again an editor of the *Mathematische Annalen*, served informally as scientific adviser to Springer, participated actively in scientific and educational reconstruction, brought a number of German mathematicians as guests to NYU.

He never considered returning permanently to Germany, although some of his friends did. Siegel assumed his old chair in Göttingen. Artin became a professor again at Hamburg and resumed his old friendship with Hasse. Max Born, after his retirement from Edinburgh, spent his final years at Bad Pyrmont, a little spa not far from Göttingen. On the other hand, many of Courant's German-born colleagues, most notably Einstein, were never able to bring themselves to go to Germany again, even for a visit.

"There is no question but that after the war Courant was absolutely committed to the United States and to his enterprise at NYU," Friedrichs told me, "but he simply couldn't resist going back to Germany as soon as possible. In spite of what had happened to him, he wanted *to help*. He was very much criticized for that by many people. But such was his nature."

TWENTY-SIX

PEACE WAS NOT peaceful. The victors separated into two opposing camps. Berlin, divided into four sectors within the city but situated entirely in the Russian-controlled sector of Germany, became a pawn in the struggle between East and West. It was a time of the Truman Doctrine, the Marshall Plan, the Berlin Airlift, the North Atlantic Treaty Organization, the civil war in China, the retreat of the government of Nationalist China to Formosa, the inevitable explosion of an atomic bomb in the USSR.

As a result of the war and the ensuing "cold war," half a dozen centers of applied mathematics had sprung up in America. The subject had experienced a kind of boom. Other universities began to look acquisitively upon Courant's group. In 1949 there were offers to take over the operation from NYU and set it up as an institute for advanced study and research in the mathematical sciences on a new campus. To Courant such a development was not only flattering but also in line with his idea of centers of scientific research to be located in various regions of the country. He was particularly gratified by an offer from the University of Maryland because he had long advocated a development in or near the nation's capital. And still—". . . after much soul-searching, I must say that, at this moment, the proposed shortcut of simply transferring our present institute from New York University to the University of Maryland does not seem to us the most suitable way to implement your general plan," he wrote to Dr. H. C. Byrd, the president at Maryland. "It is not only natural inertia but, even more, loyalty to our mission in the New York region and to our institutional background which makes us hesitate simply to accept your generous and very tempting offer."

Yet, after fifteen years, the problem of "stabilizing" the group at NYU remained. The word began to crop up with increasing frequency in Courant's letters and memorandums in spite of the fact that his secretaries red-penciled it as often as they could. There were also increasing attacks from the outside. Max Shiffman, whose promising career was ultimately to be blighted by a long and serious illness, accepted a position at Stanford. Flanders finally decided to give up the academic life and join the scientific staff at the Argonne Laboratory. Several people went into industry, where they sometimes received twice what they had been paid at NYU.

The Graduate Center of Mathematics was still a poorly supported unit of the university, inadequately housed, and with just three professors whose primary appointments were in the graduate department. Basic salaries of these and of others were paid only in part by NYU. To a large extent even money for such academic necessities as fellowships, books and journals, desks and other equipment had to be obtained by Courant from outside sources or by juggling contract funds which in principle were not to be used to meet academic needs.

The mathematicians were beginning to feel like the stepchild of fairy tales who is sent out to work but whose wages are confiscated at the end of each day. The University of Maryland proposal to take over the applied mathematics group had been turned down in a couple of weeks. A similar proposal later in 1949 from the University of California could not be refused so quickly.

In the years that Courant had been at NYU, Griffith Evans had built up at Berkeley one of the top mathematics departments in the country. He was one of the few outstanding native-born American mathematicians who worked in the field of partial differential equations as well as one of the few who had encouraged Courant's efforts at NYU in a friendly fashion. Evans had recently retired and had been succeeded by Charles Morrey, a mathematician for whom the people at NYU also had tremendous respect.

Friedrichs still recalls vividly the first time he met Morrey.

"It was one of the New York meetings of the American Mathematical Society about 1938; and this inconspicuous-looking boy came up to me and said modestly that he wanted to tell me he had been working on partial differential equations and he had solved such and such a problem. I said that was very nice, and went on. Then—wait a minute!—I suddenly turned and went back to him and said, 'Would you tell me once more which problem it was you said you had solved?' *I couldn't believe it*. It was one of those problems many of us had worked on for years and years—I just couldn't believe it. Oh yes, Morrey is powerful!"

Thus, when Courant was hiking with Morrey and Hans Lewy in the Berkeley hills one afternoon in August 1949 and Morrey wondered aloud whether the NYU group would not like to transfer its operation to Berkeley, the idea was attractive to the New Yorkers.

Things began to move very rapidly. Originally—Morrey told me—he had been thinking about getting just Courant, Friedrichs and Stoker; but the proposal quickly ballooned into the transfer of nine full and part-time faculty members plus twenty student assistants and computers, one secretary and three typists from New York to Berkeley.

Even back in the reality of the Bible House, Courant could not close off the Berkeley proposal as quickly as he had the one from Maryland. The lure of Berkeley, "general and personal," as he described it to Morrey, was too great. He saw a possibility (which he later conceded was "perhaps naive") that financial support in Berkeley could be obtained directly from the State of California and his group would then no longer have to compete for funds with other university departments or to beg for support from wealthy citizens and private foundations. He visualized, he wrote to Morrey, "an institute for advanced study"—such as he was trying to establish at NYU—on the west coast as well as on the east.

Throughout the fall and winter he debated the move with himself, with Friedrichs and Stoker. He sought the advice of friends on the Board of Trustees, of Warren Weaver and the Rockefeller Foundation, of admirals and important government officials in Washington. He wandered around the Bible House, looking helpless. Everybody, down to the most recently hired clerk, was asked for an opinion.

"What we would wish," Courant explained to Morrey, "is to find in discussions with you a way in which [the] apparent conflict can be resolved so that we can help build up the institute in Berkeley without destroying what has been developed in New York."

By the end of November, although the original proposal of moving the entire group to the west coast had not been ruled out, Morrey was suggesting as an alternative that there could be a cooperating group on each coast with two principals in New York and the third principal and Hans Lewy in Berkeley.

"It was a typically Courantian idea and naturally it appealed to Courant," Friedrichs told me, "but Stoker and I were not in favor of it."

Although Courant told Morrey that he did not feel the Berkeley offer should be used to exact concessions from New York, he had already taken the opportunity to list for Chancellor Chase fourteen specific suggestions for improving the situation of the group at NYU. These included a personal request from Courant himself, approaching his sixty-second birthday:

"Orally I [have been] assured that I would not have to retire at the age of sixty-five. In view of the fact that this age limit does not apply at the places from which I have received recent offers, I should appreciate that such an assurance be given to me in writing."

"Courant would never leave a simple situation alone," Bromberg told me. "He always liked a maximum of unstructured developments. In a state of confusion, where others were most uneasy, he was most receptive to the significant aspects of the situation; and he always felt that much more was accomplished in such a state."

He had a "fantastic" sense of timing, in Bromberg's opinion, and he would use it to throw people off balance in even the simplest situation. As an example Bromberg recalled Courant's talking to someone in his office— "particularly a younger person who was probably in the dark about why he had been called in anyway." Someone else would stick his head into the office—the door was almost always open—and, seeing that Courant was occupied, apologize for the interruption and start to back away. "Come in, come in," Courant would mumble and, turning to the person with whom he had been talking, say, "We have no secrets, have we?" Just as the newcomer started to enter, Courant would murmur apologetically, "But perhaps—" with a look across his desk—"you would prefer?" As the visitor turned a second time to leave—"No, no, don't go—we have no secrets here, have we?" to the person opposite him, and then just as the visitor once more started to enter the office—"Unless you prefer?"

This sense of timing is very evident in Courant's correspondence about the proposed Berkeley transfer; and as the academic year 1949–50 progressed and the various alternatives were discussed and rediscussed, the group in the Bible House moved step by step to a schizoid frenzy.

In New York, as late as March 1950, secretaries were being hired on the condition that they would be willing to go to California in September; but in Berkeley the developing controversy over the loyalty oath required by the regents was beginning to tear that faculty apart. By May the negotiations for the transfer were faltering, and—as Friedrichs told me— "even before then, we had definitely decided we would not leave New York."

A question still debated by those involved is whether Courant *ever* had any real intention of transferring his group to Berkeley.

There are many who feel that he merely liked to play with the scheme, stir things up, pit Berkeley against New York to improve the situation there.

Morrey feels that the transfer of the group—in whole or in part—was seriously under consideration.

De Prima, who was at Caltech and thus not personally involved, says that Courant was honestly torn: "He called me several times and asked would I please come east. He just wanted somebody to talk to about Berkeley, the pros and cons."

Stoker, who had been most enthusiastically in favor of the move to Berkeley, told me that this was the only time he seriously considered leaving the group.

"I was so disgusted with the stingy ways of NYU that I wanted to go to Berkeley," he conceded, "although now I'm not sure that it would have been the best thing. Here Courant was his own boss. He went directly to the heads of the university and I did, too, afterwards when I became director. We could never have done that at Berkeley. And I don't think Courant really wanted to go. He didn't like the idea of leaving New York altogether. He felt this was the proper base for him."

"Courant really wanted to be at both places. That was what he always wanted," Friedrichs told me. "Even when you drove with him, you were aware of that quality in his nature. When he came to a fork in the road, he would quite literally *want to go both ways*. You could actually feel that!"

One of the attractions of Berkeley for Courant may well have been the powerful position it had in physics in 1950.

"It was one of our regrets that there was a poor physics department here at NYU," said Joe Keller, who came to the group after the war and is now the director of the Division of Wave Propagation and Applied Mathematics at the Courant Institute—the successor to the Division of Electromagnetic Research, developed by Morris Kline after the war.

In this connection Keller recalled his first trip to Europe with Courant around 1950 (in the course of which Courant arranged that Keller learn to ski by outfitting him, giving him some minimal instruction, and sending him down the slope). Although the primary purpose of the trip was to find out what was being done in Europe with explosives, it was also a talent search. In Göttingen, Courant met Bruno Zumino, a young Italian who was a student of Heisenberg's. A year or two later he brought Zumino to NYU, where he eventually became head of the physics department and instrumental in building it up.

Keller feels that if the NYU group had gone to Berkeley, as proposed, it might have had more impact on mathematics in the United States than it has had.

"The point I want to make is this. Our institute has been eminently successful in its field of analysis and applied mathematics, and it is undoubtedly the leading place in the world in those two disciplines. However, it has not had a corresponding impact on mathematics in the United States. Throughout the United States applied mathematics is given short shrift. The parts of analysis that *are* emphasized verge on the pure—what we call 'soft analysis' in contrast to 'hard analysis,' which is closer to the applications. If we had gone to Berkeley in 1950, we would immediately have become part of the leading mathematical establishment in the country and we would have had about a ten or fifteen year headstart on what we have been able to do here."

By summer 1950, however, the pros and cons of transferring to Berkeley were quite definitely a thing of the past for the group at NYU.

Courant was finally bringing to a conclusion his book on *Dirichlet's principle, conformal mapping, and minimal surfaces*. Almost all of his mathematical work had been one or another variant of Dirichlet's principle; yet he felt that many attractive questions were still open. He hoped and expected that his book would provide a stimulus for further research in the field.

The book is, according to Friedrichs, the most "Courantian" of all Courant's books in subject matter and the least "Courantian" in style and treatment.

"It is very sober, very concise mathematically. There are many technical details, delicate points, attractive geometrical situations and considerations. I would say that mathematically it is probably his best book. Yet of all his books, it has had the least impact."

"It is very difficult to say why it didn't register more," Courant himself said some ten years after its publication. "It was maybe not well written. Or not written with sufficient fanfare. As one gets older, of course, things that are close to one's heart go out of fashion. At least temporarily."

The same summer (1950) that Courant was completing the book on Dirichlet's principle, the first international mathematical congress since 1936 was held in Cambridge, Massachusetts. Originally scheduled for summer 1940, it had been the special project of Birkhoff, who had seen the first official congress on American soil as an event which would signify America's arrival on the international mathematical scene. With the declaration of war in 1939, the planned congress had had to be postponed until all the mathematicians of the world would again be able to attend. Birkhoff, who died in 1944, did not live to see that time.

In 1950 there were still many difficulties in bringing together the international mathematical community. In June of that year the United States sent air and sea forces to the support of the Republic of Korea. The USSR refused to permit any of its mathematicians to go to the United States. The American State Department looked askance at some mathematicians who wished to come. A number of Americans had strong reservations about the extent to which they would associate with mathematicians whose record in relation to the Nazis could be questioned.

As a result of the Berkeley offer, Courant had managed to obtain some money from NYU to invite some of the foreign mathematicians who came to the congress to stay over and give lectures. One of those he invited to give a few talks was Heisenberg. Another was Franz Rellich, who lectured at NYU for a semester.

Rellich was the first of a number of German scientists to be brought to NYU in the coming years. A more recent visitor has been Martin Kneser, a later director of the Mathematics Institute of Göttingen, the son of Hellmuth Kneser and the son-in-law of Hasse.

The invitations to German scientists were always the subject of considerable uneasiness in the faculty. Some people objected to anyone who had remained in Hitler's Germany being invited to NYU. Others made distinctions of different degrees. In the half a decade since the end of the war, Courant himself had found it increasingly difficult to draw a sharp line between "good" and "bad" Germans.

"Courant is amazingly tolerant of flaws in character—in a way that, you know, I couldn't be," Anneli Lax told me. "He seems to look at the white and the grey and the black of a person and then put weights according to his own values. He can excuse things in people which most of us could not excuse. At the same time, somebody like Flanders, who was above reproach as far as character is concerned, is an object of great respect for Courant."

"I personally don't have the strength of character [of someone like Einstein] that I can hate people or a country so absolutely without reservation that it influences my whole existence," Courant explained to me during the last year of his life. "One cannot forget all the suffering, of course; but I was always in favor of being positive toward Germany, reestablishing contact. I felt very bad about Bieberbach once, but now it's been so long—so much time has elapsed—and I can always see that there were mitigating circumstances with all these people. I never had any contact with Bieberbach again, but if I would meet him here now I would be friendly with him."

Following the 1950 congress, Laurent Schwartz also gave a series of talks at NYU. Schwartz's theory of distributions was considered a conspicuous example of the trend toward generality and abstraction in mathematics which Courant had opposed throughout his career and from which he always tried to "shield" his students.

"There was a certain amount of provincialism at NYU, which was somewhat Göttingen-like," Peter Lax explained to me. "I guess Friedrichs told you how von Neumann was considered a flash in the pan with his operator theory there. Too abstract and all that. Well, we felt the same way about Schwartz's theory of distributions. It is one of those theories—there's no depth in it, but it's enormously useful—and it's different from the Hilbert-space approach in which Friedrichs pioneered—so we resisted it a little bit."

In spite of this resistance, Schwartz was invited to talk at NYU. A few years later, using Schwartz's theory, Lax found a way of proving some things about the solutions of differential equations and, as he said, "that certainly changed my mind and I think it changed Friedrichs's mind and Courant's mind too. In particular—I am jumping ahead now—the last big scientific effort Courant made was to write an appendix on distributions for the English edition of Courant-Hilbert."

Courant often spoke out very strongly against the trend toward purism, generality, and abstraction in mathematics. Once, when I saw a reference in a speech of his to "the blasphemous nonsense of 'mathematics for its own sake,'" I asked him if this didn't represent a considerable difference of opinion between him and Hilbert—who on occasion admiringly quoted Jacobi's rebuke to Fourier: "A philosopher like Fourier should know that the sole aim of science is the glory of the human spirit."

"With Hilbert there is really no contradiction," Courant replied promptly, "because Hilbert didn't live to see this overemphasis on abstraction and the self-emulation and self-adulation that some of these abstractionists show."

"We at NYU recognized rather tardily the achievements of the leading members of 'Bourbaki,'" Friedrichs explained in this connection. "We really objected only to the trivialities of those people whom Stoker calls *les petits Bourbakis.*"

Lax was one of the last students to take his Ph.D. with Courant. He found him "a very original guy" whose way of expression matched his way of thinking.

"The way he wrote was like nothing anybody else wrote. He hated that style of stating a theorem which goes, *Let M be a manifold, X a differential structure, Y a vector field, and so on.* He would really want to describe first what the problem was about and how one goes about attacking it. In fact, perhaps Courant went rather more in the other direction of not having enough theorems. But—as he said—'Most people have too many.' He was pretty old by the time I started working with him, but by far the most interesting work we did together was when he was sixty-eight. We generalized Huygens's principle."

I asked Lax how he and Courant had worked together.

"It came about this way. In Courant-Hilbert the propagation of discontinuities was worked out for second-order hyperbolic equations. Courant thought it should be worked out for general systems. I thought that could be easily done, and I did it—it went along the same lines as in the second order case. About the same time Courant had written a paper with Anneli on solving the initial-value problem. It was something that worked only for equations with constant coefficients, but I observed that putting together what Courant and I had done for the propagation of discontinuities with the technique described in that paper of his and Anneli's, one could get, not an exact formula for the solution of general hyperbolic equations with variable coefficients, but a formula that gives correctly the propagation of singularities. And it worked out very simply."

"But it sounds as if you did most of the work."

"No. Because, first of all, it was Courant who said the propagation of discontinuities should be worked out for the general equations and, secondly, it was he who with Anneli worked out the formula for the solution of the initial-value problem—which then could be put together."

"So you don't think it was an instance of Courant's having the first night, as it were, with one of his students?"

"No, definitely not," Lax laughed. "I would go out to New Rochelle to talk to him—in fact, on this particular problem I saved several pieces of correspondence, Courant's scribbly handwriting—it was a very natural and easy collaboration."

"Courant has always been completely generous in providing ideas for what you should do," said Jerome Berkowitz, who was a student a little later than Lax. "He was quite uninhibited—which most mathematicians are not, by the way—about telling someone else, not just a student, what he ought to do. Of course young people were especially susceptible to this; and he had really good ideas about what they should do—also their talents—and steered them into fields which certainly they would not have gone into otherwise.

"In his courses he would always give you the big picture. He had very good instincts about what could be done and how you could get from here to there; but when it came to actually doing it, his handwriting on the board would get worse and worse and every symbol would look like the same Greek letter, and things would be erased, and he would say, 'And there it is!' But he would make the idea of what you had to do to achieve it very clear, and you would then do it."

That September of 1950, when Rellich and Schwartz were delivering lectures at NYU, enrollments at all American colleges and universities were sharply down. This was partly due to the number of young men being drawn off by the armed forces, but also to the fact that the 18-year-olds of that year had been born in 1932, the worst year of the Depression, and were as a consequence relatively few in number. At NYU alone the drop in enrollment represented a loss of $1,600,000 in tuition fees. Nevertheless, as a result of the possibility that all or part of the Applied Mathematics Group might go to Berkeley, the general situation of Courant's people had improved.

"Under the prodding from various agencies"—in Courant's words—the university had become "more actively concerned with the material welfare of our enterprise." There had been a substantial increase in the budget and the university had assumed a larger share of basic salaries. But, as Courant wrote to Morrey in one of their last exchanges, the problem of *stabilizing* the group had not really been solved "and probably is not solvable in the structure of NYU."

In the fall of 1950 Courant began to ponder once again his proposal for some sort of national institute of the sciences. The need seemed even more urgent now than it had before the war. The universities of Europe were no longer what they had been. The American universities had "a major responsibility as guardians of the intellectual values of Western Civilization." The principles upon which Flexner had founded the Institute for Advanced Study should prevail, but—rather than a single and exclusive scholarly retreat, reaching only a small group of students—there should be "facilities for freely advanced learning . . . created at many places, [unhampered] by the multitude of educational services which, in a democratic society, have to be rendered on a non-selective mass scale."

During the Korean War a new series of lengthy memorandums began to go out from Courant's desk—some paragraphs carried over word for word from the original memorandum drawn up in the winter of 1939-40. The aging Flexner, eighty-five and loosely attached to the Carnegie Foundation in an advisory capacity, was still interested in Courant's ideas; and Courant tried hard to interest Robert Oppenheimer, the new director of the institute in Princeton.

"I could envisage great benefits for higher education if the Institute for Advanced Study would take the leadership in such an effort," he wrote to Oppenheimer.

But Oppenheimer did not respond.

In the middle of December 1950, President Truman declared a state of national emergency and outlined plans for placing the country again on a war footing.

Courant scurried back and forth between New York and Washington. He saw to it that there were always some of "his people" working for the ONR. He was constantly alert for connections between students at NYU and names in the daily headlines. His circle of acquaintances in government, military, and business—already large—continued to widen. Friedrichs and Stoker assumed more and more duties and responsibilities for the group. Stoker's wife, Nancy, became Courant's administrative assistant, handling with efficiency many of the multifarious problems of the growing operation. And yet—as people in the group frequently observed to me—even when Courant wasn't actually at the Bible House, he seemed to be there.

It was just at this point that even the success of Courant's efforts as an advocate of applied mathematics in the United States began to work against him. For a number of years NYU's Graduate Center for Mathematics had been receiving an annual grant from the Rockefeller Foundation. At the end of 1950, a five-year grant of $12,000 a year having expired, Courant, Friedrichs and Stoker wrote a joint letter to Weaver asking that it be renewed and, if possible, augmented.

Weaver responded regretfully.

"Ten years ago the situation in this country with respect to applied mathematics was very different from what it is now." In 1940 there had been only two places which had qualified for help from the Rockefeller Foundation to encourage applied mathematics—New York University and Brown University—and it had therefore made an exception to its program by giving some modest help to both institutions. Since then the general attitude in America toward applied mathematics had changed. There were now several effective centers for training in the field. "You and your colleagues can take a special pride in that statement, for you have without any question played a large role in making that statement true." But the period of pioneering help—as an exception to the program of the Rockefeller Foundation—was over.

At the same time that this important source of funds was being cut off, the financial position of New York University was becoming increasingly precarious. The war in Korea continued to drain off enrollment. Chase had retired. No successor had been chosen. There had been virtually no implementation by the university of the recommendation of Roberts's committee for the establishment of a mathematics institute at NYU. Yet in spite of the many disappointments, as Courant reminded himself, "moral backing" by a number of distinguished people, among them Augustus B. Kinzel, an outstanding engineer who was the vice president of Union Carbide, had persisted. In 1951, with the encouragement of Kinzel and others, Courant came to the conclusion that what he had been trying to achieve for so many years at NYU—an advanced institute of mathematical research and training—simply could not be achieved within the framework of that university. Still once again, in the spring of 1951, he brought up his idea for a national institute.

But he couldn't convince the people he wanted to convince.

"If one thinks of [your proposal] concretely in the terms of New York University and Columbia, for example, just specifically why would this combined institute offer something which is not realizable within the natural framework of the separate institutions?" Warren Weaver demanded. It was certainly true that what Courant had in mind had not been realizable "in its best form" at New York University. "But again would there not be those who would say that this leads to a criticism of New York University rather than to a proposal to start something new?"

For that, it seemed, there was no answer.

TWENTY-SEVEN

IN 1952, with the appointment of Henry Heald as chancellor of New York University, Courant—"an incurable optimist," as he admitted—returned again to the idea that perhaps an institute of science could be achieved within the framework of NYU. A memorandum of April 15 mentions "an encouraging discussion with Dr. Heald" and the fact that it now seemed "proper to suggest" certain initial steps of implementation and development.

Then, just at this time, Courant heard a significant piece of news. The Atomic Energy Commission was planning to expand its utilization of the recently developed high-speed electronic computers and was planning to install a machine in each of its laboratories. It would then locate the single computer that it had previously bought—the UNIVAC 4—in some institution in the eastern part of the United States where scientists, removed from the pressure of the laboratories, could develop more efficient methods for solving numerical problems involved in the work of the AEC.

During the past few summers Courant had frequently visited Peter Lax when that young man was a consultant at Los Alamos. He had found the big machines fascinating, although he had not been able to go as far as many mathematicians in their enthusiasm for what computers could do. He sometimes referred to them drily as "the emperor's new clothes."

"Maybe because the evidence that was offered for computers then was bad," Lax said to me. "I think there were a few instances where things were offered—oh, such and such a calculation was done on a computer which couldn't have been done otherwise—and then it turned out—I remember one such instance—that Isaacson could do it with just a little theory and a few desk machines. However, the conclusion to draw is that, well, that was a bad example; there are other good examples.

"*Computers can do marvelous things!* Computing really gives new content to scientific theories and completely alters the nature of applied mathematics. Mathematicians like *general* things, and now for the first time with high-speed electronic computers we have a chance to work on general things in the applied field."

With the news that a generously-budgeted computer was going to be available, whatever reservations Courant had about the big machines evaporated. But before he made any move to try to get the UNIVAC 4 for his group, he consulted Friedrichs and Stoker.

"Courant would never have begun an enterprise if Friedrichs and Stoker disagreed with him as to the value of it," I was told by Isaacson, who had continued to run the NYU "computing department" in the years since the war. "He would try in his own inimitable fashion, indirectly, to get them to see the light so that they would come around to his point of view; but he would never in a direct way oppose a position which they had both taken. In this case 'education' wasn't necessary. They immediately saw that it would be a very good idea."

Thus began what is still remembered by Isaacson and others as "the battle for the computer"—the turning point in the history of what was to become the Institute of Mathematical Sciences at New York University.

Besides its value as a tool of applied mathematics, there was another important reason for trying to get the UNIVAC 4 at NYU. Because of the high initial investment in the machine and its placement, the program would of necessity be a continuing source of work and support which could possibly result in the long sought stabilization of the applied mathematics group. Yet, according to Isaacson, the farsightedness of Courant, Friedrichs and Stoker was somewhat surprising; for the project as originally envisaged by the AEC was a mere service activity. A minimal staff would run it, develop the necessary programs, and maintain the machine.

During the summer of 1952, the Atomic Energy Commission scheduled a meeting at which an AEC representative would present the details of the project to the interested institutions eligible to submit proposals for the use of the computer. At the time Courant was in Europe. Stoker, who was lecturing that summer at Stanford, flew to New York. He, Isaacson, and a member of the university administration represented NYU at the meeting in the Empire State Building.

The representatives of the various institutions were seated around an oblong table. The chairman took the 12 o'clock position. The NYU contingent was assigned the 1 o'clock position at his left. After the AEC representative had explained the nature of the project, the chairman went around the table, counterclockwise, and asked a representative of each institution to comment in turn. To the amazement of Stoker and Isaacson, the response of those who spoke was, without exception, negative.

"They said they couldn't possibly think of making a proposal unless they had the use of the machine at least fifty per cent of the time for their own purposes—at the AEC's expense of course."

When a break was called for lunch, only NYU had not yet been heard from. At lunch Stoker and Isaacson freely said that it was ridiculous for the others to feel the way they did. Although the immediate objective was to run a computing center for the use of the AEC laboratories only, it was not going to persist that way. After lunch, when NYU was finally called upon, Stoker expressed his opinion that for people interested in applied mathematics having the computer would be a great opportunity.

"We then went our merry ways," Isaacson recalled, "and, lo and behold, in spite of the fact that all the other people had objected to the terms, every one of the institutions which had been represented at the meeting submitted a proposal for the machine essentially on the terms the AEC had laid out."

Since the budget for the project had to be based on the minimal staff that Remington Rand had suggested, all the institutions proposed roughly the same figures. The people from NYU were rather confident their proposal would be the one approved because of their extensive experience in applied mathematics. A month or so later, however, after Courant had returned from Europe, they learned indirectly that the AEC was planning to locate the computer at another institution—oddly enough, one which had not been represented at the meeting at the Empire State Building.

Everybody was angered and deeply disappointed, but what was there to do about it?

Stoker simply refused to give up.

"I think Stoker deserves a great deal of credit for that," Friedrichs said to me. "I remember that he told Courant, 'You can't let them do that to us!' He simply *made* Courant go back and fight to change the AEC's mind, because—as we were told—the contract had not yet been officially awarded."

"Courant was a tremendous fighter," Isaacson contributed. "He got all the scientists and the experts in Washington that knew about our work to get in touch with the administrators of the Atomic Energy Commission —scientists in scientific jobs who as government employees could speak on a par with other government employees. And then there were other people he brought in. He was absolutely inexhaustible. He was thirty years older than I, and I was worn out after an hour spent with him. But he would go on at great length, drafting letters, calling people, planning the moves that would be necessary to overturn the decision—which still had not been announced officially."

Isaacson is not sure which effort of Courant's was ultimately decisive: "Apparently the combination of all of them convinced the research division of the AEC—and in the end they awarded the contract to us!"

The AEC insisted that the building in which the UNIVAC was to be housed must be fireproof. The Bible House did not meet this last requirement; but Chancellor Heald had assured Courant that if he got the computer, a suitable building would be acquired for it.

Suddenly, after nearly twenty years of constant effort—of letters and memorandums, schemes and proposals, hopes and fears, disappointments and small successes—an institute of the mathematical sciences with all the multifarious activities to be housed under one roof, as in the old days in Göttingen, appeared to be coming at last into existence at NYU.

Another series of memorandums began to flow from Courant's office to that of Heald, but these were now concise and specific:

"The activities of (1) the present Institute for Mathematics and Mechanics (about 55 people), (2) Professor Kline's electromagnetic research group (about 28 people), (3) the AEC computing facility (initially probably about 25 people . . .), and (4) possibly in the near future a group in statistics and probability ought to be fitted into a single framework to be called 'Institute of Mathematical Sciences.' "

On the morning of January 18, 1953—ten days after Courant's sixty-fifth birthday—the *New York Times* carried the headlined story: NYU WILL EXPAND MATHEMATICS UNIT.

Although the newspaper announcement mentioned "a Division of Computing Services" as a part of the new "Institute of Mathematical Sciences," there was no mention of the AEC's UNIVAC.

This was for a very good reason.

With Chancellor Heald's assurance, Courant had had no qualms about promising that NYU would be able to provide the type of building required to house the computer; and it had been so stipulated in the contract with the AEC, Isaacson told me.

"Then, just a day or two before or after—I forgot which—when the representative of the AEC came to meet Heald and officially inform us that we had been awarded the computer, the building deal fell through. The owner, consulting his lawyer about the sale, had discovered that it would pose certain tax problems for him."

Heald moved quickly. In a very short time he had managed to sign a contract for another, larger building at 25 Waverly Place. On January 25, 1953, the *Times* was able to announce: NYU BUYS BUILDING.

The new building had housed a number of now bankrupt hat factories. Some tenants still occupied the upper floors under lease, but the mathematicians were able to take over the basement and the first and second floors immediately. They began to make the necessary preparations to install the UNIVAC.

Remington Rand had sent a template for the concrete platform which would have to be constructed for the machine. This indicated openings for various wiring conduits and air conditioning ducts. The university architect drew up the plans. The platform was poured with the holes in position as indicated. The frame for the machine was delivered. Then it was discovered that the architect had turned the template upside down and his blueprint had been the mirror image of the correct placement of the necessary openings.

An emergency meeting was called. The engineers from Remington Rand came up to New York from Philadelphia. To everybody's great relief, they concluded after inspection that with a few modifications the existing platform could be used.

"But there was still some consternation," Isaacson went on. "At that time the machine was the sole computing facility of all the AEC laboratories. It had to be kept going. So for four months we had to operate it in the factory in Pennsylvania where it was still located. I commuted quite regularly to check on the progress of our engineering and technical people and on the way in which problems were being done. The machine worked continuously 24 hours a day and every day of the week. The delay with the flooring could have been a catastrophe if the repairs had taken longer. As it was, we were able to work straight through. It was a *very* exciting time," he concluded, a little wistfully.

Bromberg, who had left the group at the end of the war to go into industry, was asked to take over the management of the new facility. He had obtained his degree with Stoker while still working at the Reeves Instrument Company, a firm that manufactured analog computers; and he had worked with numerical methods in his own research. He also had experience and interest in administration. He felt, he told me, that "it was like the family calling you home," and so he returned.

Initially the computer was a classified project under armed guard and with access only to a minimum number of people who had been cleared by the FBI. Stoker was the first senior faculty member to use the machine. Almost immediately he and Isaacson collaborated on work involving water waves and meteorology which ultimately led to Stoker's work on flood control.

Courant was always particularly proud of this work by Stoker—"a very fruitful combination of down-to-earth and purely mathematical components"—because, among other things, it perfectly illustrated his belief that although science cannot be *directed*, it can be given *opportunity*. For a number of years some members of the NYU group had been concentrating on the theory of hyperbolic partial differential equations, which describe wave propagation phenomena. Other members had directed their attention to numerical methods based on finite-difference schemes. Both of these activities were to a degree outgrowths of the work done in Göttingen by Courant, Friedrichs and Lewy. When in the 1950's Stoker was asked by the U.S. Corps of Engineers to devise methods of predicting floods, he had conveniently at hand this body of theory as well as a high-speed computing facility.

During 1953–56 Stoker took the record of one of the largest floods of the Ohio River to verify whether the flows in the river could be calculated theoretically to the accuracy desired. He found that this calculation was possible for flows in the 375 miles between Wheeling and Cincinnati and for periods up to three weeks. The calculation stretched the capacity of the UNIVAC 4 to the limit, but required only $6\frac{1}{2}$ hours of computer time. The results established that utilizing the theoretical method was not only feasible, but also much more economical than what might be called the practical method of building a large-scale model (costing as much as the UNIVAC 4) for each river or reservoir to be studied—a method used in the past.

Yet, in a way, Stoker told me, he always regretted the acquisition of the UNIVAC.

"I saw it was impossible not to try to get it because it's such an instrument for applied mathematics, but I also saw that it would mean an enormous expansion and a kind of destruction of the place the way I liked it to be. From that time on, it lost its coherence. Friedrichs and I practically never worked together on a long problem, because the senior people all had duties and so did a lot of other people. But I saw it was inevitable, so I tried to help it along."

By 1958, when Stoker succeeded Courant as director, the AEC had replaced the UNIVAC 4 with an IBM 704 computer; and the computing center had taken over four floors of a nearby building at 4 Washington

Place. As Stoker and Isaacson had predicted, the machine was generally available to the institute.

Although NYU was by that time one of the five laboratories carrying on the major portion of AEC research, its participation remained unique in respect to the role played by students and by more than token representation of faculty. In addition, more mathematicians at NYU were engaged in the theoretical aspects of Project Sherwood (the AEC's continuing effort to harness nuclear energy for peaceful purposes) than at all the other four centers combined. While the other centers were charged in general with such immediate objectives as engineering design and development, the assignment of the NYU group was to help the AEC understand in detail what was happening inside the devices, insofar as mathematical science could do so.

Neither Courant nor Friedrichs ever used the computer in connection with their mathematical work; however, the ideas of the Courant-Friedrichs-Lewy paper on numerical solutions of differential equations became basic as the use of the high-speed machines increased. In 1967, some twenty years after the wartime development of the computer and almost forty years after the publication of the original three-man paper, IBM devoted an entire issue of its journal to the impact of "C-F-L" on modern investigations in the numerical analysis of partial differential equations—"an outstanding instance of research undertaken for purely theoretical purposes turning out to be of immense practical importance."

After the installation of the first computer in 1953, the whole mathematical development at NYU began to move with a speed which "quite overwhelmed" Courant. Leaving in June 1953 to attend the 1000th anniversary of the city of Göttingen, he wrote to Harry Woodburn Chase:

"I certainly have not forgotten the sympathy you have shown to my interests from my very beginning at the university. If it hadn't been for your intervention at critical points, the Institute would never have reached its present state of development."

Courant's visit to Göttingen in 1953 was almost as significant for him as his first post-war visit. As a part of the anniversary celebration, Born, Franck, and he were to be made "honorary citizens" of the town. The decision to accept this status had not been automatic for any of the three men.

In 1933 they had all reacted to the personal repercussions of Nazism in different ways. Twenty years later, each arrived at the same decision as the other two. In a ceremony in the town hall, which still bore emblazoned on the wall of its cellar the ancient motto, *Extra Goettingen non vita est*—"Away from Göttingen there is no life"—they accepted the impressive scrolls which announced that they were *Ehrenbürger* of Göttingen.

When Courant returned to NYU that fall, preparations were being made for the move to the new quarters. Shirley Twersky, who had been working as a secretary, told me that she saw this as a good opportunity to leave and join her husband, a former member of the group, who was already on the west coast in a new position.

"I was supposed to stay around just another month or so and train another girl to take my place, because Courant required very special handling."

"What do you mean by 'very special handling'?"

"Oh, he liked to play little games. Little puppet-type games. They were used to set up a situation and see what the reaction would be. For example, when I did finally convince him that I was going to leave, he came in one morning, just ambled in, and said, 'Well, it's too bad that you're going to leave just when everybody no longer dislikes you.' Natascha, who was there, got very upset and said emphatically that she had never disliked me. But I just laughed. He wasn't really being malicious, but he was an observer of people. He liked to see what they would do."

It should be said, however, that there were others with whom I talked who felt that Courant *was* sometimes malicious in his human experiments —that when he saw two people who were close to each other, for instance, he would frequently try to sow a seed of dissension between them. He himself shrugged and said that if he could break up a relationship, it could not be a good relationship for those involved.

Those who were closest to Courant freely concede his faults.

"In his case one is quite aware, you know, that he's not perfect," Anneli Lax said to me. "He does some things that are really quite objectionable, and yet you like the man. Perhaps it is because in spite of the somewhat devious ways the kinds of things he really stands for and accomplishes are O.K. I mean, they are the things you would want to do too, although you would do them straightforwardly. I am not a person who believes that any means is justified by the end, not at all, but he seems to get away with it. Somehow."

". . . a never failing loyalty bound Courant's associates together," Neugebauer said after half a century of friendship. "He inspired an unshakable confidence in his profound desire to do what was right and what made sense under the given circumstances. His ability to create a feeling of mutual confidence in those who knew him intimately lies at the foundation of his success and influence."

The move to the former hat factory in the fall of 1953 represented for Courant the culmination of his almost twenty years of effort to bring his experience in Göttingen to bear upon the situation of science in America. Although he had earlier applied the name "institute" to his group in the Bible House, and before that to the group in the Judson Dormitory, this was the first time that an institute in the sense of the one he had had to leave behind in Göttingen would come into existence at New York University. It was also an important development in American mathematics. A few years later (1958) *Fortune*, featuring a pair of extensive articles on mathematics in this country, ran Courant's picture and that of Oswald Veblen of the Institute for Advanced Study, side by side, as "the founders of two great mathematical centers." New York University's Institute of Mathematical Sciences was described as "the national capital of applied mathematical analysis."

Curiously enough, to the people in the group the new "institute" did not seem like "such a big thing." They were going to move gradually into a larger building. All of the mathematical activities would eventually be brought together. There would be more money, a new name. But to Courant —it is clear from his correspondence and memorandums—the move was comparable to that made long ago to the new building on Bunsenstrasse. He postponed the dedication for more than a year so that Niels Bohr could deliver the principal address.

Since Harald Bohr's death in 1951, Niels Bohr "had entrusted himself a little bit more," as Courant put it, to his brother's old friend. On Courant's side there was no question but that he considered Niels Bohr, not only the most outstanding scientist, but also the most outstanding human being he had ever met.

"There is nobody in my opinion who can be rated as highly as Bohr," he told me. "Hilbert was very original, and Bohr had great respect for him; but Hilbert was a very different kind of personality. I would not even rate Einstein as highly as Bohr. And of course in Göttingen I met gradually very many outstanding people in different fields. I might talk of Poincaré —but nobody—Bohr was unique. He was on a different plane. It was not only breadth. It was his penetration of nature. He understood what went on in physical phenomena. He understood it the way birds understand the singing of other birds."

There was in any case no more appropriate choice for the speaker at the dedication than Bohr. He had always been generous in backing Courant's American efforts with his own great international prestige, and he and his brother had made the original suggestion that Courant approach the Rockefeller Foundation for assistance in erecting a mathematics institute in Göttingen.

During the first year of the existence of NYU's Institute of Mathematical Sciences, the American scientific community was aroused and divided by what came to be known as the Oppenheimer case.

In April 1954, when headlines announced the withdrawal of Oppenheimer's security clearance by the Atomic Energy Commission, Courant wrote immediately to the physicist and offered "to join the ranks of your friends who rally in defense of your character." But he did not give up his friendship with Admiral Lewis L. Strauss, the chairman of the AEC, as many of Oppenheimer's supporters felt he should. He wrote to Niels Bohr: "[The case of Robert O.] has so many sides that I personally cannot accept the prevailing simplified version and attitude."

"In general, one can say that multisidedness was an essential feature of Courant's *Weltanschauung*," Friedrichs said to me. "He hated one-sided statements. He was suspicious of them. He was suspicious of all final and sharply given statements and formulations."

After the denial of Oppenheimer's security clearance was upheld, Courant deplored the bitterness which the decision engendered among scientists. He agreed with Strauss that the scientist as such had no more claim to public influence than any other citizen, but he felt that the government could not afford to disregard "the necessary psychological conditions of the scientific mind." Something should be done to reassure scientists of the government's fairness and reasonableness in handling security problems. He wrote a long letter to Strauss setting forth his suggestions for action by the AEC to relieve the tension. He also felt, as he wrote to Niels Bohr, that Oppenheimer should modify his attitude in the affair: "It would be much healthier if he understood that, while there is practical unanimity about the outcome of the formal procedure, there is also much criticism of him in many respects. By showing conciliatory humility, he could really become a great man and make a great contribution."

There was also a "security problem" closer to home.

Donald Flanders, who had finally given up his professorship at NYU for a job at the Argonne Laboratories, had become the subject of an investigation to determine whether he was a security risk, because of his friendship with Alger Hiss, who had been accused of spying and convicted of perjury. Asked by Flanders if he would furnish an affidavit of his loyalty, Courant promptly agreed. Shortly afterwards, an FBI agent came to Courant's office to question him further about Flanders, who persisted in visiting Hiss while he was in prison. Eyes sparkling behind his glasses, Courant looked across his desk at the agent sitting in the chair opposite him.

"Ja, ja, Alger Hiss," he mumbled. "That chair," he said, not quite inaudibly, "that chair you are sitting in, when he came to my office, ja, Alger Hiss used to sit in that chair."

The FBI agent quickly came to his feet and, as Courant was subsequently to tell the story, conducted the rest of the interview standing.

When Flanders's brother Ralph, by that time a senator from Vermont, demanded the censure of Joseph R. McCarthy by his colleagues in the senate, Donald Flanders was again investigated. The month before the dedication of the institute, Walter Winchell announced in his Sunday evening broadcast that "only last Thursday" Senator Flanders's brother Donald had said that he had no intention of going back on his friendship with Alger Hiss: "Significantly, ladies and gentlemen, Alger Hiss, in prison for lying about his Red connections, will be out in about six weeks and he can count on a very, very dear, dear friend working on America's most secret weapon—the atomic bomb."

A few years later Flanders was to take his own life; but he had lived to see the dedication of the Institute of Mathematical Sciences at New York University, for which he had in a sense laid the cornerstone that long ago day in 1933 when he had approached Oswald Veblen for advice about improving the mathematical situation at NYU.

The dedication ceremony for the new institute took place in the auditorium of Vanderbilt Hall on November 29, 1954, almost to the day a quarter of a century after the dedication of the institute in Göttingen. After appropriate remarks by various officials of the university, there was music by the Moyse trio. Then came the director of the new institute. For once, he did not repeat the message that Klein had distilled from the example of the École Polytechnique and the broad universal works of Gauss, Dirichlet and Riemann—the message to which Hilbert had given living embodiment. Instead he limited himself to a few remarks about his long relationship with the Bohr brothers and the role which Abraham Flexner had played "[in stimulating] the hope that at least in the field of Mathematical Sciences something parallel to the Institute in Göttingen could be developed in New York City."

Then Niels Bohr, speaking as only he could, even more inaudibly than the director of the new institute, referring gratefully to Courant as his "other brother" and "the man who opened up mathematics to physicists," announced his subject—"to indicate," as he said, "the guidance which the mathematical sciences through the ages have offered as regards our orientation in that nature of which we ourselves are part, and especially to stress the fundamental role which mathematical abstractions and the art of calculation have played in the development of the physical sciences."

Among the men and women in the audience there were only a few who remembered the similarly cold but bright winter day when the Göttingen institute had been dedicated—the way in which the spirit of Klein had seemed to be a living presence at the ceremonies—the frail old Hilbert saying delightedly in his sharp Königsberg accent, "There will never be another institute like this! For to have another such institute, there would have to be another Courant—and there can never be another Courant!"

Just when Courant's dream of an institute of mathematical sciences in New York became a reality, his dream of Göttingen's returning to its old position as a great mathematics center under the leadership of Franz Rellich was coming to an end. The year following the dedication of the new institute, Rellich, forty-nine years old, died of a brain tumor.

"Everywhere that memories of old mathematical Göttingen are still alive, Rellich's death has caused great distress and sorrow," Courant wrote sadly. "Centers of science are fragile organisms, sensitive to the loss of the personalities who promote and unify them. When good fortune puts a man like Rellich in the right place, such centers can blossom for a short time, as we have seen with admiration in Göttingen. . . . Throughout the mathematical world it is hoped that in spite of Rellich's death the Göttingen tradition will live on."

TWENTY-EIGHT

COURANT served as director of the Institute of Mathematical Sciences at New York University from 1953 to 1958—the same number of years that he had been permitted to serve as director of the newly built institute in Göttingen.

For his seventieth birthday on January 8, 1958, his friends established a fund to endow a series of Courant Lectures to be delivered every two years. Friedrichs was chosen to present the gift.

"One may think that one of the roles mathematics plays in other sciences is that of providing law and order, rational organization, and logical consistency; but that would not correspond to Courant's ideas," Friedrichs said at that time. "In fact, within mathematics proper Courant has always fought against overemphasis of the rational, logical, legalistic aspects of this science and emphasized the inventive and constructive, esthetic and even playful on the one hand, and on the other hand those pertaining to reality. How mathematics can retain these qualities when it invades other sciences is an interesting and somewhat puzzling question. Here we hope our gift will help."

There were some other honors in the year of Courant's retirement. New York University awarded him an honorary degree. The Navy presented him with its Distinguished Public Service Award. The Federal Republic of Germany decorated him with the Knight-Commander's Cross and the Star of the Order of Merit—an honor that surpasses even the Iron Cross, First Class.

In the fall of 1958, Courant's position as director of the Institute of Mathematical Sciences passed quite naturally to Stoker, who immediately began to put into effect his own long held ideas about the way in which the financial and administrative position of the institute should be set up in relation to the university.

Intellectually, Courant believed that when one retired, he should step down completely. He was fascinated by the Rembrandt painting of young David playing the harp while Saul—with fear and hatred in his face—looks on. There was a print of this painting in his office, and he had often given prints to colleagues to remind them that another generation was coming along. Emotionally, however, he had intensely dreaded retirement; and the actual event brought on a depression, the extent of which was apparent only to the immediate members of his family.

"I think he had really felt that he could go on forever," his son-in-law Jerome Berkowitz told me. "He was stunned by the fact that everybody at the institute simply accepted the idea that he would retire at seventy. I think he felt a little bit betrayed that a delegation hadn't come to him and demanded that he continue as director."

It was impossible for Courant to give up trying to guide what he had created. In addition, he was afraid, he told me, that Stoker "[might spoil] many things by his passionate and aggressive ways, which he has sometimes."

To take over the responsibility of directing the institute, Stoker had to shut Courant out until he had established himself as the director. This was not easy.

Besides having different views on the organization and financing of their mutual enterprise, Stoker found many aspects of Courant's personality very repugnant. Stoker is, as some of his friends say, "straightforward to a fault." He said quite frankly to me:

"There were things about Courant which I found detestable. Really very bad—and I can see why he had enemies, who couldn't see him as I could, what he could do positively. He was such a calculator. That was the worst. And his utter subservience to people who were rich—even if they were utterly boring and he wouldn't have anything to do with them if they weren't rich.

"But still, on the other hand," he said, his voice changing to one of great affection and admiration, "his merits were really so great. When I say those horrible things, I feel ashamed in a way because, after all, he created this affair. Under the most terrible difficulties. And he was so helpful to people of all kinds, me too—so willing to put himself out for people who weren't rich, because he had some feeling that they were worth the bother. And there were many of them.

"I don't think there's a better place anywhere in the world than this place. There is no such diversified group which is so cohesive. It still remains that way. The people are very cooperative. There are no internal bickerings or jealousies in the place. Just none at all. It's absolutely amazing. The worst trouble we ever had was between Courant and me. We did really clash, because, I think, I was the only one who was willing to stand up to him. In retrospect—well, that I should say—I regard Courant as a very great man. An absolutely tremendous man in his accomplishments."

By the time that Stoker became director, there was at NYU a core of outstanding mathematicians, among them nine future members of the National Academy of Sciences, who have remained until the present time in spite of tempting offers from other institutions. One of the harshest criticisms of Courant—voiced to me by Saunders MacLane of the University of Chicago—was that he hung on to the people he had trained and in that way actually hindered the development of applied mathematics in the country as a whole.

Stoker snorts at this criticism.

"MacLane and the others would have taken everybody Courant had trained, and where would we have gotten replacements? Nobody else was training them!"

Nevertheless, by the time Stoker succeeded Courant as director, a decision had already been made that there should be a greater effort to bring in people from the outside. Among those who came during the late fifties and early sixties were Jacob Schwartz, Paul Garabedian, and Monroe Donsker.

After I heard MacLane's criticism, I asked some of the second-generation people at the Courant Institute why they had stayed in spite of many financially tempting offers from other institutions. Their answers can be summed up in the statement that they like the place.

"Here, of course, most of us of my generation have been very well treated by Courant, socially, scientifically, and so on," Joe Keller told me. "That's not to say that our salaries were exorbitant. In fact, it was quite the contrary. But Courant made it nice for us in whatever ways he could. Somehow or other that made up for the salaries. Then we have all been pleased that we've had hardly any of the strife that is prominent in many other mathematics departments. An accompaniment of that, which is not independent of it, is that there is lots of mutual work—collaboration—which is not common in other mathematics departments. It is possible to ask someone for his help and then use it and not have to make him a coauthor."

"There are very few places in the country where you have this spectrum of pure and applied, and people can still talk to each other," Harold Grad contributed. "I have developed a theory. I am of the group you mentioned of my generation probably the most applied of them all while Louis Nirenberg —he is in mathematics straight, he doesn't do any applied mathematics. But maybe because we were both students together and I know that he respects me and I respect him, we figure that if the other one is doing something and we don't really know what he's doing, it must still be all right. In other places—of course I have never been anyplace else except to visit—I find that there seems to be only antagonism or disrespect for someone who isn't doing what you are doing. It is frequently considered a mistake to have kept so many people who were trained here, but maybe that's the reason we can talk to each other."

Since I had heard that the Courant Institute is a place where newcomers—and sometimes those not so new—feel that they are shut out from an "in group," I asked Grad if people who had not had the common training he spoke of could also communicate.

"Yes, communicate—no question. There are very subtle differences that in a sense you have to be born here to feel. There *is* a hierarchy—it's like China or Japan in the old days, where everybody knew exactly where he stood and what was going on. Almost impossible for an outsider. But if you ignore these subtle differences, which do exist, then communication is rather easy—just because our training has been to be able to talk to other people."

During Stoker's years as director, there was a spectacular expansion in the institute's research and graduate education program in mathematics. In 1961 a grant of $2,750,000 was made by the Alfred P. Sloan Foundation —this to be matched by an equal amount from the Ford Foundation and the National Science Foundation. Of the total, $4,000,000 went toward the construction of a new building for the Institute of Mathematical Sciences on ground already owned by the university. The other $1,500,000 went to support additional pre- and post-doctoral fellowships and new programs in statistics and mathematical physics as well as to establish a special $300,000 fund for research.

Sloan was eager to honor Warren Weaver, who after his retirement from the Rockefeller Foundation had gone to the Sloan Foundation. It was natural that Sloan would select the group at NYU for his tribute to Weaver's efforts on behalf of applied mathematics in the United States. For a few years after the war there had been plans at several important locations for substantial developments in applied mathematics; but, as Weaver himself wrote later in his autobiography, "it is distressing to have to record that in general these brave starts were not sustained." Since the war there had been only one "truly significant development"—and that had been at New York University.

Stoker and the other mathematicians at NYU did not object to their new building's being named Warren Weaver Hall, as Sloan suggested; but they did feel that the building should then also carry Courant's name. It had always been "Courant's institute" anyway. Although Courant could have objected—as Friedrichs pointed out to me—he did not.

"But I think that in a sense he was vaguely embarrassed by the whole thing," said Jane Richtmyer, an administrative assistant who acted as liaison between the faculty and the architects in the planning of the new building. "He loved the idea of his castle, but he also saw that it might lead to a breaking up or a loss of some of the things he valued. Certainly he was ambivalent."

Jane Richtmyer, the wife of Robert Richtmyer, the scientific director of the computing center after 1954, was another of Courant's "flames."

"He did like the ladies," she laughed when I talked to her in Boulder, Colorado, where she and her husband now live. "I used to think that he was a very 'human' human being with sort of an excess of everything. I never thought he was especially one way or another. He was both. Everything was just a little bit more than with everybody else. But basically he was just a very balanced, reasonable man with sort of excessive ways."

"Let me tell you about the time Papa felt that Jane was abandoning him," Lucile Wolff said to me during our interview. "He and I went for a long walk out in New Rochelle, very slow, inch by inch—he was quite old by then—and he told me how she was so cool to him and so on. 'Well, Richard,' I said, 'you know that this is not the first time you have been disappointed in love.' 'Ja,' he said, 'but maybe it will be the last time.'

"He dreaded the thought, you know—he loved the state of being infatuated. It was such a painful thing. When he was really in love, he was such a pest—you couldn't move without him being there. He was really a remarkable man, Papa—to be so distinguished and still to be so glad to humble himself. It made him feel again that he was true. It's hard to say. He was ambitious and humble. The more he wanted to be great, the humbler he got. And that was very charming."

In the planning of the new building, Jane Richtmyer worked with a committee of faculty members, all of whom had definite ideas about the needs of mathematicians. Friedrichs, for instance, wanted some *Gemütlichkeit*—in the library "a talking room" but also "a silence room" with easy chairs and paintings on the wall. Outside the auditorium and lecture halls on the main floor, he thought that there should be a sort of "temporary office" for preparation and relaxation before a talk.

Jane Richtmyer saw her job as translating these ideas and others to the architects "and trying to get them to value such things and make them practical."

On November 20, 1962, ground was broken for the new building. Courant and Weaver took turns holding the shovel for the newspaper photographers. Two years later the fourteen-story building was completed. By 1965 the mathematicians were in their new quarters.

I first met Courant shortly after the move to Warren Weaver Hall. I had recently returned to the United States from Germany, where I had visited the old institute on Bunsenstrasse, and I remember I could not help thinking that he had come a long way from Göttingen.

Visitors from Europe who have seen pictures of the building often say with confidence (although without success) to their New York taxi drivers: "Take me to the Courant Institute." It is a modest but elegant skyscraper, constructed of bronze-colored brick and glass. Powerful vertical columns rise thirteen floors to an imposing brick facade. A girdle of large bay windows around the twelfth floor provides the necessary horizontal contrast. Inside, nine floors of offices and lecture halls are carefully arranged on Courant's principle that "everything should be all mixed up." As an example, the thirteenth floor, which contains the office Courant occupied during his lifetime, also contains a large office for another senior professor, two lecture halls, a series of small offices for visitors and graduate fellows, and a long handsome lounge with couches and easy chairs where faculty members and students can enjoy coffee or tea throughout most of the day. The "second generation" professors sometimes tell stories of what it was like in the old days "at the Judson" and "at the Bible House" and, looking at the faces of the young people listening, suddenly see themselves again as students listening to Courant tell stories of Göttingen.

At the time I first met Courant, although he was seventy-seven years old, he was still very active and had only recently given up skiing. In addition to being director emeritus of the Courant Institute and a member of its board of governors, he served as scientific consultant for a number of large firms, most importantly IBM. Two years before, he had headed a delegation from the National Academy to the USSR's "academic city" near Novosibirsk. While I set up my tape recorder, he glanced constantly at his watch, rather like the White Rabbit, and murmured that he had to be in Washington that afternoon. I began our interview concerning his recollections of Hilbert by complimenting him on the handsome new building that bore his name.

"Ja," he mumbled, shuffling the pile of papers on his desk and glancing away. "Ja, ja, it is nice," he agreed. "But it does not guarantee achievement."

That same year Courant was asked by an interviewer what in mathematics he saw as most likely to be productive for the scientist.

"I don't want to emphasize or advocate future mathematical activities because of my personal taste," he said. "It is the same as in music. Even now I have difficulty appreciating Bartok or more modern music to any extent. Yet my grandchildren sit at the piano and play such pieces as a matter of course. They don't know there is any difference, and so will it be with respect to the attitude of the younger generation toward scientific subjects, such as computing and computers, or outlying fields of topology or logic. It may well be that older people are just no longer able to adequately absorb new material so I don't want to prophesy anything. I think as long as the attitude with which science is pursued is honest and not dominated by commercialism and as long as people are honestly dedicated, then one must have confidence that something valuable will result."

He still maintained a lively interest in writing. The year I met him (1965), he and Fritz John published in collaboration an "Americanization" of the first volume of the Courant calculus.

He himself had always intended to write a third volume of Courant-Hilbert after he retired. One of the subjects to be treated in it would be the application to practical numerical problems of the theoretical methods which had been developed in the earlier volumes. The other subject would be the existence of the solutions of elliptic partial differential equations. Already, in the English edition of the second volume (1962), he had omitted the seventh chapter on this subject, over which he and Friedrichs had struggled so long and valiantly in 1937. Now, in the third volume, he planned to rewrite the chapter and treat the subject afresh.

Over the years several young men were brought as visitors to work on the project. Among the first of these was De Prima. He found himself sadly unable to help Courant.

"The material for the third volume came from the old Courant-Hilbert, but it had changed and developed over the years," he explained to me. "Courant had really not kept up with the new results except in a general way. He needed to be educated so that he could understand the developments within his own framework. That occasioned some difficulties, in that Courant may have got beyond the point where he *could* understand. The questions he was asking were frequently naive and repetitive. Of course, we all know that in mathematics the naive questions are sometimes the most penetrating; and maybe we, the people trying to help him, myself included, did not quite sufficiently appreciate that.

"You see, whether it's abstract mathematics or concrete mathematics, you often push ahead by successive abstraction and generalization. Possibly what Courant was saying is that significant progress occurs only when you return again to ground level with new understanding. In earlier times it seems that Courant had a rare talent for posing simple and natural questions which often led to fruitful penetration of the how and the why of apparently abstruse and technically complicated mathematical situations. But now, it seemed, Courant was not really asking *those* questions. Maybe that was the trouble—perhaps he didn't know how to formulate them in the present context anymore—or perhaps he was too distracted in those last years."

In 1966, after eight years as director, Stoker resigned. During that time, with the assistance of Bromberg, his friend and former student, he had managed to achieve at last the long-desired recognition of the mathematics group as an integral part of the university. After Stoker resigned, Friedrichs accepted the directorship for a year. It was arranged that in the future the position would be assumed by different members of the staff on a rotating basis. Friedrichs's successor under this plan was Jürgen Moser, a pure mathematician whose work has nevertheless had a great deal of influence on the understanding of the phenomena of the real world.

Moser, who is the husband of Courant's daughter Gertrude, was born in Königsberg, the son of a neurologist. He attended the gymnasium at which Hilbert received the *Abitur*, and studied with Rellich and later with Siegel in Göttingen. Courant's other son-in-law, Jerome Berkowitz, Lori's husband, is also a mathematician and the chairman of the Graduate Division at the Courant Institute. He is typical of the students attending NYU when Courant came there. He is Jewish, a native New Yorker, a graduate of CCNY, the only son of a widow who had to work hard to support him after his father's early death. He came from a family who, as he says, "did not have quite the imagination for a career in mathematics, or in the university at all." Bromberg, for whom he was working at the Reeves Instrument Company after graduation, encouraged him to go and see Courant about continuing his education. Courant offered him a fellowship. Berkowitz's mother was a little upset at his giving up a job that paid $300 a month for one that paid $100, but Berkowitz knew somehow that it was all right. Today he says, "I think that very much of my education occurred right here, and I think that many of the people feel that way. The education was not just mathematical. It was political and social and musical, too."

The balancing and contrasting quality of Courant's sons-in-law is also exemplified by their wives, and by his sons and their wives. Courant's daughters carried on their careers with marriage. Gertrude became a biologist; and Lori, a professional musician. Not unexpectedly, Courant's sons became physicists rather than mathematicians. Ernest married Sara Paul, a young woman whose grandfather had been a rabbi. Hans married Maggie Spaulding, who had what Courant would describe as "a 100 per cent American background."

I came in contact with Courant again at the end of 1968 when he was reading the finished manuscript of *Hilbert*. He was eighty then, but he was trying hard to understand the alienation of the young whites, the anger of the young blacks. Because I came from the west coast, he questioned me closely about the Black Panthers. When I gave him Eldridge Cleaver's *Soul on Ice*, he would not let it out of his hands until he had finished reading it. But he was dreadfully unprepared and shaken in May 1970, during the Cambodian crisis, when a group of students took over Warren Weaver Hall and—announcing that they were holding the AEC computer for $100,000 ransom—covered the walls with obscenities and demands "to get the government out of the university."

All during the time that Moser was director, Courant was again tempted to meddle a little in the affairs of the institute. He made a confidant of young Monroe Donsker, a relative newcomer to the faculty. At lunches in the Village—insisting always on sitting against the wall "to avoid stabs in the back"—he expressed to Donsker his fears about the future of the institute and the weaknesses of his possible successors among the young people. He was most concerned that the institute was slipping away from its mission. In 1968—when he was eighty—he submitted to his colleagues "some remarks about the substance and continuity of our enterprise."

The role attributed to the institute as a center of applied mathematics must not be repudiated but cheerfully accepted. "It is, realistically seen, the brightest hope and justification for us in expecting... support from society, as our historical development clearly indicates, and it is a challenge that we must meet." There was a clear warning. "Before our eyes the process of more and more mathematical elements encroaching on more and more fields of scientific, technological, and other endeavors has been developing without finding in our institute the necessary degree of interested attention... the danger of losing contact with live developments has become real indeed."

During Courant's last years, the institute was in fact sliding somewhat toward pure mathematics. Louis Nirenberg, the "purest" of the second generation, had succeeded Moser as director. There was a feeling that the offerings should be broadened, more fields should be represented. The best students seemed to be attracted to the pure aspects of mathematics.

The struggle to write the third volume of Courant–Hilbert continued. The last young men drawn into it were Stefan Hildebrandt, a German visitor, and Don Ludwig, one of Courant's last Ph.D.'s. At one point Peter Lax suggested that Courant do the book in two small volumes, one with Hildebrandt and the other with Ludwig.

"I told Courant he wouldn't get it done any other way, and indeed he didn't," Lax said to me. "I don't know why he wouldn't do it. Maybe he thought he would be exploiting these people. He wouldn't have. I think it would have been a good opportunity for them, and something good would have come out of it."

Curiously, even in Courant's old age, when he and Hildebrandt began to work together, he conveyed to his young collaborator the same optimism in approaching mathematical problems which he had conveyed long ago in Göttingen.

"Technical difficulties—even if he was quite ignorant—never frightened him," Hildebrandt recalled. "He made one feel that any mathematical task is solvable provided that it is based on a sound and convincing idea.

Courant

After having spoken with him about a problem, I found myself always very confident that—in the end—the solution would be within reach."

In May 1971, in one of the last extended conversations I had with Courant, we returned again to the subject of the institute he had built up at NYU. It had been, he told me—psychologically as well as professionally—"a natural and intense" development for him:

"I felt so dedicated to doing something here for the benefit of the country and the best I could imagine that I might be able to do was to transfer my Göttingen experience and the skills and abilities I had acquired there to the situation here—this is what the institute downtown originated from. I also wanted to do something about the unity of science, including the mathematical sciences in this country. Of course, I consider that what we have done at our institute, we have achieved something; but we have not really won over people, have not achieved the continuation of some of the Hilbert tradition and the general tradition in Germany that was so close to my heart."

All during 1971 he was deeply depressed. The younger people at the institute continued to keep him informed about developments and to consider his wishes, but those who remembered the tremendous optimism and energy which had characterized him in the old days dreaded visiting him.

He no longer played the piano, and sometimes it seemed he could hardly bear the happy activity of others. Nina's passionate devotion to music continued. Her quartet evenings were regular events. Courant withdrew morosely to his study upstairs.

He became inordinately concerned with his financial situation. Yet even during his last years, in addition to his savings and investments, he had a more than generous yearly income which included consulting fees, royalties from his many books, a pension from the German government as well as the one from NYU.

"Look, Courant—I know—you *must* be a millionaire," Stoker remembers saying to him one time when he was fretting again about money.

"But it is so precarious, Jim," Courant said, "it is so precarious."

At NYU and other places, in 1971, the days of the Sputnik-inspired emphasis on science and the days of affluence were coming to an end. The last of the post-war babies who had thronged the colleges in the 1960's were being graduated. New York University, even more than most institutions, was feeling a financial squeeze which would shortly result in the forced sale of the entire University Heights campus to the City University of New York. As a result of the demand for "relevance," the claim of mathematics to be a fundamental academic discipline was being questioned in a way which was a throwback to the situation in America in the 1920's and 1930's.

The 83-year-old Courant worried incessantly about the future of the institute and of mathematics:

"I should go to Washington once more," he said wearily. "The new director of research for the National Science Foundation is a close friend of mine. But I cannot make up my mind to go and talk to him. You see, to do what I think should be done, everybody must be enthusiastic in the same direction. But gradually things have changed. It has become more and more difficult to deeply convince people of the need to do certain things. The situation of the world has changed too, and what should be done—in mathematics anyway—is no longer as clear as it was when I started. And so it is very difficult. Also there is a lot of unfriendliness and coolness toward us. I have had to make a long fight."

"You mean—here—in the beginning?"

"Always."

The last time I saw Courant was at the Courant Institute—the day before he suffered the stroke which resulted in his death in 1972.

A little more than three years later, I visited the Courant Institute again. Fritz John, who had been the last "non-Aryan" mathematician to obtain his Ph.D. in Göttingen in 1933, occupied Courant's old office. Peter Lax had succeeded Louis Nirenberg as director. In nominating Lax as a member of the National Academy in 1962, Courant had described him as "[embodying] as few others do . . . the unity of abstract mathematical analysis with the most concrete power in solving individual problems and with a well-balanced scientific philosophy." At the institute there was now the hope, often expressed to me, that perhaps—after a succession of directors who did not want the position but had assumed it out of a sense of duty—*Peter is the one.*

Lax feels that mathematics is flourishing today at the Courant Institute. True, there are the difficulties common to all universities and in all university mathematics departments—less money, fewer students—and it is no secret that New York University's financial situation is particularly desperate. There are also problems unique to the institute itself. The distinguished faculty members—the young people trained during the war years—are middle-aged now; and no "third generation" has appeared. On the other hand, the current economic situation—the disillusionment with knowledge that has no value in the job market—has worked to the advantage of a mathematics group as applied as the one at NYU.

Lax also feels that the institute is carrying on its mission—"to bring mathematics to applications—new applications—and to relate mathematics to science."

"This does not mean that we are continuing in exactly the same way," he told me. "We have modified to a certain extent, and we are continuing to modify. The institute today is more 'applied' than it has been for a long time."

He picked up a copy of the catalogue for the year 1974–75. On the cover there was a schlieren photograph of a shockless airfoil designed by Paul Garabedian, the principle of which has since been applied by one of his own students to the design of compressor blades and turbines for jet engines. The catalogue expresses, virtually unchanged, the philosophy which Courant espoused for so many years in so many memorandums.

"I am going to take this home and rewrite it a little," Lax said, fanning the pages under his thumb.

As I left Lax's office that day and entered a crowded elevator in which the other passengers—faculty members as well as students—were continuing a discussion begun upstairs in the lounge, I remembered the last time I had seen Courant—the last time he was in the institute that bears his name, the day before his stroke.

I had long since realized that the time for working with him on his reminiscences had passed, and I had given up trying to extract from him specific memories of people he had known and events he had experienced. That day I had been going over old letters and papers in his files. I had not realized that he was in the building, and I almost did not see him as I was leaving; for he was standing near his office, drawn back into a slight alcove so that students rushing past, most of whom no longer knew who he was, would not upset his precarious balance. Wearing a big fur-collared coat and a curly cossack's hat which he had bought in Moscow, he stood quietly, a small old man with a face like a miserable troll.

I stopped to greet him, shocked at how much he had failed since I had seen him last.

"I am a little tired and do not feel very enterprising," he murmured in an attempt at apology for his lack of response to what I told him about my work that day.

Seeking for something to say that would be cheering, I commented again how struck I was every time I entered the institute by the spirit, human and scientific, that I always felt there.

His eyes, which had been dull, brightened a little.

"Ja, ja," he nodded. "It is Göttingen. Göttingen is here."

Göttingen
and
New York

AN ALBUM

I would like to express my appreciation to Natascha Brunswick and Nina Courant, who both furnished a number of pictures, and to each of the following for individual pictures: the Archiv für Kunst und Geschicht, Prof. Garrett Birkhoff, Elizabeth Schoenberg Brody, Nellie Friedrichs, Prof. Konrad Jacobs, Prof. Fritz John, Michael Lewy, *Life* Magazine (for the copyrighted photograph of Warren Weaver), Elisabeth Franck Lisco, the M.I.T. Historical Collections (for the picture of Norbert Weiner and Max Born), Prof. Jürgen Moser, Prof. George Pólya, Margaret Pryce-Born, Brigitte Rellich, Rockefeller University, Prof. Peter Swinnerton-Dyer and the Trinity College Library, Caroline Underwood and the School of Mathematics at the Institute for Advanced Study, Warren Weaver, F. Joachim Weyl, and Eva Toeplitz Wohl.

Top: Felix Klein / *Bottom:* Alfred Haar, Franz Hilbert, Minkowski, ... , the Hilberts, Ernst Hellinger

Top: Courant as a student / *Bottom left:* Hermann Weyl / *Bottom right:* Nina Runge, later Courant

Top: Edmund Landau and daughter Dolli / *Bottom:* Otto Toeplitz

Top: G. H. Hardy / *Bottom left:* Carl T. Runge / *Bottom right:* Harald Bohr

Courant in the trenches, 1915

Top: James Franck (with Gustav Born) and Max Born / *Bottom left:* K. O. Friedrichs / *Bottom right:* Hans Lewy

Top: Courant in his first car, 1929 / *Bottom:* Otto Neugebauer

Top: David Hilbert from painted portrait / *Bottom left:* Weyl / *Bottom right:* Gustav Herglotz

XXXVIII

Emmy Noether

Top: Norbert Wiener with Born / *Bottom left:* George D. Birkhoff / *Bottom right:* Oswald Veblen

Top: Courant, Landau, and Weyl / *Bottom left:* Carl Ludwig Siegel / *Bottom right:* Emil Artin

Abraham Flexner

XLII

Top: Courant and J. J. Stoker / *Bottom left:* Donald Flanders / *Bottom right:* Franz Rellich

Top: Franck and Born back in Göttingen / *Bottom left:* Fritz John / *Bottom right:* Warren Weaver

Top: Groundbreaking for Warren Weaver Hall / *Bottom:* Friedrichs and Courant

Courant in front of the Courant Institute of Mathematical Sciences, 1965

Index of Names

Note: Photographs in the HILBERT ALBUM are numbered I-XXVIII and appear following page 220; those in the COURANT ALBUM are numbered XXIX-XLVI and appear following page 533.

Abraham, Max (1875–1922), XIII
Ackermann, Wilhem (1896–1962), 173, 189
Adler, Auguste, *see* Minkowski, Auguste
Ahlfors, L.V. (1907), 411
Alexander III, The Great (B.C. 356–323), 14, 33, 216
Alexandroff, P.S. (1896–1983), 165–67, 320, 330–31
Althoff, Friedrich (1839–1908), 39, 40, 43, 44, 46, 52, 90, 96, 100, 202
Andrae, Albert, 87
Archibald, R. C. (1875–1955), 358
Archimedes (B.C. 287?–212), 47
Arco, Georg Wilhelm Alexander Hans, Count, 279, 288
Arendt, Otto, 288
Aristobulus (fl. c. B.C. 160), 33
Aristotle (B.C. 384–322), 98, 177
Artin, Emil (1898–1962), 166, 191, 199, 219, 311, 315, 316, 324 348–49, 351, 375, 411, 412, 426–28, 489, 493, 494, XLI
Artin, Natascha, *see* Brunswick, Natascha

Baade, Walter (1893–1960), 358
Becker, Richard (1887–1955), 439
Beethoven, Ludwig van (1770–1827), 137
Behnke, Heinrich (1898–1979), 297
Bell, E.T. (1883–1960), 107, 253, 338
Bell, Winthrop, 272, 298, 487
Benfey, Ilse, *see* Berk, Ilse
Berk, Ilse Benfey (1903), 387, 388, 391, 401–02
Berkowitz, Jerome (1928), 491, 503–04, 521, 527–28
Berkowitz, Lenore (Lori) Courant (1928), 392, 422, 527–28
Berliner, Arnold (1862–1942), 293
Bernays, Paul (1888–1977), 151–53, 172–75, 190, 198, 200, 203, 204, 206, XXII, 302, 368
Bernoulli, Daniel (1700–1782), 274, 276
Bernoulli, Jakob (1654–1705), 68
Bernoulli, Johann (1667–1748), 68, 75
Bernstein, Felix (1878–1956), 118, XIII, XXII, 268, 367
Bernstein, S.N. (1880–1968), 118
Bers, Lipman (1914), 229, 494

535

Index of Names

Bessel, Friedrich Wilhelm (1784–1846), 59
Bieberbach, Ludwig (1886–1982), 189, 209, 347, 383, 385–86, 393, 423, 487, 501–02
Birkhoff, George D. (1884–1944), 180, 269–70, 327–29, 331–32, 353, 357, 358, 432, 435, 436–37, 500, XL
Bismarck–Schönhausen, Otto von (1815–1898), 2, 3, 368
Blackett, P.M.S. (1897–1974), 171
Blaschke, Wilhelm (1885–1962), 375, 393, 483
Bliss, G.A. (1876–1951), 358, 360
Blumenthal, George (1858–1941), 399
Blumenthal, Otto (1876–1944), 48, 49, 51, 86, 87, 97, 100, 111, 114, 150, 158, 187, 199, 208–13, 215, XIII
Bohr, Harald (1887–1951), 119, 122, 134, 135, 163, 168, 189, 212, 268–69, 271, 301, 320, 326–27, 346–47, 367, 371, 377, 378, 385–86, 393, 395, 404, 412, 423, 432, 438, 439, 443, 472, 515, 516, 518, XXXIV
Bohr, Niels (1885–1962), 134, 135, 161, 171, 189, 212, 268–69, 313, 326, 337, 377, 439, 443, 472, 486, 488, 515–16, 517, 518
Boileau, Nicholas (1636–1711), 11
Bolyai, Johann (1802–1860), 58, 106, 126
Bolyai, Wolfgang (1775–1856), 106
Bolzano, Bernhard (1781–1845), 26
Bonnet, (Pierre) Ossian (1819–1892), 23
Bormann, E., 310
Born, Gustav, XXXVI
Born, Max (1882–1970), 95, 103, 105, 109, 112, 113, 123, 129, 141, 154, 167, 171, 172, 180–82, 184, 191, 203, 215, 219, XV, XXV, 233, 248, 266, 268, 269, 271, 306–07, 310, 337, 349, 355, 365, 366–67, 371, 379, 381, 391, 395, 397, 428, 439, 443, 494, 513–14, XXXVI, XL, XLIV
Brauer, Alfred (1894–1965), 437
Breit, Gregory (1899), 388, 397

Bromberg, Eleazer (1913), 459, 460, 498, 511, 527–28
Brouwer, L.E.J. (1881–1966), 148–50, 154–57, 175, 176, 184–88, 198, 324, 346–48
Bruell, Nellie, *see* Friedrichs, Nellie
Brüning, Heinrich (1855–1970), 359
Brunswick, Natascha Artin (1909), 349, 427–28, 483–85, 488–89, 491, 493, 514
Bryan, William Jennings (1860–1925), 84
Burchard, Adelheid, *see* Courant, Adelheid
Burchard, Joseph Jehuda (1785–1874), 230
Busemann, Herbert (1905), 330, 351, 356–57, 377, 471
Bush, Vannevar (1890–1974), 459, 467
Byrd, H.C. (1889–1970), 495

Cantor, Georg (1845–1918), 10, 26, 27, 50, 82, 99, 149, 154, 156, 175–77, 185, 212
Carathéodory, Constantin (1873–1950), 138, 146, 152, 159, 187, 210, 213, 214, XXI, 266–67, 271, 297, 376
Carleman, Torsten (1892–1949), 211, 212
Caruso, Enrico (1873–1921), 134
Cauchy, Augustin-Louis (1789–1857), 26, 80
Cayley, Arthur (1821–1895), 16, 32, 33, 37, 38, 59
Chase, Harry Woodburn (1883–1955), 395, 399, 408, 409, 427, 478, 497, 513
Chisholm, Grace, *see* Young, Grace
Clausius, R.J.E. (1822–1888), 306
Cleaver, Eldridge (1935), 528
Clebsch, Alfred (1833–1872), 38, 325
Comfort, Edwin, 176
Compton, K.T. (1887–1954), 171
Comte, Auguste (1798–1857), 196
Condon, E.U. (1902–1974), 180, 182
Confort, Tom, *see* Comfort, Edwin
Cook, Walter W.S. (1888–1962), 413

Courant, Adelheid Burchard (1824–1883), 230
Courant, Ernest (Ernst) (1920), 303, 311, 381, 392, 441, 447, 528
Courant, Ernst (1891–1916), 231, 235, 275–76, 281, 285, 288, 372
Courant, Eugen, 259
Courant, Fritz (1889), 231, 235, 401, 435, 438, 440
Courant, Gertrude (Gertrud), *see* Moser, Gertrude Courant
Courant, Hans (1924), 392, 528
Courant, Jakob, 231–33
Courant, Leonore (Lori), *see* Berkowitz, Leonore Courant
Courant, Margaret (Maggie) Spaulding (1928), 528
Courant, Martha Freund (1863–1935), 231–35, 259, 285, 299–300
Courant, Nelly Neumann, *see* Neumann, Nelly
Courant, Nerina (Nina) Runge (1891), XX, XXII, 226, 227, 279, 280, 284–85, 289, 290, 291, 293, 299, 300, 303, 311, 333, 349, 356, 357, 359, 361, 381, 387, 388, 389, 390, 391, 392, 397, 402, 403, 405–06, 411, 413, 415, 423, 428, 438, 440, 441, 442, 466, 472, 487, 492, 530, XXXII
Courant, Richard (1888–1972), 100, 107, 109, 111, 123, 124, 132, 139, 152, 159–62, 164, 165, 167, 168, 170, 171, 175, 178, 179, 187, 191, 203, 204, 209, 216, 219, 220, XVII, XX, XXII, XXIV, page references not given for COURANT, XXXII, XXXV, XXXVII, XLI, XLIII, XLV, XLVI
Courant, Salomon (1815–1896), 230, 231, 232, 488
Courant, Sara Paul (1924), 473, 528
Courant, Siegmund (1853–1925), 230–31, 232, 233, 234, 235, 259, 285, 299–300
Curie, Marie (1867–1934), 127

Curie, Pierre (1859–1906), 127
Curry, Haskell B. (1900–1982), 190

Darboux, Gaston (1842–1917), 23, 106, 145, 203
Davenport, Harold (1907–1969), 393, 402–03
Dawney, XIII
Debye, Peter (1884–1966), 135, 136, 138, 140, 141, 152, 170, 184, 219, XXII, 267, 279–80, 306
Dedekind, Richard (1831–1916), 11, 26, 35, 42, 44, 76, 98, 121, 126
Dehmel, Richard (1863–1920), 147
Dehn, Max (1878–1952), 63, 85, 219, 266, 404, 432, 437–38
De Prima, Charles (1918), 429–31, 444, 458–59, 479, 499, 527
Descartes, René (1596–1650), 15
Deuring, Max (1907–1984), 403
Diestel, XIII
Dirac, P.A.M. (1902–1984), 171
Dirichlet, P.G. Lejeune (1805–1859), 19, 65, 68, 213, 259–60, 311–12, 373, 518
D'Ocagne, Philibert Maurice (1862–1938), 23, 24
Donsker, Monroe D. (1924), 522, 528–29
Douglas, Jesse (1897–1965), 397–98, 304–05, 407–08, 411, 414
DuBois-Reymond, Emil (1818–1896), 13, 72, 196, 299

Eberhard, Victor (1862), 39
Ehrenfest, Paul (1880–1933), 381
Ehrlich, Marianne, *see* Landau, Marianne
Ehrlich, Paul (1854–1915), 138, 180, 250, 435
Einstein, Albert (1879–1955), 6, 92, 105, 112, 138, 141, 142, 148, 171, 187, 191, 194, 205, 208, 219, 269, 295, 296, 314, 363–64, 371, 443, 494, 501, 516

Eötvös, Roland, Baron (1848–1919), 271
Epstein, Paul Sophus (1883–1966), XIII
Erdtmann, Karl, 2
Erdtmann, Maria Therese, *see* Hilbert, Maria
Esslen, Elisabet, *see* Springer, Elisabet
Euclid (fl. B.C. 300), 17, 57–61, 63, 127, 213
Euler, Leonhard (1707–1783), 3
Evans, Griffith C. (1887–1973), 449, 461, 495
Ewald, Ella Phillipson, 267
Ewald, Paul P. (1888–1985), 108, 109, 129–33, 139, 171, 212, 219, 267, 322

Fanla, XIII
Feller, William (Willy) (1906–1970), 159, 219, 335, 449, 451, 471
Fenchel, Werner (1905), 250
Ferdinand, Archduke of Austria, 136
Ferdinand, Duke of Brunswick (1721–1792), 47
Fermat, Pierre (1601–1665), 120, 164
Fischer, Emil (1852–1919), 138
Fischer, Ernst (1875–1954), 555
Flanders, Donald (1900–1958), 382, 388–89, 397, 400, 402, 406, 412, 425, 426, 427, 440, 450, 465, 469, 476, 477–78, 483, 495, 501, 517, XLIII
Flanders, Ralph (1880–1970), 440, 450, 517
Fleisher, XIII
Flexner, Abraham (1866–1959), 205, 320, 359, 360–61, 367, 370, 377, 378, 379–80, 393–94, 395, 397, 399, 402, 409, 412, 439, 450, 459, 504–05, 518, XLII
Fourier, Jean Baptiste Joseph (1768–1830), 196, 502
Franck, Hertha Sponer (1895), XXII
Franck, Ingrid, XXII
Franck, James (1882–1964), 167, 171, 191, 203, 204, 219, 269, 306–07, 310, 326–27, 331, 333, 334, 349, 355, 362, 363–65, 368, 371, 378, 380, 381, 382, 395, 396, 397, 406–07, 412, 439, 443, 513–14, XXXVI, XLIV
Frank, Philipp (1884–1966), 138
Frankfurter, Felix (1882–1965), 450
Frankfurter, Magda, *see* Frei, Magda
Franklin, Fabian (1853–1939), 34, 35, 102
Frederick the Great (1712–1786), 1, 204
Fredholm, Erik Ivar (1866–1927), 85, 86, 93, 100, 124, 126
Frege, Gottlob (1848–1925), 98
Frei, Magda Frankfurter (1885–1972), 234, 265
Frenzel, Elise Hilbert (1868–1897), 3, 54, 55
Freund, Louis (1825–1895), 231
Freund, Martha, *see* Courant, Martha
Friedman, Bernard (1915–1966), 463, 481
Friedrichs, Kurt O. (1901–1983), 159, 219, 225–27, 228, 229, 231n., 249, 251–52, 294, 305, 315–16, 318, 319, 321–22, 329, 334–35, 336, 339–41, 344, 346, 351, 352, 357, 362, 373–76, 380, 386, 396, 400–01, 405, 406, 410–11, 412, 414–15, 419–21, 422–23, 425–26, 427, 431, 438, 442, 444, 450–51, 452, 458, 460, 462, 465–66, 467, 468–69, 471, 479–80, 481, 491, 494, 496, 497, 498, 499, 500, 502, 505, 508, 509, 512, 513, 520, 524, 525, 526, 527, XXXVI, XLV
Friedrichs, Nellie Bruell (1908), 225, 228, 400, 421–22
Frobenius, Ferdinand Georg (1849–1917), 86
Fry, Thornton C. (1892), 461
Fuchs, Lazarus (1833–1902), 10, 31, 89
Fueter, Rudolf (1880–1950), 200, 266
Funk, Paul (1886–1969), 183, 193
Furtwängler, Philipp (1869–1940), 200

Index of Names

Gaisman, Henry, 416–17
Galilei, Galileo (1564–1642), 92
Galois, Evariste (1811–1832), 77, 208
Garabedian, Paul (1927), 522, 532
Gardner, Clifford S. (1924), 492
Gardner, Lucile, *see* Wolff, Lucile
Gauss, Carl Friedrich (1777–1855), 9, 17, 19, 41, 42, 45, 47, 55, 56, 58, 59, 61, 65, 90, 126, 163, 192, 195, 196, 199, 208, 253, 308, 311–12, 323, 346, 366, 373, 387, 518
Gelfond, A.O. (1906–1968), 164, 330
Gentzen, Gerhard (1909–1945), 212, 217, 218
George II, Elector of Hannover (1683–1760), 47, 111
Giannini, A.P. (1870–1949), 359
Gilberg, David (1918), 447
Goebbels, Joseph Paul (1897–1945), 206
Gödel, Kurt (1906–1978), 197–99, 217, 219
Goethe, Johann Wolfgang von (1749–1832), 6, 137, 175, 210
Goldman, Henry, 397, 399, 402
Gordan, Paul (1837–1912), 20, 23, 24, 27–38, 125, 142, 165, 166, 202, XVI
Grad, Harold (1923), 464, 522–23
Graves, L.M. (1896–1973), 461
Grommer, Jakob, 131, 143
Grotrian, Eva Merkel XXII
Grotrian, Walter, XXII

Haar, Alfred (1885–1933), 136, 240, 242, 243–45, 248, 266, 268, XXXI
Haber, Fritz (1868–1934), 310
Hadamard, Jacques S. (1865–1963), 339
Haeckel, Ernst Heinrich (1834–1919), 201
Hahn, Kurt (1886–1974), 257, 258, 259, 296, 333
Halphen, Georges (1844–1899), 23–25
Hamburger, Hans Ludwig (1889–1956), 437
Hamel, Georg (1877–1954), XIII, 266
Hamilton, William Rowan (1805–1865), 181
Hansen, XIII
Hardy, G.H. (1877–1947), 114, 118, 119, 163, 184, 206, 320, 346–47, 357, 386, 413, 414, 432, 483, XXXIV
Haskell, M.W. (1863–1948), 358
Hasse, Helmut (1898–1979), 154, 199, 200, 203, 207, 211, 212, 300–01, 376, 384–86, 393, 402–03, 426–27, 439, 474–75, 487, 494, 501
Heald, Henry (1904–1975), 507, 510–11
Hecke, Erich (1887–1947), 129, 130, 164, 165, 187, 210, XXII, 237, 262, 263–64, 266, 268, 269, 297, 301, 302, 310, 376, 483
Hegel, Georg Wilhelm Friedrich (1770–1831), 49, 150
Heinz, Erhard, 486–87
Heisenberg, Werner (1901–1976), 167, 170, 171, 180–82, 337, 338, 376, 381, 485–86, 499, 501
Hellinger, Ernest (Ernst) (1883–1950), 95, 109, 219, 233, 237, 238, 239, 242, 252, 260–61, 266, 268, 271, 404, 435, 437, 443, XXXI
Helly, Eduard (1884–1943), 241–42, 438
Helmholtz, Hermann Ludwig Ferdinand von (1821–1894), 10, 31, 35, 126, 195
Hensel, Kurt (1861–1941), 260–61
Herglotz, Gustav (1881–1953), 179, 213, 311, 324, 351, 352, 369, 376, 381, 470, 474, XXXVIII
Hermite, Charles (1822–1901), 24, 30, 41, 121
Hertz, Heinrich (1857–1894), 35, 40, 100, 127, 224
Hilbert, Christian David, 1, 2, 209
Hilbert, David (1862–1943), Frontis., page references not given for HILBERT, VIII, XII, XIII, XIX, XXII, XXVII, XXVIII, 225, 226,

539

Index of Names

Hilbert, David (*cont.*)
    228, 229, 237, 238–43, 244, 245–46, 247–49, 250, 251–52, 254–55, 256–59, 260–61, 262, 263–64, 268, 271, 272, 288–89, 290–91, 294–96, 297, 300, 301, 302, 305, 306–07, 310, 311, 312, 313–14, 317, 321–22, 324, 336, 339, 344, 347–48, 350, 351, 352, 353, 358, 364, 373, 376, 377, 381, 383, 387, 395, 397, 408, 412, 415, 422, 430, 438, 444, 451, 458, 465, 491, 502, 515–16, 527, 530, XXXI, XXXVIII
Hilbert, David Fürchtegott Leberecht (1782?–1858), 2
Hilbert, Elise, *see* Frenzel, Elise
Hilbert, Franz (1893–1969), 43, 44, 53, 91, 92, 104, 123, 124, 138, 139, 151, 210, 215, XII, 240, XXXI
Hilbert, Ilse, 132
Hilbert, Johann Christian, 1
Hilbert, Käthe Jerosch (1864–1945), 36, 39, 40, 43, 46, 49, 52–55, 91, 95, 100, 119, 120, 123, 130, 132, 139, 140, 151, 152, 190, 202, 207, 211, 212, 214, 215, VIII, XVI, XXII, 240, 247, 250, 262, 264, 382, 492, XXXI
Hilbert, Maria Erdtmann (1830–1905), 1–4, 88
Hilbert, Otto (1826–1907), 1–3, 6, 12, 14, 88, 204, III
Hildebrandt, Stefan (1936), 529–30
Hille, Einar (1894–1980), 322, 358, 407
Hindenburg, Paul von (1847–1934), 203, 207, 333, 357–58, 359, 361
Hirschland, Franz, 388–89, 396, 402
Hiss, Alger (1904), 483, 517
Hitler, Adolf (1889–1945), 203, 207, 354, 358, 359, 361, 363, 364, 377, 378, 401, 408, 411, 440, 474–75, 477, 485, 501
Hoffmann, Banesh (1906), 338
Hofmann, Mrs., XXII
Homer, 175

Hopf, Heinz (1894–1971), XX, 233, 331, 410, 418–19, 458
Houtermans, F.G. (1903), 171
Humbert, Georges (1859–1921), 25
Hurwitz, Adolf (1859–1919), 13, 14, 18, 20, 21, 23, 27, 28, 35, 40–42, 46, 54, 55, 69, 70, 72, 81, 113–15, 117, 121, 144, 145, 150, 153, 174, 202, 210, 213, VII, 237, 317–18
Hurwitz, Ida Samuels, 39
Husserl, Edmund (1859–1938), 91, 122, 145, 245, 257–58, 265, 271, 272–73, 314
Husserl, Elisabeth, *see* Rosenberg, Elisabeth
Husserl, Gerhart (1893–1973), 272
Husserl, Malvine Steinschneider (1860–1950), 265
Husserl, Wolfgang (1895–1916), 272–73

Isaacson, Eugene (1919), 466–67, 507–11, 512, 513

Jacobi, C.G.J. (1804–1851), 9, 11, 15, 76, 192, 196, 502
Jammer, Max (1915), 338
Jaspers, Karl (1883–1969), 484
Jerosch, Käthe, *see* Hilbert, Käthe
John, Charlotte Woellmer (1910), 351, 355–56, 377, 383, 402
John, Fritz (1910), 355–56, 377–78, 381, 383, 402, 406, 479, 480, 526, 531, XLIV
Joos, Jakob Christoph Georg (1894), 439
Jordan, Camille (1838–1922), 12, 23, 24
Jordan, Pascual (1902), 171, 181, 182, 317, 337, 483
Joseph, Helene, *see* Weyl, Helene

Kaluza, Theodor (1885–1954), 403, 439, 474

Kant, Immanuel (1724–1804), 3, 5, 16, 17, 62, 63, 137, 192, 194, 195
Kármán, Theodore (Theodor) von (1881–1963), 122, 128, 219, XXII, 268, 269, 341, 349, 358, 426, 450–51, 461–62, 485
Keller, Gottfried (1819–1890), 201
Keller, Joseph B. (1923), 499–500, 522
Kellogg, Oliver D. (1878–1932), 180, 266, 317–18, 358
Keppel, Frederick (1875–1943), 408, 417
Khinchin, A.Y. (1894–1959), 114
Killing, Wilhelm (1843–1923), 302
Kinzel, August B. (1900), 506
"Klärchen," XXII
Klein, Anna Hegel, 20, 49
Klein, Elisabeth, *see* Staiger, Elisabeth
Klein, Felix (1849–1925), 14, 18–25, 29, 31, 33, 37, 43–46, 48–50, 52, 55, 59, 61, 65, 67, 76, 86, 88–91, 95, 96, 100, 103, 104, 106, 111, 116–20, 136–38, 140, 142, 143, 145, 146, 152, 159–62, 165, 170, 178, 179, 189–91, 202, 203, 209, 212, 214, IX, XIII, XVI, 226–27, 228, 238, 243, 245, 246, 249, 251–52, 257, 258, 260, 263, 266–67, 270, 271, 272, 297, 300, 301, 302, 306–07, 308, 311–12, 313, 315, 318, 323–25, 326, 332, 333, 334, 341–42, 349, 353, 358, 359, 372, 373, 375, 377, 382, 383, 385, 387, 395, 405, 413, 428, 431, 446, 450, 465, 488, 518, XXXI
Kline, Morris (1908), 390, 394, 434, 499, 510
Kneser, Adolf (1862–1930), 237, 309, 333
Kneser, Hellmuth (1898–1973), 162, 309, 315, 316, 368, 369, 372–73, 375, 501
Kneser, Martin (1928), 309, 501
Knopf, Alfred A. (1892), 454
Knopp, Konrad (1882–1957), 386
Koebe, Paul (1882–1945), 255, 256, 257
König, Julius (1849–1913), 106
König, Robert (1885), 200
Koenigsberger, Leo (1837–1921), 100
Kollwitz, Käthe Schmidt (1867–1945), 6, 7, 147, 174
Kolmogoroff, A. N. (1903), 330
Kotkin, Bella Manel, 412–13, 416, 462
Kowalewski, Sonja (1850–1891), 142, 143, 166
Kratzer, Adolf (1893), 153, 297
Kronecker, Leopold (1823–1891), 10, 25–27, 31, 32, 34–37, 39, 42, 44, 50, 53, 56, 71, 76, 98, 99, 148, 149, 155, 157, 173, 175, 186, 187, 196, 198, 202, 260, 261
Kummer, Ernst Edward (1810–1893), 10, 26, 42, 44, 56, 75, 80, 126

Lagrange, Joseph Louis (1736–1813), 306
Landau, Edmund (1877–1938), 117–19, 138, 163, 164, 167, 202–04, 206, 211, XVI, XXII, 249–50, 268, 271, 284, 307, 310, 350–51, 352, 368, 379–80, 381, 383, 385, 388, 404, XXXIII, XLI
Landau, Marianne Ehrlich, 180, XXII, 250, 262, 435
Landé, Alfred (1888), 133, 134, 138, 140, 141, 219
Landsberg, Georg (1865–1912), 237
Laplace, Pierre Simon, Marquis de (1749–1827), 253
Laue, Max von (1879–1960), 68, 130, 267, 376
Lax, Anneli Leopold (1922), 491, 501, 503, 514
Lax, Peter D. (1926), 458, 473, 491, 502–03, 507, 529, 531–32
Lefschetz, Solomon (1884–1972), 358, 359, 407, 424–25
Legendre, Adrien Marie (1752–1833), 55
Leisler, Jacob (1640–1691), 391
Lenin, V.I. (1870–1924), 193
Levi, F.W. (1888–1966), 135

Levi-Civita, Tullio (1873–1941), 432
Lewy, Hans (1904), 159, 184, 204, 219, 316, 320, 332, 336, 340, 341, 344, 345–46, 351, 352, 357, 360, 369, 387, 388, 396, 415, 420, 451, 465–66, 471, 496, 497, 512, 513, XXXVI
Lichtenberg, Georg Christoph (1742–1799), 173
Lichtenstein, Leon (1878–1933), 329
Lie, Sophus (1842–1899), 21, 89, 126
Lietzmann, Walther (1880–1959), 89, 127, 212, 213
Lindemann, Ferdinand (1852–1939), 13, 15, 16, 26, 27, 42, 43, 202
Littlewood, J.E. (1885–1977), 114
Lobatchewski, N.I. (1793–1856), 58
Lohse, Walter (–1917), 273, 274, 275, 276, 278, 282, 292
Lorentz, H.A. (1853–1928), 100, 135, 303–04
Ludendorff, Erich (1865–1937), 299
Ludwig, Don (1933), 529
Lummer, Otto (1884–1925), 236
Lüneburg, Rudolf, 400
Lusternik, L.A. (1899), 330
Luzin, N.N. (1883–1950), 320

McCarthy, Joseph R. (1908–1957), 517
MacDonald, J.K.L., 463
McKinley, William (1843–1901), 84
MacLane, Saunders (1909), 353–54, 355, 361, 378, 522
McShane, Edward J. (1904), 360, 361, 378, 424
Mäder, Hanna, see Schwerdtfeger, Hanna
Manel, Bella, see Kotkin, Bella
Mann, Thomas (1875–1955), 453–54
Mannheim, V.M. Amédée (1831–1906), 23
Maschke, Erich, 232–33, 247
Maschke, Heinrich (1853–1908), 233
Mason, Max (1877–1961), 358, 417

Maximilian (Max) of Baden, Prince (1867–1929), 295, 296
Maxwell, James Clerk (1831–1879), 35, 127, 140
Mayer, Maria Goeppert (1906–1972), 358
Meissner, Ernst (1883–1939), 242, 419
Mersenne, Marin (1588–1648), 75
Meunier, Constantin, 174
Meyer, Martha, 355, 381
Mie, Gustav (1868–1957), 141, 378
Minkowski, Auguste Adler (1875–1944), 55, 95, 119, 123, 214, XXII
Minkowski, Fanny (1863–1954), 4–6, XXII
Minkowski, Hermann (1864–1909), 4, 5, 7, 11–14, 16–18, 24, 26–29, 34, 35, 37–41, 43–46, 50, 51, 53–55, 62, 69–72, 80, 81, 85, 88–97, 100, 102–05, 108, 111–22, 138, 141, 145, 146, 165, 167, 171, 202, 203, 208, 213–15, VI, XVIII, 238–39, 241, 242, 243, 245–46, 248, 249, 250, 268, 314, 324, 358, 387, 438, XXXI
Minkowski, Lily, see Rüdenberg, Lily
Minkowski, Max, 4, 11
Minkowski, Oskar (1858–1931), 4
Minot, G.R. (1885–1950), 179, 180
Mises, Richard von (1883–1953), 342, 347, 452
Mittag-Leffler, Magnus Gösta (1846–1927), 211
Morawetz, Cathleen Synge (1923), 479–80
Morrey, Charles B. (1907–1984), 496–98, 504
Morse, Marston (1892–1977), 432, 447, 452, 461
Moser, Gertrude (Gertrud) Courant (1922), 388, 392, 440, 479, 527–28
Moser, Jürgen (1928), 486–87, 527, 528, 529
Mühlestein, 301
Müller, Claus (1920), 489
Müller, Conrad (1878–1953), XIII

Index of Names

Müller, Hans, XIII
Mugdan, Bertha, *see* Stenzel, Bertha
Mugdan, Käthe (–1944), 236
Mussolini, Benito (1883–1945), 401

Nelson, Leonard (1882–1927), 121, 144, 145, 151, XXII, 245, 263, 314
Nernst, Walther (1864–1941), 52, 87, 90, 138
Neugebauer, Otto (1899), 159, 204, 219, 315, 318, 320, 324, 332, 334, 345, 348, 353, 357, 358, 359, 360, 366–67, 368–69, 370, 371, 373–76, 377, 431–32, 434, 440, 451, 471, 515, XXXVII
Neumann, Carl (1832–1925), 66, 67
Neumann, Franz (1798–1895), 9, 66
Neumann, John von (1903–1957), 172, 183, 189, 195, 219, 244, 335–36, 407, 422–23, 460, 502
Neumann, Justizrat, 234, 271, 272
Neumann, Nelly (1886–194?), 234, 237, 239–43, 246, 252, 259–60, 261–62, 263, 264, 265–66, 271–72, 279, 280, 284, 285, 288, 299
Newson, Mary Winston, 74
Newton, Isaac (1642–1727), 47, 68
Neyman, Jerzy (1894–1981), 441, 449
Nirenberg, Louis (1925), 473, 491, 522, 529, 531
Noether, Emmy (1882–1935), 142, 143, 165, 167, 199, 203, 204, 207, 208, 219, XX, 297, 352, 367, 371, 381, 411, 425, 437, XXXIX
Noether, Fritz (1884), 142
Noether, Max (1844–1921), 23, 142
Nordheim, Lothar (1899–1985), 172, 181–83, 192, 219

Oppenheimer, J. Robert (1904–1967), 171, 474, 505, 516–17
Ore, Oystein (1899–1968), 195, 219
Osgood, William Fogg (1864–1943), 190, 358

Ostrowski, Alexander (1893), 145, 146, 153, 169, 297
Ostwald, Wilhelm (1853–1932), 318

Painlevé, Paul (1863–1933), 346
Papen, Franz von (1879–1969), 361
Pascal, Blaise (1623–1662), 75
Pasch, Moritz (1843–1930), 59–61
Paul, Sara, *see* Courant, Sara
Pauli, Wolfgang (1900–1958), 167, 171, 483
Pauling, Linus (1901), 171
Peano, Giuseppe (1858–1932), 59–61, 84
Peary, Robert (1856–1920), 97
Peirce, Benjamin (1809–1880), 435
Perron, Oskar (1880–1975), 117, 118, 249
Phillips, H.B. (1881–1972), 344
Phillipson, Ella, *see* Ewald, Ella
Picard, Émile (1856–1941), 23, 118, 211, 327
Pick, Georg (1859–1943?), 19, 20
Pick, Helen (1891), 233, 234
Pick, Hilde (1899–1962), 355, 388, 391, 415
Planck, Max (1858–1947), 127, 138, 236, 376
Plateau, J.A.F. (1801–1883), 404, 413, 414
Plücker, Julius (1801–1868), 100, 325
Plutarch (c.46–c.120), 33
Pohl, Robert (1884), 306, 334, 381
Poincaré, Henri (1854–1912), 19, 20, 22–24, 41, 55, 62, 63, 69, 72, 76, 82, 84, 99, 106, 120, 125, 133, 141, 179, 185, 186, 195, 250, 266, 269, 325, 516
Pólya, George (1887–1985), 132, 149–51, 211, 219
Pontryagin, L.S. (1908), 336
Prager, William (Willy) (1903–1980), 451
Prandtl, Ludwig (1875–1953), 97, 122, 138, 249, 334, 369, 375, 376, 439, 486
Prandtl, Mrs., XXII
Pringsheim, Ernst (1859–1917), 236

543

Proskauer, Eric (Erich) S. (1903), 424–25, 448, 493

Rados, Gustav (1862–1941), 106
Rayleigh, John William Strutt (1842–1919), 172, 294–95, 303, 322
Rees, Mina S. (1902), 462, 468
Reid, Legh Wilber (1867), 93
Reidemeister, Elisabeth, 210
Reidemeister, Kurt (1893–1971), 193, 196, 484–85, 488
Reinach, Adolf, 271, 273
Rellich, Franz (1906–1955), 159, 205, 335, 362, 377, 384, 385, 470, 471–72, 473, 474–75, 486–87, 488, 494, 501, 504, 518–19, 527, XLIII
Remak, Robert (1888–194?), 437
Richardson, R.G.D. (1878–1949), 358, 394, 451–52, 457, 459
Richtmyer, Jane Seely, 524–25
Richtmyer, Robert D. (1910), 524
Riecke, Eduard (1845–1915), 97, 245–46, 306
Riemann, Bernhard (1826–1866), 11, 19, 65–67, 82, 85, 159, 163, 164, 178, 208, 213, 217, 250, 251–52, 253, 254–55, 312, 317, 346, 373, 387, 398, 460, 518
Ritter, Irving F. (1902), 399
Ritz, Walther (1878–1909), 67, 248–49, 338
Robbins, Herbert (1915), 447–49, 454–56
Roberts, E.S., 467, 478, 506
Robscheit-Robbins, F.S., 179
Rockefeller, John D., Jr. (1874–1960), 320
Roentgen, Wilhelm Konrad (1845–1923), 127–29, 138
Roosevelt, Franklin D. (1882–1945), 361, 398, 459, 470
Rosanes, Jacob (1842–1922), 237
Rosenberg, Elisabeth (Elli) Husserl (1892), 265, 266, 299
Rosenthal, Arthur (Artur) (1887–1959), 437

Rüdenberg, Lily Minkowski (1898–1984), XXII, 243
Runge, Aimée DuBois-Reymond, 279, 284, 289, 290, 299, 387, 391, 405, 415, 438, 441
Runge, Bernhard (1897–1914), 272, 273, 279, 289
Runge, Carl (1856–1927), 96, 97, 107, 108, 116, 179, XIV, 249, 271, 272, 273, 279, 280, 284, 289, 290, 297, 299, 300, 324, 334, 341–42, 355, 387, 389, 392, 420, 466, XXXIV
Runge, Nerina, see Courant, Nina
Runge, Wilhelm (1895), 272, 273, 289, 290, 487–88
Rupprecht, Crown Prince of Bavaria, 288
Russell, Bertrand (1872–1970), 98, 144, 150, 186
Rust, Bernhard, 373, 374

Samuels, Ida, see Hurwitz, Ida
Sarnoff, David (1891–1971), 412
Schaefer, Clemens (1878), 237
Scherrer, Paul (1890–1969), 134, 135, 140, 279, 280
Schiller, Friedrich von (1759–1805), 6, 212
Schilling, Friedrich (1868–1950), XIII
Schleicher, Kurt von (1882–1934), 361
Schmidt, Arnold (1902–1967), 183, 205–07, 210
Schmidt, Erhard (1876–1959), 86, 87, 106, 109, 110, 152, 172, 183, XIII, 266, 347
Schmidt, Karl, 6
Schmidt, Käthe, see Kollwitz, Käthe
Schnirelman, L.G. (1905–1938), 330
Schoenberg, Dolli Landau, 388, XXXIII
Schreier, Otto (1901–1929), 483
Schrödinger, Erwin (1887–1961), 182, 183, 337, 338, 376, 379
Schubert, Hermann Hannibal (1848–1911), 13, 14

Schütte, Kurt (1909), 205
Schwartz, Jacob T. (1930), 522
Schwartz, Laurent (1915), 502, 504
Schwarz, Hermann Amandus (1843–1921), 31, 44, 86, 212
Schwarzschild, Karl (1873–1916), 97, XIII, 249
Schwerdtfeger, Hanna Mader, 351
Schwerdtfeger, Hans (1902), 351
Seeger, Raymond J. (1906), 460, 476
Segré, Corrado (1860–1924), 25
Shakespeare, William (1564–1616), 6
Shapiro, Harold N. (1922), 489–90, 492
Shaw, George Bernard (1856–1950), 180
Shiffman, Max (1914), 399, 400, 413, 463, 481, 495
Siegel, Carl Ludwig (1896–1981), 164, 165, 206, 210–12, 310–11, 315, 316, 375, 397, 404, 405, 407, 435, 437, 439, 443, 475, 490, 494, 527, XLI
Simon, H.T. (1870–1918), 97
Sloan, Alfred P., Jr. (1875–1966), 523–24
Smith, Henry (1826–1883), 12
Smith, T.L., 418–19
Snyder, Virgil (1869–1950), 358
Sommerfeld, Arnold (1868–1951), 5, 36, 52, 103, 128–30, 133, 135, 141, 167, 172, 202, 213, 267–68, 337, 376
Spaulding, Margaret (Maggie), *see* Courant, Margaret
Sponer, Hertha, *see* Franck, Hertha
Springer, Elisabet, XXII
Springer, Ferdinand (1881–1965), 159, 160, 201, XXII, 293–94, 296, 308–09, 333, 355, 384, 411, 424–25, 431–32, 476–77, 484, 488, 493, 494
Springer, Julius (1817–1877), 293, 411
Staiger, Elisabeth (Putti) Klein (1888–1968), 272, 273, 488
Staiger, Robert (–1914), 272, 273
Stalin, Joseph (1879–1953), 485
Stein, Auguste Courant (1848–1936), 235, 265–66

Stein, Edith (Sister Terese Bendicta of the Cross) (1891–1942), 230, 231n., 233, 234, 235, 261, 265–66, 271–72, 488
Stein, Rose (1883–1942), 266, 488
Steinhaus, Hugo (1887–1972), 102, 103, 118, 119
Stenzel, Anne K. (Anna), 411, 412, 415–16
Stenzel, Bertha Mugdan, 236, 403–04
Stenzel, Julius (1883–1935), 236, 403–04, 411
Stern, Otto (1888–1969), 233
Sternberg, Grete, 233
Sternberg, Wolfgang (1887–1953), 233, 440–41
Stieltjes, Thomas Jean (1856–1894), 25
Still, Carl, 306, 313, 317, 319–20, 334, 359, 383, 385
Still, Hanna, 320
Stoker, James J. (1905), 410, 418–19, 425–26, 442, 444, 458, 460, 474, 479, 481, 496, 497, 499, 502, 505, 508–09, 511, 512–13, 520–24, 527, 530, XLIII
Stoker, Nancy, 419, 505
Straus, Percy (1876–1944), 416
Strauss, Lewis L. (1896–1974), 439, 478, 516
Study, Eduard (1862–1930), 20, 22–25, 27, 30, 39
Sturm, Rudolf (1863–1919), 237
Sylvester, James Joseph (1814–1897), 16, 24, 36–38, 325, 382
Synge, Cathleen, *see* Morawetz, Cathleen
Synge, J.L. (1897), 479
Szegö, Gabor (1895–1985), 193, 196, 219

Takagi, Teiji (1875–1960), 86, 199
Tamarkin, J.D. (1888), 400, 432, 451
Tarski, Alfred (1902–1983), 218–19
Taussky-Todd, Olga (1906), 201, 202, 219

Index of Names

Teichmüller, Oswald (1913–1943), 362, 380, 386, 393, 403
Teubner, B.G., 308, 309
Thompson, Dorothy (1894–1961), 450
Thomson, J.J. (1856–1940), 35, 127
Toeplitz, Otto (1881–1940), 95, 109, 131, 181, 237, 238, 239, 242–43, 244–45, 252, 268, 300–01, 313, 415, 435, 437, 443, XXXIII
Tornier, Wilmar Hermann Erhard (1894), 385, 393, 402
Trowbridge, Augustus (1870–1934), 325–26, 331–32, 336, 342–43
Truman, Harry S. (1884–1972), 505
Twersky, Shirley, 514

Urysohn, P.S. (1898–1924), 320, 331

Vahlen, Theodor (1869–1945), 372–73, 386, 402
van der Waerden, B.L. (1903), 162, 166, 167, 187, 199, 207, 209, XX, 309, 341, 376
Veblen, Oswald (1880–1960), 212, 357, 381, 383, 404, 407, 432, 461, 515, 517, XL
Veronese, Giuseppe (1854–1917), 25
Victoria, Queen (1819–1901), 84
Vieweg, Gotthold Richard (1896), 484
Viviani, Vincenze (1622–1703), 75
Voigt, Woldemar (1850–1919), 97, 123, 245–46, 257–58, 306
Volterra, Vito (1860–1940), 133

Walther, Alwin (1898–1967), 403
Waring, Edward (1734–1798), 113, 126
Wassermann, August von (1866–1925), 138
Weaver, Warren (1894–1878), 397, 417–18, 459–62, 476, 484, 489, 497, 505–06, 523–24, 526, XLIV, XLV

Weber, Heinrich (1842–1913), 11, 13, 35, 44, 45, 48, 208
Weber, Werner (1906–194?), 361, 380, 381, 404
Weber, Wilhelm (1804–1891), 61, 323, 366
Weierstrass, Karl (1815–1897), 9, 10, 26, 31, 65–68, 76, 78, 82, 145, 175, 176, 254, 317, 414
Weinstein, Alexander (1897–1979), 322
Weyl, F. Joachim (1915–1977), 381
Weyl, Helene (Hella) Joseph (1893–1948), 268, 313–14, 351, 352, 361, 377
Weyl, Hermann (1885–1955), 94, 95, 104, 123, 143, 148–51, 155–57, 160, 161, 170, 171, 173, 175, 186, 187, 191, 192, 200, 201, 204, 205, 211, 214, 216, 219, XXIII, XXVII, 244–45, 266, 268, 269, 271, 303–05, 307, 313–15, 349, 351–52, 353, 360–61, 369, 371–72, 376, 377, 378, 381, 384, 405, 407, 411, 427–28, 431, 453, 465, XXXII, XXXVIII, XLI
Whipple, G.H., 179
Whitehead, Alfred North (1861–1947), 144, 150, 186
Whitney, Hassler (1907), 254, 255
Wiechert, Emil (1861–1928), 16, 36, 61, 62, 97
Wien, Max (1866–1938), 5
Wien, Willi (1864–1928), 5, 138
Wiener, Hermann (1857–1939), 57
Wiener, Leo, 328–30
Wiener, Norbert (1894–1964), 164, 169, 170, 328–31, 340, 344, 411, XL
Wigner, Eugene (1902), 183, 219
Wilhelm I (1797–1888), 2, 3
Wilhelm II (1859–1941), 137, 144, 147, 272, 295, 297, 474
Wilks, Samuel S. (1906–1964), 461
Winchell, Walter (1897–1972), 517
Witt, Ernst (1911), 403

546

Woellmer, Charlotte, *see* John, Charlotte
Wolff, Lucile Gardner (1927), 492–93, 524
Wolfskehl, Paul (1856–1906), 120
Wright, Orville (1871–1948), 97
Wright, Wilbur (1867–1912), 97

Yoshiye, XIII
Young, Grace Chisholm (1868–1944), 48, XIII

Zermelo, Ernst (1871–1953), 97, 98, 135, 150, XIII, 249, 250
Zumino, Bruno (1923), 499